AF788991

Bone Markers: Biochemical and Clinical Perspectives

Edited by

Richard Eastell MD FRCP FMedSci FRCPath
Professor of Bone Metabolism
Director of the Division of Clinical Sciences
Bone Metabolism Group
University of Sheffield
Sheffield, UK

Matthias Baumann MD
Integrated Health Care Solutions
F Hoffmann-La Roche Ltd.
Basel
Switzerland

Nicholas R Hoyle PhD
Director of Bone Metabolism R & D
Roche Diagnostics GmBH
Penzberg
Germany

Lothar Wieczorek PhD
Integrated Health Care Solutions
F Hoffmann-La Roche Ltd.
Basel
Switzerland

MARTIN DUNITZ

© 2001 Martin Dunitz Ltd, a member of the Taylor & Francis group

First published in the United Kingdom in 2001
by Martin Dunitz Ltd, The Livery House, 7–9 Pratt Street, London
NW1 0AE

Tel.: +44 (0)20 7482 2202
Fax.: +44 (0)20 7267 0159
E-mail: info.dunitz@tandf.co.uk
Website: http://www.dunitz.co.uk

All rights reserved. No part of this publication may be reproduced, stored in a retrieval system, or transmitted, in any form or by any means, electronic, mechanical, photocopying, recording, or otherwise, without the prior permission of the publisher or in accordance with the provisions of the Copyright, Designs and Patents Act 1988 or under the terms of any licence permitting limited copying issued by the Copyright Licensing Agency, 90 Tottenham Court Road, London W1P 0LP.

The Author has asserted his right under the Copyright, Designs and Patents Act 1988 to be identified as the Author of this Work.

Although every effort has been made to ensure that all owners of copyright material have been acknowledged in this publication, we would be glad to acknowledge in subsequent reprints or editions any omissions brought to our attention.

Although every effort has been made to ensure that drug doses and other information are presented accurately in this publication, the ultimate responsibility rests with the prescribing physician. Neither the publishers nor the authors can be held responsible for errors or for any consequences arising from the use of information contained herein. For detailed prescribing information or instructions on the use of any product or procedure discussed herein, please consult the prescribing information or instructional material issued by the manufacturer.

A CIP record for this book is available from the British Library.

Distributed in the United States by:
Blackwell Science Inc.
Commerce Place, 350 Main Street
Malden, MA 02148, USA
Tel: 1-800-215-1000

Distributed in Canada by:
Login Brothers Book Company
324 Salteaux Crescent
Winnipeg, Manitoba, R3J 3T2
Canada
Tel: 204-224-4068

Distributed in Brazil by:
Ernesto Reichmann Distribuidora de Livros, Ltda
Rua Coronel Marques 335, Tatuape 03440-000
São Paulo
Brazil

ISBN: 1-84184-023-8

Composition by Wearset, Boldon, Tyne and Wear

Printed and bound in Spain by Grafos S.A. Arte sobre papel

Contents

Faculty of Bone Markers: Biochemical and Clinical Perspectives Workshop vi

Chairmen and principal authors vii

Foreword x
Lawrence G Raisz

Preface xi
Richard Eastell, Matthias Baumann, Nicholas R Hoyle and Lothar Wieczorek

1 Introduction: bone metabolism and its regulation 1
Graham Russell

Section A: Potential candidates

2 Potential candidates for bone turnover markers—N-telopeptide cross-links of type I collagen (NTX) 27
Peter R Ebeling

3 C-telopeptides 39
Elizabeth T Leary

4 Pyridinium cross-links as bone resorption markers 49
Simon P Robins

5 Bone-specific alkaline phosphatase 57
Laurence M Demers

6 Osteocalcin 65
Caren M Gundberg

7 The N- and C-terminal propeptides of human procollagen type I (PINP and PICP): molecular heterogeneity and assay technology 73
Jette Brandt, Jette Klose Frederiksen, Charlotte Harken Jensen and Børge Teisner

8 Bone sialoprotein 83
Berthold Fohr, Henning W Woitge and Markus J Seibel

Discussion paper A 91

Section B: Sources of biological variation

9 Biochemical markers of bone turnover: age, gender and race as sources of biological variability 95
Yvette M Henry and Richard Eastell

10 Circadian rhythm studies of serum bone resorption markers: implications for optimal sample timing and clinical utility 107
William D Fraser, Mark Anderson, Christine Chesters, Brian Durham, Aftab M Ahmad, Paula Chattington, Jiten Vora, Christine R Squire and Michael J Diver

11 Sources of biological bone marker variability 119
Charles H Chesnut III

12 The effects of fracture or disease on biochemical markers of bone remodelling 123
Michael Kleerekoper

13 Variation in biochemical markers of bone turnover: sources, quantification, minimization 131
G Russell Warnick

Discussion paper B 141

Section C: Osteoporosis; monitoring

14 The use of biochemical markers of bone turnover for monitoring treatment of osteoporosis 149
Pierre D Delmas

15 The use of biochemical markers of bone turnover to monitor response to therapy: statistics and logic 159
Aubrey Blumsohn

16 Prediction of response in bone mass by biochemical markers of bone turnover during antiresorptive therapy for prevention of osteoporosis 167
Pernille Ravn

17 Biochemical markers of bone turnover as predictors of bone loss and response to therapy 179
Clifford J Rosen

Discussion paper C 187

Section D: Osteoporosis; risk prediction: who, when, how

18 Assessment of fracture risk from bone mineral density and bone markers 197
Olof Johnell, Anders Odén, Alison Dawson and John A Kanis

19 Markers of bone turnover, endogenous hormones, rate of bone loss and fracture risk in the OFELY study 203
Patrick Garnero, Elisabeth Sornay-Rendu and Pierre D Delmas

20 Prediction of hip fracture with markers of bone turnover in the EPIDOS study 211
Roland D Chapurlat, Patrick Garnero, Pierre J Meunier, Gerard Bréart and Pierre D Delmas

21 Biochemical markers of bone turnover: the Study of Osteoporotic Fracture 219
Douglas C Bauer

22 The association between bone turnover and fracture risk (Sheffield Osteoporosis Study) 225
Diana M Greenfield, Rosemary A Hannon and Richard Eastell

Discussion paper D 237

Index 245

Faculty of *Bone Markers: Biochemical and Clinical Perspectives Workshop*
Hotel President Wilson, Geneva, Switzerland
Thursday March 23rd to Saturday March 25th, 2000

Chairmen and principal authors

Douglas C Bauer MD
Associate Professor of Medicine, Epidemiology and Biostatistics
University of California, San Francisco
School of Medicine
San Francisco, CA
USA

Aubrey Blumsohn MBChB PhD MRCPath
Sheffield University
Bone Metabolism Group
Section of Medicine
Clinical Sciences Centre
Northern General Hospital
Herries Road
Sheffield, UK

Jette Brandt
Department of Immunology and Microbiology,
University of Southern Denmark
Odense, Denmark

Roland D Chapurlat MD PhD
Prevention Science Group
University of California San Francisco
San Francisco CA
USA

Charles H Chesnut III MD
Professor of Radiology, Medicine, and Nutritional Sciences
Adjunct Professor of Orthopaedics
Director, Osteoporosis Research Group
University of Washington Medical Center
Seattle, WA
USA

Pierre D Delmas MD PhD
Professor of Medicine
Department of Rheumatology and Bone Diseases
Hôpital Edouard Herriot
Lyon
France

Laurence M Demers PhD DABCC
Distinguished Professor of Pathology and Medicine
The Penn State University College of Medicine
Departments of Pathology and Medicine
Hershey, PA
USA

Richard Eastell MD FRCP FRCPath FMedSci
Professor of Bone Metabolism
Division of Clinical Sciences
Director of the Bone Metabolism Group
University of Sheffield
Sheffield, UK

Peter R Ebeling MBBS MD FRACP
Associate Professor
Research and Development Laboratory,
Department of Diabetes and Endocrinology
Royal Melbourne Hospital
Victoria, Australia

Solomon Epstein MD
Medical Director
Roche Laboratories Inc.
Nutley, NJ, USA

Berthold Fohr MD
Department of Internal Medicine
University of Heidelberg
Heidelberg, Germany

William D Fraser MD
Reader/Honorary Consultant
Department of Clinical Chemistry
Royal Liverpool University Hospital
Liverpool, UK

Patrick Garnero PhD
INSERM Unit 403 and SYNARC
Lyon, France

Susan L Greenspan MD
Professor of Medicine
Division of Endocrinology
University of Pittsburgh Medical Center
Pittsburgh, PA, USA

Diana M Greenfield BSc PhD
Bone Metabolism Group
Clinical Sciences Centre
University of Sheffield
Sheffield, UK

Caren M Gundberg PhD
Department of Orthopaedics and Rehabilitation
Yale University School of Medicine
New Haven, CT USA

Yvette Henry BSc
Bone Metabolism Group
Clinical Sciences Centre
University of Sheffield
Sheffield, UK

Olof Johnell MD PhD
Department of Orthopaedics
Malmö University Hospital
Malmö, Sweden

John A Kanis MD FRCPath
WHO Collaborating Centre for Metabolic Bone Diseases
University of Sheffield Medical School
Sheffield, UK

Sundeep Khosla MD
Division of Endocrinology, Metabolism, Nutrition and Internal Medicine
Mayo Clinic
Rochester, MN
USA

Michael Kleerekoper MD
Professor of Internal Medicine
Wayne State University School of Medicine
Detroit, MI
USA

Elizabeth T Leary PhD
Chief Scientific Officer
Pacific Biometrics, Inc.
Seattle, WA
USA

Michael R McClung MD FACE
Director
Oregon Osteoporosis Center
Portland, OR, USA

Christopher P Price MA PhD FRCPath
Professor of Clinical Biochemistry
St Bartholomew's and the Royal London School of Medicine and Dentistry
London, UK

Lawrence G Raisz MD
Professor of Medicine
Program Director
Lowell P Weicker, Jr General Clinical Research Center
University of Connecticut Medical Center
Farmington, CT, USA

Pernille Ravn MD
Center for Clinical and Basic Research Ballerup
Ballerup, Denmark

Juha Risteli MD PhD
Department of Clinical Chemistry
University of Oulu
Oulu, Finland

René Rizzoli MD
Division of Bone Diseases
WHO Collaborating Center for Osteoporosis and Bone Diseases
Department of Internal Medicine
University Hospital
Geneva, Switzerland

Simon P Robins PhD DSc
Skeletal Research Unit
Rowett Research Institute
Aberdeen, UK

Clifford J Rosen MD
Maine Center for Osteoporosis Research and Education
Bangor, ME
USA

Graham Russell PhD DM FRCP FRCPath
Institute of Musculoskeletal Sciences
University of Oxford
Nuffield Orthopaedic Centre
Oxford, UK

Markus J Seibel MD PhD
Department of Medicine
University of Heidelberg
Heidelberg, Germany

Børge Teisner MD
Department of Immunology and Microbiology
University of Southern Denmark
Odense, Denmark

G Russell Warnick MS MBA
President
Pacific Biometrics Research Foundation
Seattle, WA
USA

Nelson B Watts MD
Professor of Medicine
Emory University School of Medicine
Director, Osteoporosis and Bone Health Programs
Emory Clinic
Atlanta, GA
USA

(Discussion A)
Panel: Laurence M Demers, Peter R Ebeling, Solomon Epstein, Caren M Gundberg, Elizabeth T Leary, Christopher P Price, Juha Risteli, Graham Russell, Simon P Robins, Markus J Seibel and Børge Teisner

(Discussion B)
Panel: Charles H Chesnut III, Richard Eastell, William D Fraser, Yvette Henry, Sundeep Khosla, Michael Kleerekoper, Lawrence G Raisz, René Rizzoli and G Russell Warnick

(Discussion C)
Panel: Aubrey Blumsohn, Pierre D Delmas, Susan L Greenspan, Michael R McClung, Pernille Ravn, Clifford J Rosen, Markus J Seibel and Nelson B Watts

(Discussion D)
Panel: Douglas C Bauer, Roland D Chapurlat, Pierre D Delmas, Peter R Ebeling, Patrick Garnero, Diana Greenfield, Olof Johnell, John A Kanis and Michael R McClung

Foreword

The concept of using biochemical markers to assess bone formation and resorption in skeletal disorders is an old one. Total serum alkaline phosphatase has been used to assess bone formation in Paget's disease, hyperparathyroidism and osteomalacia for many decades. Measurements of total urinary hydroxyproline can reflect collagen breakdown, most of which is due to bone resorption, if the patient is on a gelatin-free diet. While total serum alkaline phosphatase and urinary hydroxyproline measurements are useful in patients with high rates of bone turnover, particularly in Paget's disease, they are not sufficiently accurate or precise to assess bone turnover in osteoporosis.

Over the last three decades new sensitive and specific biochemical markers of bone formation and resorption have been developed which have had a remarkable impact on the study of human skeletal physiology and disease. The discovery of osteocalcin as a bone specific protein almost 30 years ago and the subsequent discovery of collagen crosslinks which are relatively specific for bone have made it possible to assess rates of bone formation and resorption with much greater accuracy.

Nevertheless each of the major markers currently in use has limitations as well as advantages. Most importantly, while biochemical markers have been of enormous utility in research, their clinical use in individual patients has been less successful. This is due in part to the variability of these measurements. Moreover little is known about the metabolic degradation of proteins released from bone which may be another source of variation. Thus markers do not provide actual quantitation of the rates of resorption or formation, but only reflect relative differences during progression or treatment of disease. In clinical research, conclusions concerning changes in resorption or formation can be strengthened by the use of multiple markers. This approach would be strengthened still further by the development of new markers which assess these processes in different ways.

Biochemical markers have been particularly useful in clinical studies evaluating antiresorptive drugs. They provide a rapid indication of drug efficacy and can be used to compare different agents, doses and schedules of administration. There are data suggesting that one can predict the effect of an antiresorptive agent on bone mineral density (BMD) based on the early marker response. Prospective population studies have shown an association between high bone turnover and increased fracture risk, independent of differences in BMD. Other studies show an inverse relation between markers and BMD.

This monograph explores that impact of bone markers on skeletal research especially in osteoporosis. There are up-to-date chapters by recognized experts which describe the chemistry of the various markers as well as their analytic precision and sources of biologic variation. The uses of markers in monitoring therapy in osteoporosis as well as predicting bone loss and fractures are thoughtfully presented from several different points of view. The material presented will also enable the reader to keep pace with and understand the rapid developments that are likely to occur in this field in the near future.

Lawrence G Raisz, MD
Professor of Medicine
Program Director
University of Connecticut Health Center
Connecticut
USA

Preface

This book represents the proceedings of a meeting 'Bone Markers: Biochemical and Clinical Perspectives' held at the Hotel President Wilson in Geneva, Switzerland on 23rd and 25th March, 2000. The meeting was jointly sponsored by Hoffmann-La Roche Ltd and the University of Sheffield. The speakers, discussants and audience comprised a large proportion of the scientists currently active in research in the biochemical markers of bone turnover.

We felt a single-topic meeting was necessary because there have been a number of developments in this field. The introduction of automated immunoassay analysers makes these tests more widely available. There is now a wealth of potentially useful bone markers and we felt it would be helpful to compare these. There are many publications on the sources of variability of bone markers and these need to be summarised so that we can use the markers optimally. There are differing views on whether the markers are best used for prediction of the risk of fracture or in the monitoring of response to therapy, and we felt that to bring together the proponents of the different views would allow individuals to draw their own conclusions.

The meeting was structured to match these issues and so after an overview of bone remodelling by Graham Russell it began with a consideration of the potential candidate markers, followed by the sources of biological variability, and then the use of these markers for monitoring osteoporosis and for the prediction of fractures. The following day included a discussion on the use of bone markers in malignant bone disease chaired by Rob Coleman and those proceedings will be published separately in *Cancer Treatment Reviews*.

The session on potential candidates (Discussion paper A) was chaired by Caren Gundberg and Juha Risteli and the other discussants included Sol Epstein and Chris Price. The session considered in detail pyridinium crosslinks and their related telopeptides, the collagen propeptides, alkaline phosphatase, bone sialoprotein and osteocalcin the assays available, their performance and the characteristics of each marker.

The session on sources of biological variability (Discussion paper B) was chaired by Richard Eastell and Sundeep Khosla and the other discussants included René Rizzoli and Larry Raisz. The speakers described the uncontrollable sources of variation such as age, gender, race, diseases and fracture as well as controllable sources of variation such as the circadian rhythm, diet and exercise. A clinical chemistry perspective was presented by Russ Warnick on approaches used in other areas to minimize variability when using these tests in practice.

The session on the use of bone markers to monitor osteoporosis (Discussion paper C) was chaired by Mike McClung and Susan Greenspan and the other discussants included Markus Seibel and Nelson Watts. An overview of monitoring was given by Pierre Delmas and different approaches were described by others. The evidence for an association between change in markers and change in bone density was summarised by Cliff Rosen.

The session on the use of markers as risk predictors (Discussion paper D) was chaired by

Pierre Delmas and John Kanis and the other discussants included Olof Johnell and Peter Ebeling. Olof Johnell begin by giving an overview of the prediction of fractures by any method so as to set the scene for the place of bone markers. The evidence was examined side by side from the EPIDOS, OFELY, Study of Osteoporotic Fractures and Sheffield Osteoporosis Study.

We believe that by having these speakers with different perspectives attend and discuss their results in the same forum allowed a deeper understanding of the issues concerning the clinical application of bone markers in the practice of osteoporosis. The discussion was taped and the transcript checked by each of the contributors. This provides a lively commentary on the presentations and reflects the lively debate that was encouraged at this meeting.

We are very grateful to the speakers, chairpersons and discussants for their contributions to this monograph. We hope that the lively debate has been captured and that this stimulates further research in this important area. We are grateful to Gill Higginbottom and Monique Kruithof and to Phocus for their contibution to the logistics of the meeting. We appreciated the importance of having these proceedings published within a year of the meeting and we are grateful to the publishers Martin Dunitz for making this possible and particularly to Pete Stevenson and Dan Edwards.

Richard Eastell, MD, FRCP
Sheffield
Matthias Baumann, MD
Basel
Nicholas R Hoyle PhD
Penzberg
Lothar Wieczorek PhD
Basel

November 2000

1

Introduction: bone metabolism and its regulation

Graham Russell

Summary • General introduction • Bone differentiation and function • The cells of bone • Potential regulatory factors • Factors affecting the action of cytokines and growth factors • Prospects for the future • Clinical assessment of bone status • The major diseases of bone resorption • Current therapy and drugs in development • References

SUMMARY

The physiological control of calcium metabolism and of skeletal remodelling is under the regulation of systemic hormones, especially the calcium-regulating hormones, parathyroid hormone, 1,25-dihydroxyvitamin D (calcitriol) and calcitonin. Other hormones, including thyroid and pituitary hormones, and adrenal and gonadal steroids, also have major effects on the skeleton, as seen in clinical disorders in which their secretion is abnormally high or low. Many additional factors, notably cytokines and growth factors, and non-peptide mediators such as nitric oxide, must also play a role in these processes, in many cases by interacting locally with systemic hormones. Mechanical loading of the skeleton is also a major influence on bone remodelling. Bone is metabolically active throughout life. After skeletal growth is complete, remodelling of both cortical and trabecular bone continues and results in an annual turnover of about 10% of the adult skeleton. The remodelling of both cortical and trabecular bone requires the sequential and coordinated actions of osteoclasts to remove bone, and osteoblasts to replace it. Osteoclasts differentiate from haematopoietic stem cell precursors under the direction of factors that include cytokines and other mediators. Osteoblasts within trabecular bone probably differentiate from stromal cell precursors in bone marrow, and manufacture a complex extracellular matrix, which subsequently mineralizes. The genetic factors that regulate skeletal development and function are gradually being identified, and recent examples include the *cbfa1* gene for osteoblast differentiation and the RANK system for osteoclasts. Many cytokines and growth factors are involved in the induction of new bone formation, and the activation and modulation of remodelling. These and other mediators contribute not only to the physiological regulation of bone metabolism but also to the pathogenesis of skeletal diseases. The commonest clinical disorder of bone metabolism is osteoporosis, which affects one in three women over 50 years, and about 1 in 10 men. Its pathophysiological basis includes genetic predisposition and subtle alterations in systemic and local hormones, coupled with environmental influences. Treatment depends mainly on drugs that inhibit bone resorption, either directly or indirectly. These include bisphosphonates, oestrogens, synthetic oestrogen-related compounds (SERMs – selective oestrogen receptor modulators), and calcitonin. There are several other clinical disorders in which bone resorption is increased, and these include Paget's disease of bone, and the bone changes secondary to cancer, such as occur in myeloma and metastases from breast cancer. Monitoring of bone metabolism by biochemical means depends upon measurement of enzymes and proteins released during bone formation (such as alkaline phosphatase, osteocalcin and collagen propeptides), and of degradation products produced during bone resorption. The most useful markers of bone resorption are degradation products derived from the enzymatic hydrolysis of type I collagen, particularly peptides related to regions of cross-linking with the pyridinolines. Tissue-specific processes in bone can thereby be monitored by appropriate assays, as described in this book.

GENERAL INTRODUCTION

The physiological regulation of calcium homeostasis involves three main organs, the gut, the kidney, and the skeleton. The fluxes of calcium and phosphate through these organs contribute to the integration of calcium metabolism throughout growth and adult life.

The hormonal control of calcium metabolism can be attributed mainly to the effects of systemic hormones, especially the calcium-regulating hormones, parathyroid hormone (PTH), 1,25-dihydroxyvitamin D (calcitriol) and calcitonin (CT). The regulation of plasma calcium concentrations is determined mainly by the renal tubular reabsorption of calcium and the effects of PTH on this process (Fig. 1.1).

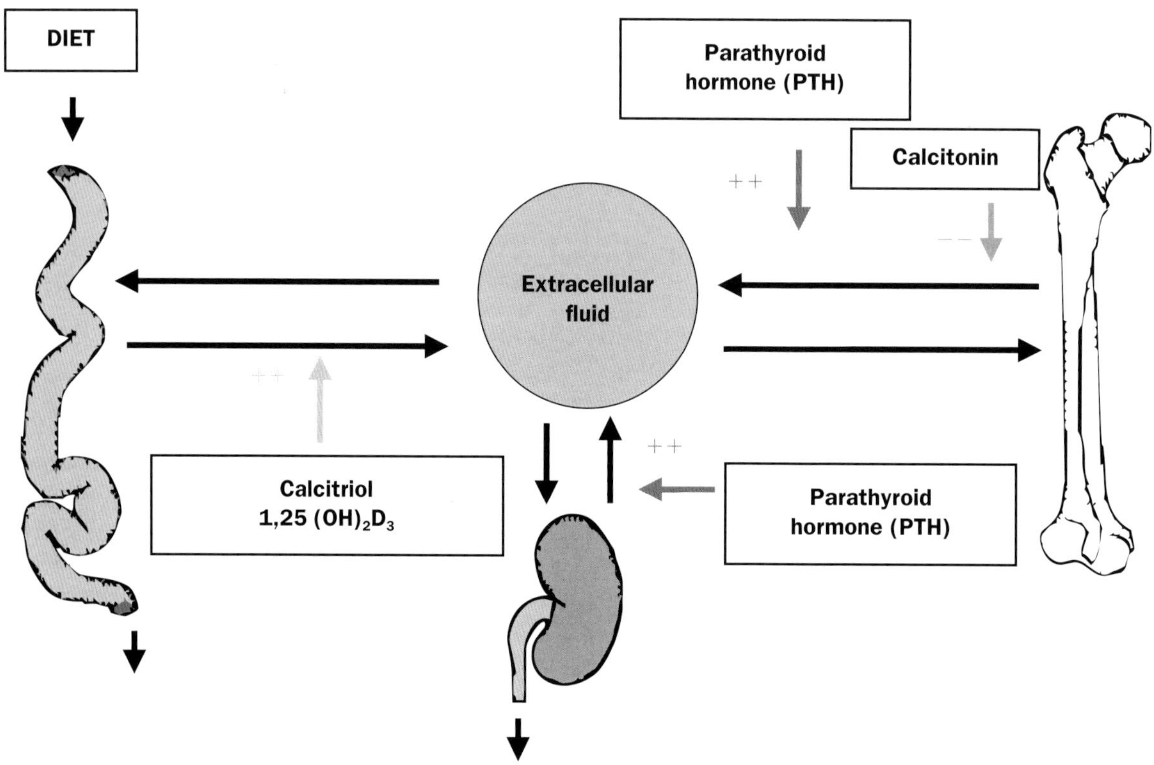

Fig. 1.1 The major physiological regulatory mechanisms in calcium metabolism.

Other hormones, including thyroid and pituitary hormones, and adrenal and gonadal steroids, also have major effects on the skeleton, as seen in clinical disorders in which their secretion is abnormally high or low. Many additional factors, notably cytokines and growth factors, also play a role in skeletal metabolism, in many cases by interacting locally with systemic hormones. Mechanical loading of the skeleton also has a major influence over bone remodelling.

The genetic factors that regulate skeletal development and function are gradually being identified, and recent examples include the *cbfa1* gene for osteoblast differentiation[27] and the RANK system for osteoclasts. Many cytokines and growth factors are involved in the induction of new bone formation, and the activation and modulation of remodelling. These and other mediators contribute not only to the physiological regulation of bone metabolism but also to the pathogenesis of skeletal diseases.

Monitoring of bone metabolism by biochemical means depends upon measurement of enzymes and proteins released during bone formation (such as alkaline phosphatase, osteocalcin and collagen propeptides), and of degradation products produced during bone resorption. Tissue-specific processes can be measured by appropriate assays, such as pyridinoline cross-links, to reflect bone resorption.

An excellent and comprehensive review of basic and clinical aspects of metabolic bone disease can be found in Favus.[1]

BONE DIFFERENTIATION AND FUNCTION

Bone growth and development

During growth, the skeleton enlarges. In long bones this is achieved by the epiphysial growth plates, which produce increases in length, while increases in diameter result from deposition of new bone on the periosteal surfaces, accompanied by resorption from the endosteal surfaces.

During development and growth, bone is produced by two main processes, intramembranous ossification, as in skull bones, and by endochondral ossification involving the growth plate, as occurs in limb bones. Many clinical defects of skeletal growth can be traced to genetic abnormalities in matrix production and cellular signalling mechanisms (Table 1.1).

Thus, parathyroid-related peptide (PTH-RP) and fibroblast growth factors (FGFs) are involved in cartilage differentiation, and defects in their production or action via defective receptors are associated with chondrodysplasias.

After skeletal growth is complete, remodelling of both cortical and trabecular bone continues. Cortical bone remodels at a lower rate (~2% per annum) than trabecular bone (~10% per annum) and results in an annual turnover of about 10% of the adult skeleton. The purpose of remodelling is to allow the bone to adapt to changes in distribution of mechanical forces, and to repair microdamage. The remodelling process requires the coordinated actions of osteoclasts to remove bone, and osteoblasts to replace it, and these processes may be monitored by histological means. Changes in the quality or amount of bone arise from disorders of bone modelling during growth or remodelling during adult life.

THE CELLS OF BONE

Osteoblasts and bone formation

Osteoblasts within trabecular bone differentiate from stromal cell precursors in bone marrow, and manufacture a complex extracellular matrix, which subsequently mineralizes (Figs. 1.2 and 1.3). Important events for induction of new bone formation include the activation of specific 'master' genes for skeletal development, notably *cbfa1*.[2] Many growth factors affect bone formation.[3] These include insulin-like growth factors (IGFs), FGFs, and especially members of the TGF-β (transforming growth factor beta) family, such as the bone morphogenetic proteins (BMPs).[4,5] Other members of this family can induce cartilage formation, e.g. the cartilage morphogenetic factors (CMPs).

Osteoblasts themselves, as well as marrow stromal cells, are capable of producing a wide array of factors that can potentially act as

Table 1.1 Bone matrix proteins, cytokines and other factors. Some molecular and genetic defects that result in a skeletal phenotype.

	Bone phenotype	
Affected genes	**Mice**	**Human**
Matrix proteins		
Type I collagen	Osteogenesis imperfecta	Osteogenesis imperfecta
Decorin	No bone phenotype	–
Biglycan	Osteoporosis	–
Fibrillin 1	–	Marfan syndrome
Thrombospondin	Increased bone density and cortical thickness	–
Osteopontin	Protection against ovariectomy-induced bone loss	–
Osteocalcin	Increased bone formation	–
Matrix Gla protein (MGP)	Extraskeletal calcification (blood vessels	–
Cytokines		
Macrophage colony stimulating factor (M-CSF)	Osteopetrosis	–
Osteoclast differentiation factor (ODF) (also known as RANK Ligand)	Osteopetrosis	–
Interleukin 4	Osteoporosis	–
Osteoprotegerin (OPG)	Over expression leads to osteopetrosis Knockout leads to osteoporosis	–
Signalling molecules		
Core binding factorα1 (Cbfαl)	Lack of osteoblasts and ossification	Cleidocranial dysplasia
TNF receptor associated factors (TRAF6)	Osteopetrosis	–
Nuclear factors (NF–κB1 and NF–κB2)	Osteopetrosis	–
C-src	Osteopetrosis	–
Enzymes		
Alkaline phosphatase	Impaired bone mineralization	Hypophosphatasia
Tartrate resistant acid Phosphatase (TRAP)	Defects in endochondral ossification	–
Carbonic anhydrase II	No bone phenotype	Osteopetrosis
Cathepsin K	Osteopetrosis	Pycnodysostosis
Matrix metalloproteinase 9 (MMP–9)	Delayed endochondral ossification	–
MT1–MMP	Defective ossification and osteopenia	–

– = information not available

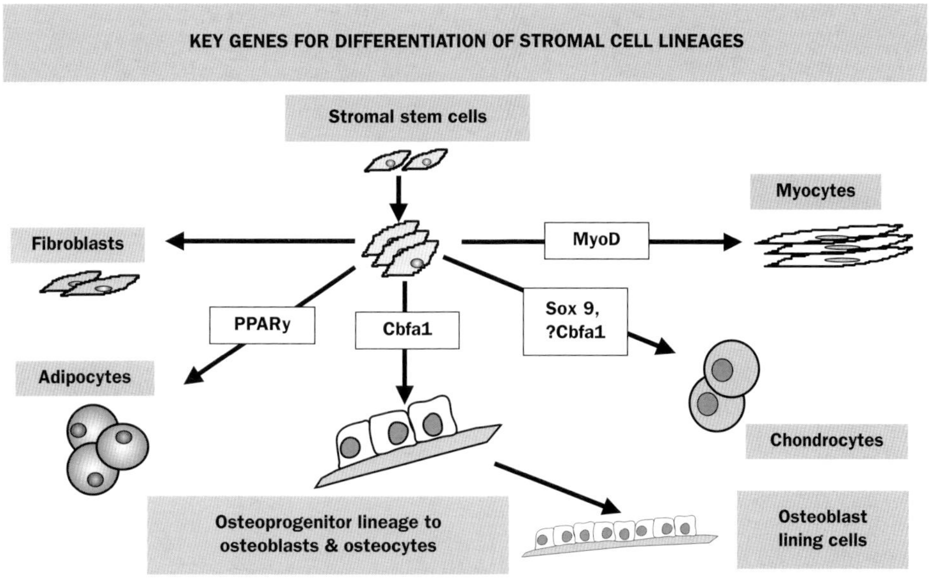

Fig. 1.2 Scheme to show the distinct cellular lineages of osteoblasts and related cells, and some of the key regulatory genes.

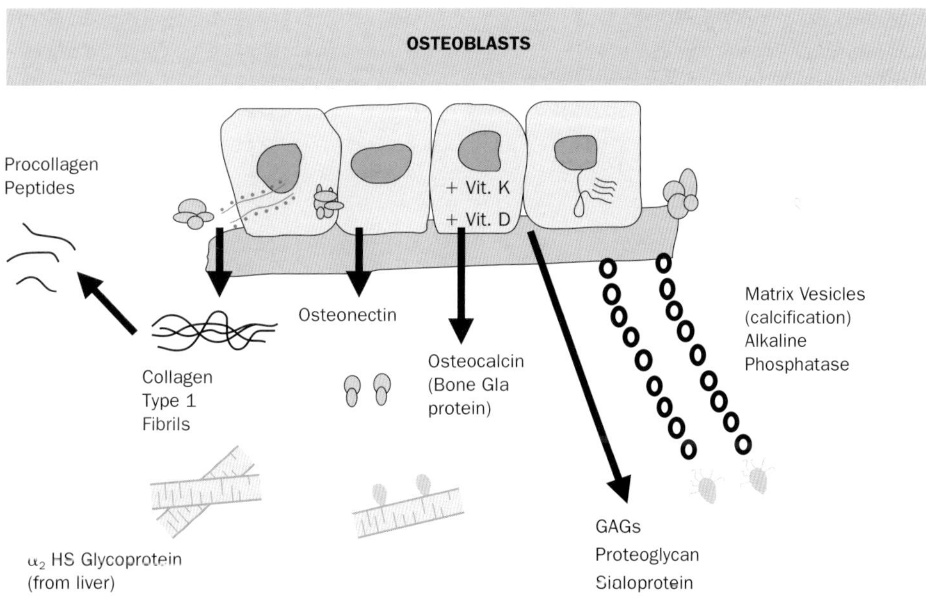

Fig. 1.3 Osteoblasts: mineralization of extracellular matrix.

autocrine and paracrine regulators of bone cell function. Several cytokines that can be produced by stromal cells and osteoblasts, such as RANK-ligand, colony-stimulating factors (CSFs) and interleukin-6 and interleukin-11, affect osteoclast recuitment In some cases, systemic hormones such as PTH have been shown to be capable of influencing the production of these local factors (e.g. IL-6, prostaglandins and IGFs) from bone. Products of osteoblasts are used to monitor bone formation (Fig. 1.4).

Osteoclasts and bone resorption

Osteoclasts are multinucleated cells with a number of distinctive features (Fig. 1.5) and are the major cells involved in bone resorption.

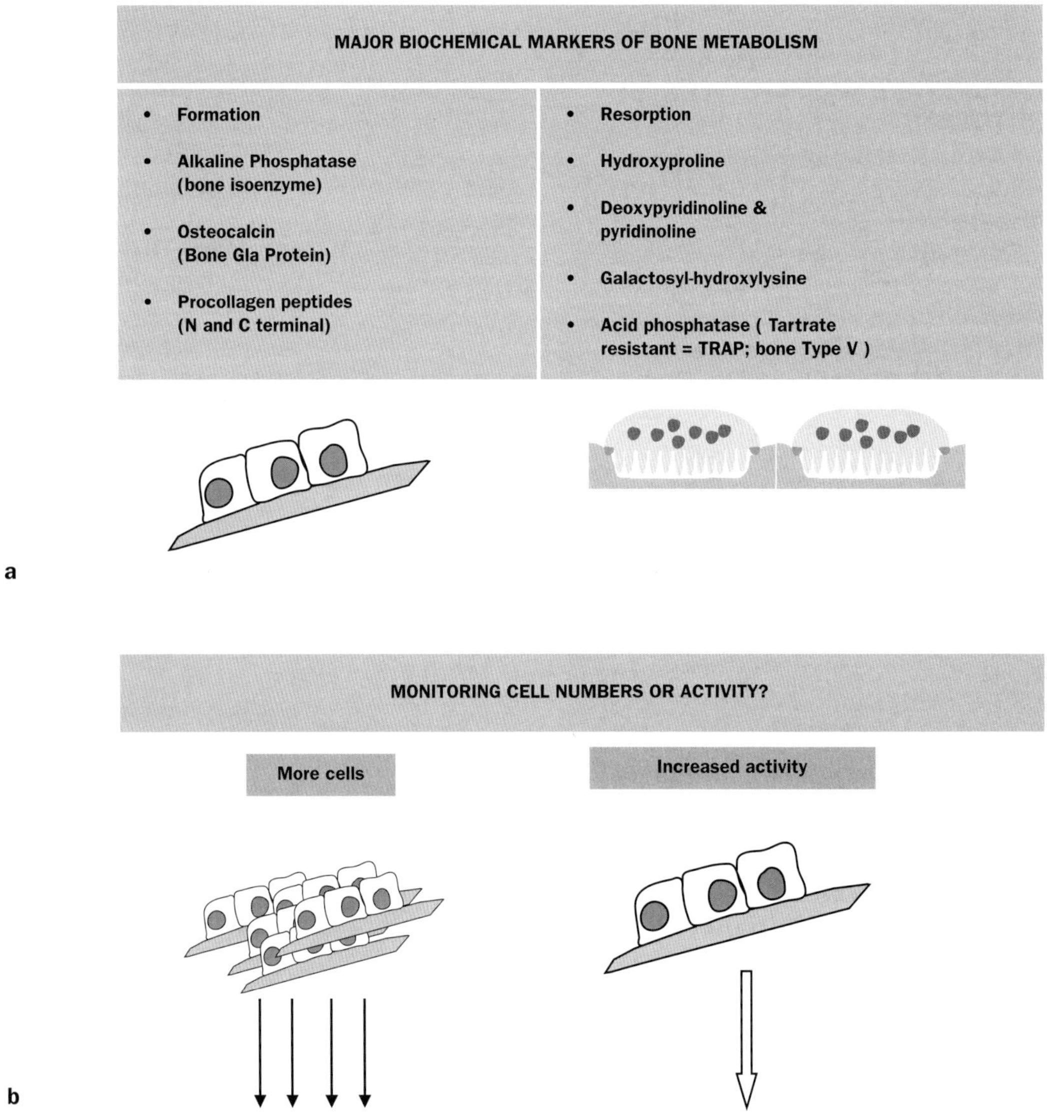

Fig. 1.4a,b Osteoblasts are used to monitor bone formation.

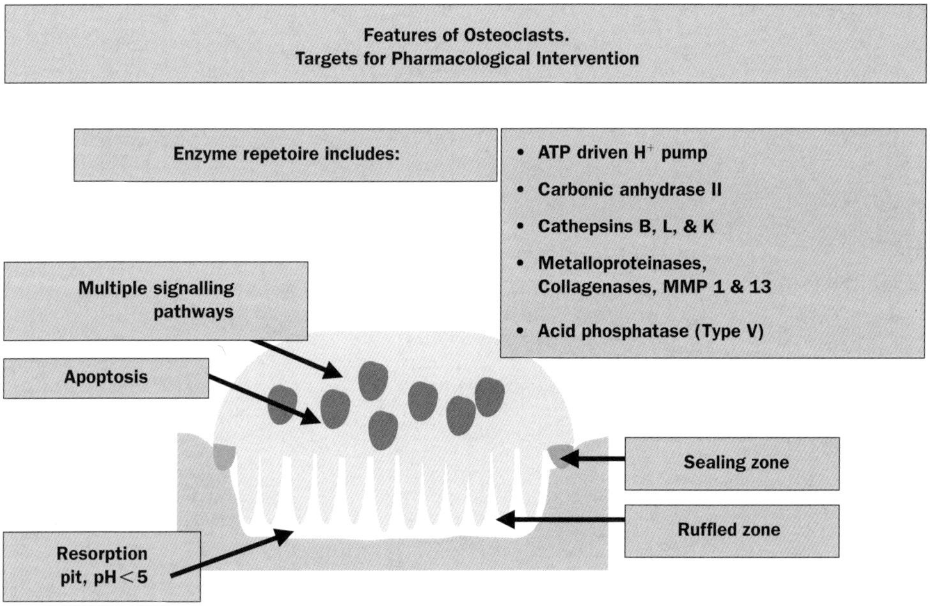

Fig. 1.5 The main features of osteoclasts involved in bone resorption.

Osteoclasts differentiate from haematopoietic stem cell precursors under the direction of factors that include cytokines such as CSFs (especially m-CSF), interleukins (e.g. IL-1, IL-11), and other factors (Fig. 1.6).

The recent discovery of the RANK/RANK-ligand system has revealed that this is one of the major regulatory systems for osteoclast recruitment and action. RANK and RANKL are members of the TNF and TNF-receptor families respectively.[6]

Many cytokines can affect the differentiation and activity of osteoclasts and activate bone resorption. In pathological states, such as the bone destruction that occurs in rheumatoid arthritis or periodontal disease, the proinflammatory cytokines such as IL-1 and IL-6 and tumour necrosis factors play a prominent role, as well as the RANK/RANK legend system.[28–31]

Nitric oxide (NO) is another endogenous mediator that appears to have complex effects on osteoclast function. It is likely that some of the effects of cytokines, such as those of IL-1 and interferon-gamma (IFN-γ) on bone resorption, may be mediated by changes in the production of NO. There are several isoforms of nitric oxide synthase (NOS). One of these (iNOS) is inducible by cytokines and can be inhibited by glucocorticosteroids.

Osteocytes and the response to mechanical effects on bone

Osteocytes lie embedded within individual lacunae in mineralized bone, and connect with each other via the canicular system. Osteocytes thus form a cellular network, much like a neural network, and are believed to be the cellular system that responds to mechanical deformation and loading in bone. Mechanical forces exert strong influences on bone shape and modelling.[7] Early biochemical responses to mechanical loading may include induction of prostacyclin synthesis and of NO production, and later there are increases in IGFs, changes in amino acid transporters, and eventually increases in new bone formation.[8,9]

The role of apoptosis in the regulation of bone turnover

Apoptosis is emerging as a major means of regulating the lifespan of bone cells of all lineages, osteoclasts, osteoblasts and osteocytes.[10] This may contribute to changes in bone turnover under physiological and pathological conditions. Drugs with adverse effects on bone, such as glucocorticoids, may induce osteocyte apoptosis, while therapeutic agents that inhibit bone resorption, including oestrogens and bisphosphonates, may shorten the lifespan of osteoclasts in this way.

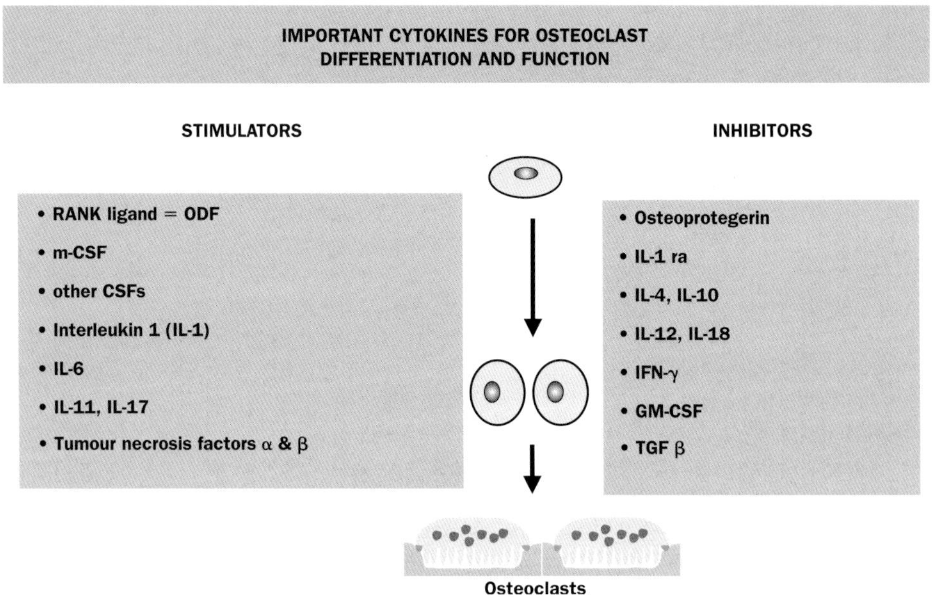

a

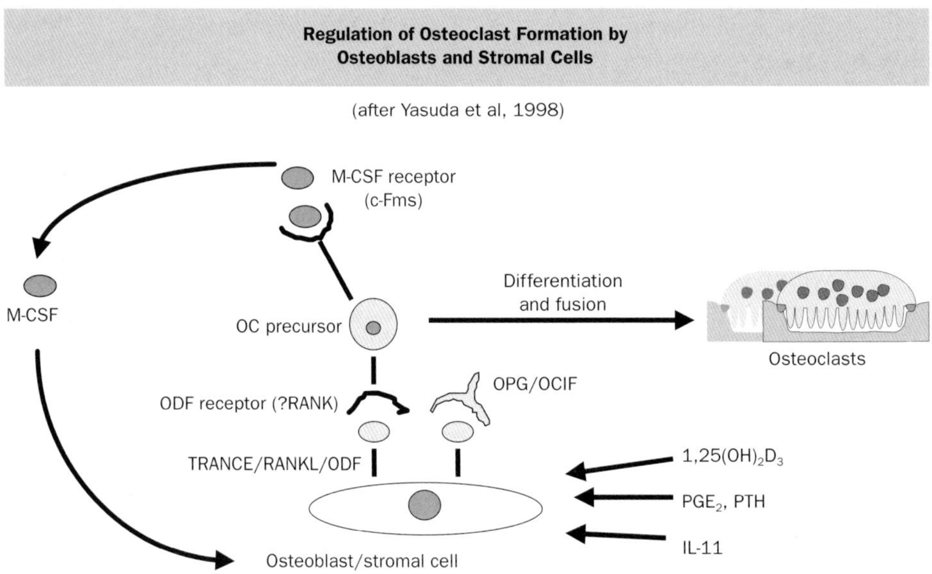

b

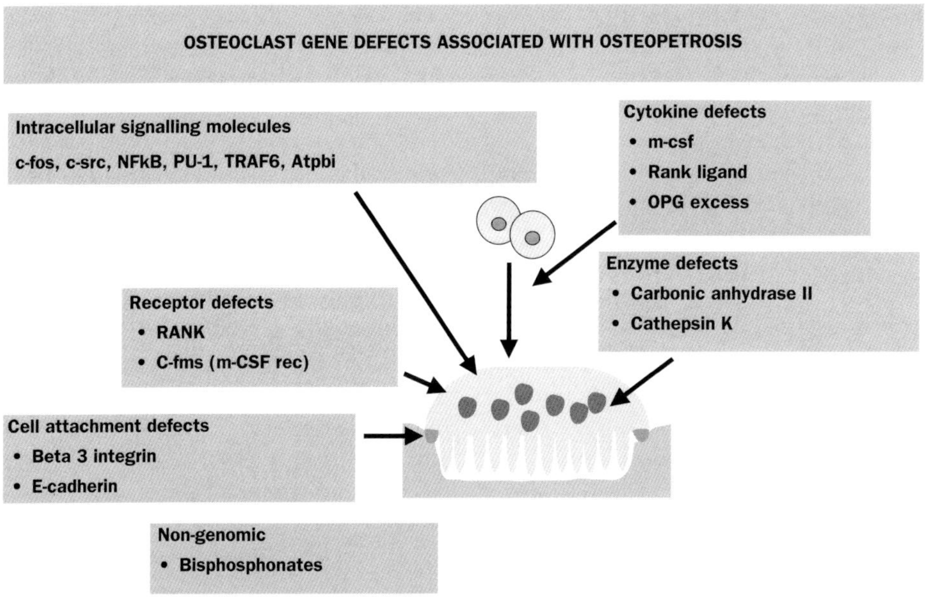

c

Fig. 1.6 The regulation of osteoclast differentiation and action. (a) Activating and inhibitory cytokines; (b) the importance of the RANK/RANK ligand system and its interaction with systemic hormones acting via stromal cells and osteoblasts; (c) various genetic and other factors that inhibit osteclast development and function, and lead to osteopetrosis.

Bone remodelling

During adult life, bone remodelling continues. The remodelling cycle within bone involves a similar sequence of cellular activity at both cortical and trabecular sites. An initial phase of osteoclastic resorption is followed by a more prolonged phase of bone formation mediated by osteoblasts. The amount of bone made under normal conditions corresponds very closely to the amount removed, so that in any remodelling cycle within bone, the total amount of bone tends to remain constant. Even in conditions such as Paget's disease, where there seems to be a primary acquired abnormality of osteoclasts, the subsequent formative and reparative phase of bone deposition is still closely matched to the preceding resorption. The nature of these coupling mechanisms is still poorly understood but they are very important, since minor disturbances in them are likely to contribute to osteopenic or osteosclerotic states. Furthermore, any therapeutic attempts to increase bone mass, in osteoporosis for example, may be difficult to achieve unless these regulatory mechanisms can be circumvented.

In both cortical and trabecular bone, the remodelling process is initiated by the activation of osteoclasts, followed by the filling in of the resorption cavity by bone deposited by osteoblasts. The way in which these events are regulated is still only partially understood, but cytokines and growth factors are likely to be involved.

The processes of bone formation and resorption may be influenced by the exposure to matrix-derived growth factors. Regulation that is achieved by factors attached to matrix, e.g. FGF attached to heparin-like glycosaminoglycans, may be a mechanism for limiting cellular responses to specific sites within bone.

A further way in which the activities of osteoblasts and osteoclasts are coordinated may be through the generation of enzymes such as plasminogen activator (PA) from osteoblasts in response to bone-resorbing agents. Proteolysis of the surface matrix of bone may be an essential step in preparing it for subsequent resorption by incoming osteoclasts. Many of the agents which stimulate bone resorption (e.g. retinoids, PTH, 1,25-dihydroxyvitamin D (1,25D), IL-1) can stimulate the production of PA by osteoblast-like cells. Another mechanism may involve contraction of cells of the lining osteoblast layer in response to resorbing agents such as PTH, thereby allowing access by osteoclasts.

Intrinsic bone matrix proteins may themselves be regulators of cell function. Thus, propeptides of collagen may act as endogenous regulators of procollagen synthesis within connective tissues. Osteocalcin, and matrix Gla protein, and the sialoproteins, e.g. osteopontin, may be chemoattractants, for recruiting cells to the bone surface. Several matrix proteins, including collagen, osteonectin and osteopontin, have the cell receptor binding domain characterized by the Arg–Gly–Asp (RGD) amino acid sequence, and can bind to integrins, which are transmembrane receptors on cell surfaces.

POTENTIAL REGULATORY FACTORS

Some of the regulatory factors that may have important effects on bone and cartilage are described below.

The Interleukin-1 (IL-1) family

There are two major forms of IL-1, IL-1α and IL-1β, derived by complex proteolytic cleavage from larger precursors. IL-1 can be produced in large amounts by monocytes and macrophages and is a major mediator of inflammatory responses. Its many actions include the induction of pyrogenic responses, and the stimulation of an array of other cytokines, e.g. IL-2 and IL-6.

IL-1 also stimulates the synthesis of prostaglandins, particularly via the inducible form of cyclooxygenase (Cox II), which in turn is suppressible by glucocorticosteroids.

IL-1 can stimulate several types of cell, e.g. synovial cells, chondrocytes and fibroblasts, to secrete proteinases, including metalloproteinases (MMPs) such as collagenase (MMP-1) and stromelysin (MMP-3), as well as PAs. Collectively, these enzymes can contribute to the breakdown of connective tissue matrices. IL-1 is one of the most potent known inducers of bone resorption and was one of the first osteoclast-activating factors (OAFs) to be characterized. IL-1 causes hypercalcaemia when injected, and may account for bone resorption associated with monocytic leukaemias and inflammatory erosive conditions in bone, e.g. osteomyelitis, rheumatoid arthritis, and periodontal disease. IL-1 also has complex effects, often inhibitory, on the production of matrix components, including collagen, osteocalcin and proteoglycans.

A third member of the IL-1 gene family located in the same gene cluster on chromosome 2 is the IL-1 receptor antagonist (IL-1ra), which acts as a naturally occurring inhibitor of IL-1 by blocking binding of IL-1 to its receptor. IL-1ra can inhibit the bone loss occurring after ovariectomy in rats, suggesting that IL-1 mediates the bone loss seen after oestrogen deprivation. IL-1ra has also been studied as a therapeutic agent in humans against septic shock and in rheumatoid arthritis (RA).

Tumour necrosis factors (TNFs)

TNFs also exist in two forms, TNF-α and TNF-β, which display extensive homology and have a very similar spectrum of activity to each other and to IL-1.

TNF-α was originally isolated on the basis of activity termed cachectin, thought to be responsible for weight loss in tumour-bearing animals. TNF-β, also known as lymphotoxin, is derived from T lymphocytes and may mediate their cytotoxicity.

The effects of TNFs on connective tissues are very similar to those of IL-1 and include induction of prostaglandin and MMP synthesis, the stimulation of bone and cartilage resorption, and mitogenesis. On a molar basis, IL-1 is more potent than TNF-α, but the two can act in a synergistic fashion under certain circumstances. They can also induce the production of each other and of other cytokines.

The TNF molecules exist as trimers, and their biological activities can be neutralized by soluble shed receptors, of which there are two forms (p55 and p75). Therapeutic neutralization of TNF activity can be achieved with antibodies or by soluble receptor constructs, and has shown encouraging clinical responses in patients with RA.

Interleukin-6 (IL-6)

IL-6 can be produced by many types of connective tissue cells, including human bone and cartilage cells, and by monocytes. IL-6 is now known to be responsible for several biological activities previously ascribed to separate factors, and has some important overlapping activities with IL-1 and TNF.

IL-1, TNF and lipopolysaccharides (endotoxin) are strong inducers of the production of IL-6 from several cell types, including bone cells, so that some biological effects attributable to IL-1 and TNF may be mediated via IL-6. The most important of these may be the augmentation of hepatic acute-phase protein synthesis, and IL-6 is probably an important mediator of the acute-phase response.

The function of the large amount of IL-6 that can be produced in bone is not yet clarified. In mice, IL-6 can cause hypercalcaemia and may stimulate bone resorption, but its role in humans is less clear. It appears to be an important growth factor for myeloma cells.

Mice with the IL-6 gene ablated show less bone loss after ovariectomy, suggesting that oestrogens may act via IL-6.

Colony-stimulating factors

Colony-stimulating factors play a key role in haemopoietic differentiation, as does IL-1, and may be induced by other cytokines, e.g. TNF and IL-1.

GM-CSF and M-CSF can also stimulate bone resorption in vitro, and this may be attributable to their effects on osteoclast generation, since both can be shown to stimulate the generation of osteoclast-like cells from bone marrow cultures. Both can also be produced by stromal cells and perhaps osteoblasts, which raises the interesting possibility that bone cells can influence haematopoiesis in the bone microenvironment. Lack of biologically active M-CSF due to a gene defect is responsible for defective macrophage and osteoclast differentiation in the op/op mouse, which has an osteopetrotic phenotype. The bone lesions can be reversed by administration of M-CSF.

Interferon-gamma

Interferons were originally defined by their ability to inhibit viral replication, but are known to have many other effects, particularly on cellular proliferation and differentiation, which has led to their therapeutic use as anti-tumour agents and as immunomodulators.

IFN-γ while sharing some properties with other interferons, appears to differ from IFN-α and IFN-β in terms of its effects on connective tissues. Thus, IFN-γ inhibits bone resorption induced by IL-1 or TNF, whereas it has less effect on resorption stimulated by the classical calciotropic hormones, PTH hormone or 1,25 D.

IFN-γ also opposes other actions of IL-1 and TNF, for example on cell proliferation, on cartilage resorption and on MMP production by chondrocytes. In these respects, it can be viewed as a potential natural antagonist to IL-1 and TNF. IL-4 may have similar effects.

INF-γ also induces the expression of MHC class 2 (HLA-DR) antigens on connective tissue cells, including synovial cells, bone cells and chondrocytes, as it does on macrophages. These changes may allow these connective tissue cells to present antigens and to participate in immune responses, or in other forms of intercellular communication.

Fibroblast growth factors (FGFs)

FGFs are members of a large family of related peptides (15–16 kDa), which are mitogenic for many cell types, including fibroblasts and osteoblasts.

There are acidic and basic forms of FGF based on isoelectric points, and they share some homology with IL-1. Acidic FGF is derived mainly from neural tissue, but basic FGF is one of the heparin-binding growth factors; it appears to be produced by many cell types and is present in bone matrix. The products of at least three oncogenes (including *hst* and *int-2*) appear to be related to basic FGF, and these may be important in the vascularization of tumours. FGFs are powerful angiogenic factors and are important during embryological development, e.g. during limb bud development, during endochondral ossification and in various pathological states.

The importance of FGF in the development of the growth plate has been dramatically illustrated by the identification of the genetic defect underlying achondroplasia as a point mutation causing an amino acid substitution in the transmembrane domain of the type 3 form of the FGF receptor. This is a common form of dominantly inherited dwarfism that occurs by new mutations, and it is therefore remarkable that the genotype seems to be the same in all patients studied.

Insulin-like growth factors (IGFs) and their binding factors

IGFs, as the name implies, are growth factors which share some structural and biological properties with insulin, and were originally isolated from serum.

Insulin-like growth factor I (IGF-I), formerly known as somatomedin C, stimulates the replication of bone cells and chondrocytes, and increases production of matrix constituents.

Both IGF-I and IGF-II are produced within bone itself, and their activity is modulated by specific binding proteins, of which there are at least six, some of which are also under endocrine control.

There may be important species differences in the regulation by IGFs, with IGF-II dominating in human bone, while IGF-I is the major endogenous IGF in rat bone.

Transforming growth factor betas (TGF-β)

TGF-βs are members of a much larger gene family and are produced by a variety of cells and tissues which include bone. The TGF-βs are homodimeric polypeptides (25 kDa) which exists in at least three isoforms (1, 2, and 3) with different primary sequences but similar biological activities. TGF-βs are thought to be important in embryological development and differentiation as well as in connective tissue repair and fibrosis. They comprise one of the most important families of regulatory cytokines in connective tissues and bone.

In vitro, TGF-β stimulates the synthesis of extracellular matrix components, such as type I collagen and fibronectin. In the presence of epidermal growth factor (EGF), TGF-β stimulates the proliferation of a variety of connective tissue cell types, including bone cells. When injected in vivo, TGF-β induces a wound repair response which includes fibrosis and angiogenesis. TGF-β may be involved in inducing pathological fibrosis, e.g. in liver cirrhosis.

When TGF-β is injected in or around existing bone, it produces a remarkable induction of new bone.

TGF-β is one of the few agents known to inhibit the production from fibroblasts of proteinases such as PA and stromelysin (MMP-3), which may be involved in the degradation of extracellular matrix proteins. TGF-β also stimulates the production of inhibitors of PAs (the latter being known as PAIs) and of MMPs (tissue inhibitors of MMPs, TIMPs). These actions of TGF-β may be important in the maintenance of the integrity of extracellular matrix and, in concert with the effects of TGF-β on matrix synthesis, may promote tissue repair.

TGF-β is produced in latent form and can be activated by exposure to an acid environment. The latency is due to the existence of pro-forms, to binding to α2-macroglobulin, and to association with a large protein (135 kDa) denoted LAP, latency-associated protein. It has been suggested, therefore, that active TGF-β may be released from bone matrix during resorption, and thence be available to affect both osteoblastic and osteoclastic activity.

The TGF-β class of growth factors also include the bone-inductive factors known as bone morphogenetic proteins.

Bone morphogenetic proteins (BMPs) and cartilage-derived morphogenetic proteins (CDMPs)

The so-called bone morphogenetic proteins (BMPs) are factors present in demineralized bone matrix which can induce the formation of new cartilage and bone when implanted into various non-skeletal sites in vivo. The characterization of BMPs was a difficult task, partly because of the limitations imposed by the laborious bioassay. BMPs are now known to be important in regulating the differentiation of skeletal and other tissues, and may eventually be of therapeutic use, e.g. to promote repair of fractures and skeletal defects.

Apart from BMP-1, which, curiously, turns out to be a proteinase capable of cleaving the C-terminal propeptide from type 1 collagen, the remaining dozen or so BMPs described are members of the TGF-β family.

The TGF-β family includes the CDMPs, of which three have been described. These are involved in the differentiation of cartilage. Abnormalities in CDMP-1 are associated with brachypodism in mice.

Parathyroid hormone-related peptides (PTHrP)

The discovery of PTHrP has provided a rational basis for understanding the pathophysiological basis of hypercalcaemia of malignancy, in which it had been recognized for a long time that factors resembling PTH itself might be involved (Fig. 1.8).

PTHrP peptides are produced by non-parathyroid tissues such as breast and skin, and may also be a fetal form of PTH. In hypercalcaemia associated with malignant disease, PTHrP has been implicated particularly in breast tumours and squamous cell carcinomas.

Several closely related PTHrP peptides have now been identified, as single-chain peptides containing about 140 amino acids. The N-terminal portion shows considerable homology with PTH itself, and, at present, only a single receptor for PTH and PTHrP has been identified.

An unexpected finding has been that mice with either the PTHrP gene or the PTH-PTHrP receptor gene deleted show defects in the growth plate. Moreover, a rare inherited form of dwarfism in humans, Jansen-type metaphyseal dysplasia, is associated with a substitution of histidine by arginine at a critical position 223 in the PTH-PTHrP receptor.

These observations, together with the evidence that PTHrPs are produced by bone and cartilage cells, indicate that PTHrP may be an important autocrine and paracrine regulator within bone and cartilage.

Prostaglandins and eicosanoids

It has been known for a very long time that prostaglandins have effects on bone under experimental conditions, but their physiological and pathological significance is still not fully resolved. There have been two major recent developments in this field that are likely to have a significant impact on future research. The first is the discovery of a second form of cyclooxygenase (Cox-2), inducible by cytokines and inhibitable by glucocorticoids. Non-steroidal anti-inflammatory drugs (NSAIDs) can now be reclassified based on their different inhibitory potencies for Cox 1 and 2. The second is the characterization of the prostanoid receptors, of which at least four major subtypes exist.

Prostaglandin E_2 is one of the most potent of the prostanoids for induction of bone resorption. The ability of several agents to stimulate bone resorption (e.g. TGF-β, complement components, thrombin, IL-1, TNF) may be mediated, in part at least, by increased prostaglandin synthesis, in many cases probably by inducing synthesis of cyclooxygenase-2. In turn, production of several of these cytokines may be influenced by prostaglandins. In contrast, low concentrations of some prostanoids may inhibit osteoclast actions by directly altering cell mobility.

Prostaglandins, such as prostacyclin (PGI_2), may be involved in the response of bone to mechanical stress, and help to mediate the bone loss associated with immobilization. They may be involved in the localized bone resorption associated with periodontal disease, neoplasia and inflammation. The chronic administration of prostaglandins in vivo may lead to enhanced periosteal and endosteal apposition of bone, e.g. in children with patent ductus arteriosus, and in experimental animals. These potential anabolic effects of prostaglandins have not yet been exploited clinically.

Products of the lipoxygenase pathways, such as leukotrienes, also appear to have marked effects on bone cell function, and may activate osteoclasts directly. Some of these products may be generated in response to proinflammatory cytokines and other agents.

Nitric oxide

NO is another potentially very important endogenous mediator of many tissue responses, e.g. vasodilatation. There are several isoforms of NOS, one of which is inducible by cytokines and inhibitable by glucocorticosteroids. NO appears to have complex effects on osteoclast function, and it is likely that some of the effects of cytokines such as IL-1 and IFN-γ on bone resorption may be mediated by changes in production of NO.

FACTORS AFFECTING THE ACTION OF CYTOKINES AND GROWTH FACTORS

Cytokine and growth factor receptors

Following the identication of the many cytokines and growth factors, their receptors have now been extensively characterized. These also fall into several classes, and in many cases there is more than one receptor isoform for each agonist.

The cellular distribution of receptors helps to determine tissue responsivenes to these factors, and receptor expression itself is modulated by exposure to hormones and cytokines.

In many cases, receptors can be shed by proteolytic action from cell surfaces, and shed receptors may have important roles in neutralizing the activity of extracellular cytokines.

Role of cytokine gene polymorphisms

There is increasing evidence for cytokine gene polymorphisms that may be linked to the occurrence or severity of inflammatory on infectious disease. There are examples of such polymorphisms in the genes encoding IL-1α and IL-1β, IL-1ra, and TNF-α. These genetic variants may result in differences in cytokine production.

Other cellular interactions in bone

There are many potential interactions between the haematopoietic, stromal and immune systems and bone that may be important within the microenvironment of bone in terms of osteoclast recruitment, bone resorption, osteoblastic activity, and haematopoiesis. Many different mediators and mechanisms are likely to be involved in the interactions between these accessory cells and bone, and may therefore play a role in the physiology and pathology of bone. In cancer, there may be bidirectional interactions between tumour cells and bone.

Systemic or local effects of cytokines

Several of the cytokines may have systemic as well as local effects at their site of production. Appropriate immunoassays and bioassays demonstrate that cytokines such as IL-1, TNF, IGF-I, IFNs and IL-6 are present in the systemic circulation, particularly in disease states, e.g. infections, burns, shock, and hepatic failure. In spite of this, not all potential target tissues display a response. For example, changes in bone and cartilage are not seen in all these clinical situations.

The responsiveness of each tissue is likely to depend upon many variables, including the concentration of individual cytokines present, the relative amounts of different cytokines, the potential synergisms and antagonisms among them, the degree of receptor modulation, e.g. by glucocorticoids, and the presence of specific inhibitors, such as shed receptors and autoantibodies.

Lessons from transgenic animals

Since there are so many factors that can potentially act on bone, a major task in contemporary research is to determine how these agents interact and which are the most important under physiological conditions and in different disease states. This task is only just beginning. As already illustrated by M-CSF and PTHrP, gene knockouts or genetic defects can be very informative (Table 1.1).

Some of the transgenic models do not show the expected phenotype, e.g. the lack of a skeletal phenotype in a TGF-β knockout, while overexpression of the IL-4 gene in mice appears to be associated with osteopenia.

Interactions between oestrogens and cytokines, and their relevance to osteoporosis

Osteoporosis is clearly a multifactorial disorder, and much remains to be learnt about the many pathogenic processes that eventually contribute to the bone loss that leads to osteopenia. Although the majority of patients with osteoporosis do not display florid disturbances in immune function, abnormalities in immunoregulation, arising from oestrogen deficiency, may contribute to bone loss in osteoporosis.

The effects of oestrogen in bone are of particular interest in relation to the loss of bone after the menopause in women and the therapeutic use of oestrogen to prevent this. Earlier explanations for the bone-sparing effect of oestrogen include a stimulation of calcitonin production and a resulting decrease in bone resorption; or enhanced availability of $1,25D_3$ leading to increased intestinal calcium absorption. More important effects may include actions on the immune system, culminating, for example, in an inhibition of the release of promoters of bone resorption, as well as direct effects on bone cells through recently discovered oestrogen receptors. Oestrogen receptors exist in bone, although in low concentrations. In osteoporosis, changes in IL-1 production seem to occur. The production of several cytokines (IL-1, TNF, GM-CSF) from monocytes is increased after the menopause and can be suppressed by administration of oestrogens (Fig. 1.8).

There is considerable experimental as well as clinical evidence to suggest that at least some of the effects of oestrogen on bone (and other tissues) may be mediated by oestrogen-dependent changes in the production of cytokines and other mediators. The production of cytokines such as IL-1, TNF-α, IL-6 and GM-CSF, which can all potentially enhance production of osteoclasts and bone resorption, can be suppressed in monocytes by physiological doses of oestrogen (Fig. 1.7). In mice, ovariectomy does not lead to normal rates of bone loss in mice with an IL-6 gene knockout, while in rats, administration of

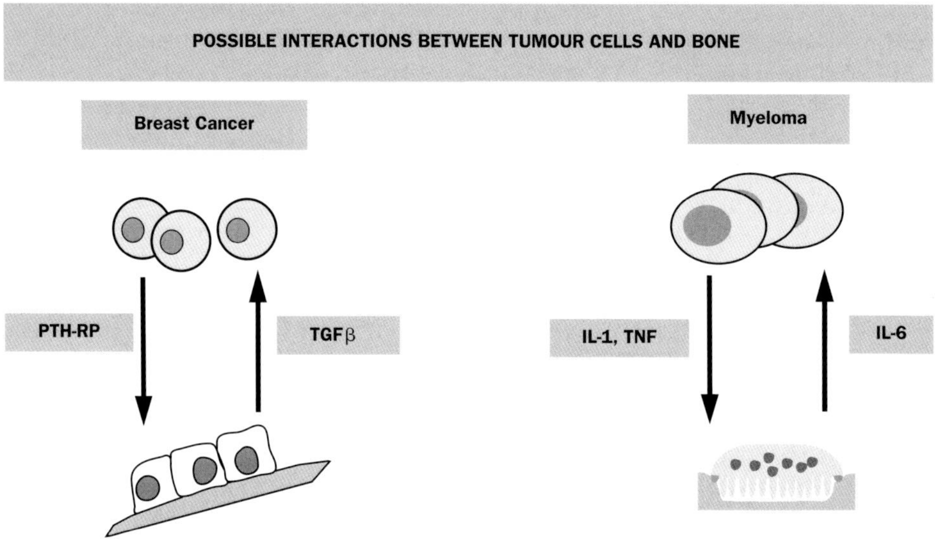

Fig. 1.7 Some of the regulatory mechanisms involved in bone resorption in cancers.

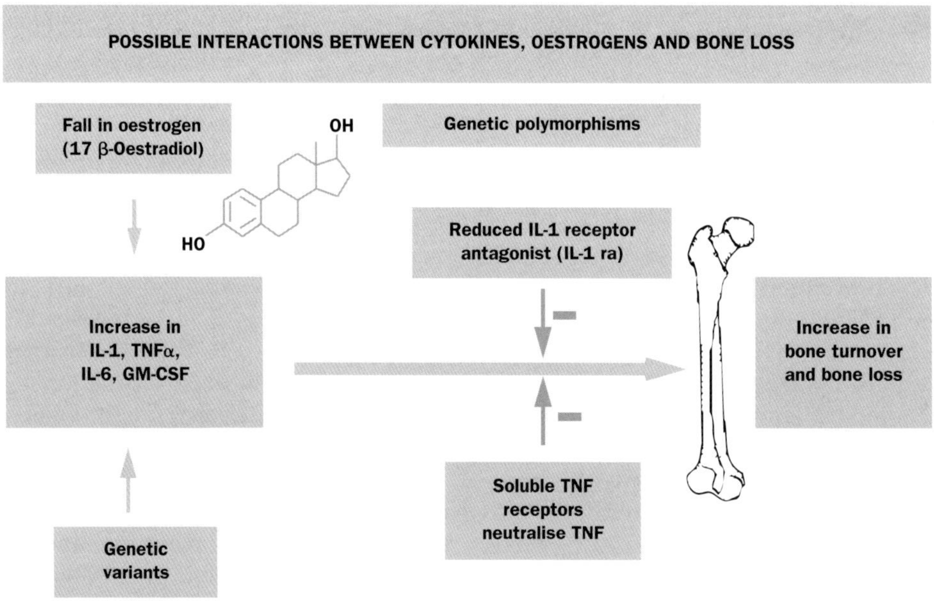

Fig. 1.8 Possible interactions between cytokines, oestrogens and bone loss.

the IL-1 receptor antagonist reduces bone loss after ovariectomy.

It is also possible that oestrogens have significant anabolic effects on bone. Oestrogens can affect the proliferation of osteoprogenitor cells and osteoblast-like cells. Part of this effect may be mediated by oestrogen affecting the production by osteoblasts of a number of potentially locally acting growth factors such as IGFs, and their binding proteins, and TGF-β by bone cells.

Oestrogens may also influence the production of prostaglandin E_2.

The pathogenesis of osteoporosis in men is less well studied than in women but is clinically important, with secondary causes, e.g. hypogonadism being common. The effects of androgens on cytokines are less well characterized.

PROSPECTS FOR THE FUTURE

The rapidly accumulating new knowledge about the multiple possible regulatory mechanisms within bone should aid the understanding of physiological bone remodelling and also offer potential explanations for the changes in bone turnover seen in a variety of disease states. This knowledge will be important in devising new therapeutic strategies to control bone formation and resorption based on these novel regulatory mechanisms.

CLINICAL ASSESSMENT OF BONE STATUS

Bone mass can now be measured very precisely by physical methods such as DXA (dual-energy X-ray absorptiometry), and turnover itself can be assessed using biochemical markers of bone resorption and formation.

Monitoring of bone metabolism by biochemical means depends upon measurement of enzymes and proteins released during bone formation (such as alkaline phosphatase, osteocalcin and collagen propeptides), and of degradation products produced during bone resorption. The most useful markers of bone resorption are degradation products derived from the enzymatic hydrolysis of type I collagen, particularly peptides related to regions of cross-linking with the pyridinolines.

THE MAJOR DISEASES OF BONE RESORPTION

Diseases of diminished bone resorption

The most florid examples of these are the various osteopetroses. There are now many examples of these in rodents, where the gene defects are known and in most cases have been generated in transgenic animals. Examples include mice lacking RANK or RANKL, or with defective intracellular signalling pathways, such as occur after gene knockouts for *src*, PU-1 and other factors.[11–14] Osteopetrosis can be due to lack of differentiation of osteoclasts, or to their failure to function normally. A human disorder in which the defect is characterized is pycnodyostosis, a form of osteopetrosis that the French painter, Toulouse-Lautrec, is thought to have suffered from, and which is associated with defective cathepsin K.

Diseases of increased bone resorption

Paget's disease

Paget's disease of bone is characterized by enlargement and deformity of bones as a result of markedly increased rates of remodelling. It is common and affects men and women with approximately equal frequency. There are marked geographical differences in prevalence, and it can affect up to 5% of people over 50 years of age, in certain parts of the UK, such as Lancashire.

The primary defect in Paget's disease appears to reside in the osteoclast. These cells are bigger than normal, more plentiful, and have more nuclei. The reasons for these changes are incompletely understood. Inclusion bodies occur in the osteoclasts from Paget's disease, and also in some other disorders of bone resorption. This has led to the notion that Paget's disease has a viral origin, with members of the paramyxovirus family as causative agents. Measles and respiratory syncytial viruses have been implicated, and there has been much speculation about the potential involvement of dog distemper virus. The case is unproven, and current work is focusing on the role of genetic factors. In many patients there is a positive family history, and several gene loci have been associated with susceptibility. Recently, a mutation in the receptor, RANK, has been found in one Paget's family, and also in a large kindred with the rare resorption disorder called familial expansile osteolysis (FEO). These mutations may cause constitutive activation of RANK, which could contribute to the increased differentiation and activity of osteoclasts.

Myeloma

Myeloma is a B-cell neoplasm which produces characteristic and substantial destruction of bone, with complications that include hypercalcaemia, and pain and fractures. Myeloma represents a paradigm of the 'seed-soil' concept of the bi-directional interaction between tumour cells and the bone cells of the host. Thus myeloma cells generally have OAFs, including IL-1 and TNF, which stimulate resorption, while host bone cells produce factors such as IL-6, which stimulate the growth of myeloma cells.

Bone metastases

Bone metastases commonly occur from dissemination from breast cancer and other tumours, including lung and kidney. In breast cancers, bone resorption may be stimulated by factors such as PTHrp, which also acts as a hormonal mediator of hypercalcaemia of malignancy.

Inflammatory bone loss

Local bone loss is a feature of several inflammatory diseases. Thus, erosive conditions occur in bone in osteomyelitis, rheumatoid arthritis, and periodontal disease. The pathogenic mechanisms are gradually being elucidated. Inflammatory cytokines such as IL-1 and TNF are probably key, and recent studies show that synovial cells may express RANKL. Reactive oxygen species (ROS), such as hydrogen peroxide and oxygen radicals, have been implicated as resorption mediators in the anoxic environment of rheumatoid joints.

Osteoporosis

The past decade has witnessed a remarkedly greater awareness of osteoporosis as a major health problem that is associated with profound socio-economic consequences. There have been impressive advances in understanding the epidemiology and pathogenesis of osteoporosis and its associated fractures, in the application of physical and biochemical methods to its diagnosis and evaluation, and in the therapeutic approaches to prevention and treatment of postmenopausal and other forms of osteoporosis. There are several recent good reviews.[15–17] Despite these advances, much remains to be done, and the development of better and more cost-effective methods of treatment must remain a high priority.

The essential features are bone loss, particularly due to lack of oestrogen after the menopause in women. Although osteoporosis is usually thought of as a woman's disease, it also occurs in men at about one-quarter of the rate in women (Fig. 1.9).

CURRENT THERAPY AND DRUGS IN DEVELOPMENT

Table 1.2 provides a list of current therapies and those likely to become available in the near future.

Current therapies[18] are based on drugs that are inhibitors of bone resorption and remodelling rather than stimulators of bone formation. The major current therapies include vitamin D and calcium supplements, oestrogens and related compounds, and the bisphosphonates.

Oestrogens and SERMs

Recent advances in the uses of oestrogens

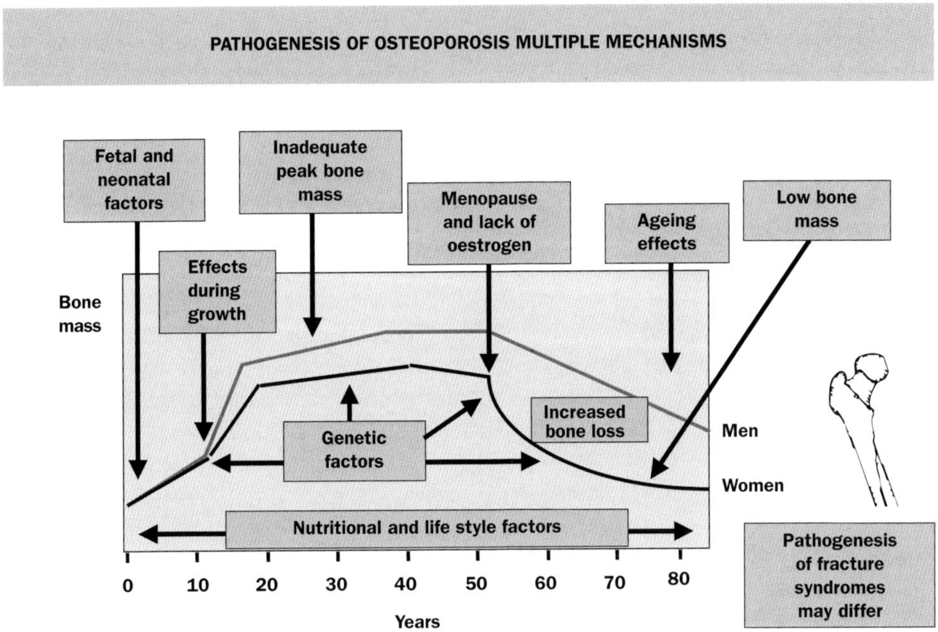

Fig. 1.9 Some of the multiple pathogenic mechanisms that contribute to osteoporosis.

include the development of novel delivery systems, particularly by the transdermal route. Future developments may include additional novel routes of administration, e.g. pulmonary inhalation.

One of the more interesting recent achievements has been the development of more tissue-selective oestrogens or SERMs (selective oestrogen receptor modulators). The ideal compound of this type would possess all the good properties of oestrogens but none of the bad.[19] Such an agent would therefore be effective in osteoporosis, ischaemic heart disease, and Alzheimer's disease, without adverse effects on the breast or uterus in terms of increasing cancer risk, and with no risk of inducing venous thromboembolism. The first compound in this class is raloxifene. Although raloxifene does not yet fulfil all these criteria, it is effective in reducing bone loss and vertebral fracture, and also the incidence of newly diagnosed breast cancers. Other SERMs may follow relatively soon (Fig. 1.10a).

It is encouraging that the opportunities for chemical innovation in this area are enormous. The biological basis for tissue selectivity is beginning to be understoood. Although there are now known to be at least two functional oestrogen receptors, α and β isoforms, the differential tissue distribution of these receptors does not account for the differences in action

Table 1.2 Some of the genes that may contribute to osteoporosis by influencing bone mass and rates of bone loss (Ralston, 1997).
Vitamin D receptor
Oestrogen receptor
Sp1 site in Alpha 1 chain of Type 1 collagen
TGF–beta
Parathyroid hormone receptor
Interleukin 1 receptor antagonist
Others (mice etc)

Table 1.3 Current and future drugs for osteoporosis.

Current treatments	Future treatments
• ESTROGENS (HRT) – Several types, with or without progestogens • SELECTIVE OESTROGEN RECEPTOR MODULATORS (SERMs) – Raloxifene • BISPHOSPHONATES – Etidronate (Didronel) as cyclical therapy – Alendronate or Risedronate as continuous therapy • CALCIUM, often plus Vitamin D • CALCITRIOL, 1,25 dihydroxy vitamin D • CALCITONINS – Nasal rather than injectable formulations may be preferred • Other drugs not specifically registered for osteoporosis (fluoride, anabolic steroids, testosterone in men, etc)	• NEW ESTROGEN FORMULATIONS – e.g. pulmonary • NEW ESTROGEN ANALOGUES (SERMs) – Lasofoxifene • 'NEW' BISPHOSPHONATES – Zoledronate, Ibandronate • PARATHYROID HORMONE and analogues (PTH–RP peptides) • CALCITONINS – New formulations and analogues, e.g. oral or nasal formulations • OTHERS e.g. Strontium salts, Ipriflavone, Growth Factors • NEW DISCOVERIES

among the known SERMs. It is likely that the differential tissue effects are related to altered conformations of the oestrogen receptors, followed by different associations with other transcriptional regulatory proteins in individual cell types. In theory, it may be possible to develop drugs with the ideal properties if the basic biology allows it.

Bisphosphonates

The most widely used drugs for all disorders of increased bone resorption, including osteoporosis, are the bisphosphonates (BPs). Recent elucidation of their mode of action, together with the rapidly increasing knowledge of regulatory mechanisms in bone biology, offer many opportunities for developing new therapeutic agents.

BPs are now well established as successful antiresorptive agents for the prevention and treatment of osteoporosis.[20,21] In particular, etidronate, alendronate and, more recently, risedronate are approved therapies in many countries, and can increase bone mass and reduce fracture rates at the spine, hip and other sites by up to 50% in postmenopausal women.

The use of BPs in osteoporosis is relatively recent compared with the many years of experience in other diseases such as Paget's disease of bone, and bone metastases, for which compounds such as pamidronate and clodronate have been used extensively.

The clinical pharmacology of BPs is characterized by low intestinal absorption, but highly selective localization and retention in bone.

In osteoporosis, the effects of BPs on bone mass may account for some of their action in

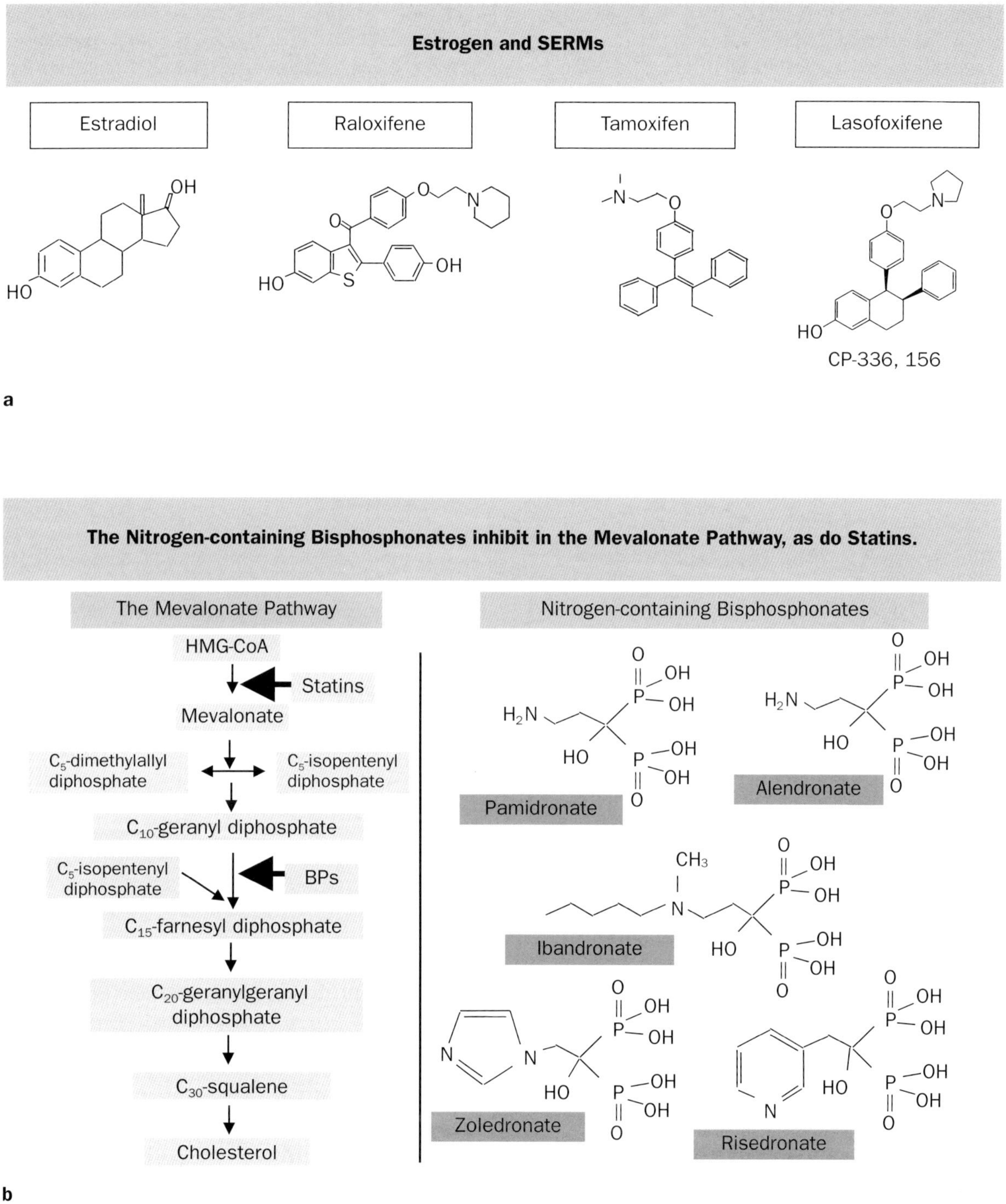

Fig. 1.10 Chemical structures of drugs relevant to osteoporosis. (a)The SERMs raloxifene, tamoxifen, and lasofoxiphene are shown alongside the natural oestrogen, oestradiol. (b)The two bisphosphonates shown here are the nitogen-containing group that appear to act by inhibiting the metabolism of mevalonate to isoprenoids.

reducing fractures, but it is likely that their ability to reduce activation frequency, and possibly to enhance osteon mineralization, may also be related to the reduction in fractures.

Current issues with BPs include the choice of therapeutic regimen, e.g. the use of intermittent dosing rather than continuous, intravenous versus oral therapy, the optimal duration of therapy, the combination with other drugs such as oestrogens, and their extended use in related indications, e.g. glucocorticosteroid-associated osteoporosis, male osteoporosis, childhood osteopenic disorders, and arthritis. There is therefore much that needs to be done to improve the way in which existing drugs can be used as well as introducing new ones.

BPs inhibit bone resorption by being selectively taken up and adsorbed to mineral surfaces in bone, where they interfere with the action of osteoclasts. It is likely that BPs are internalized by osteoclasts and interfere with specific biochemical processes. Recent mechanistic studies show that BPs can be classified into at least two groups with different modes of action. Those that most closely resemble inorganic pyrophosphate (PPi), e.g. clodronate and probably etidronate, can be incorporated into toxic ATP analogues, whereas more potent nitrogen-containing BPs, such as alendronate and risedronate, interfere with other reactions, e.g. in the mevalonate pathway, and may affect cellular activity such as apoptosis by interfering with protein prenylation, and therefore the intracellular trafficking of key regulatory proteins.[22,23] There may therefore be subtle differences between compounds in terms of their clinical effects (Fig. 1.10b).

Calcitonin

Despite many years of use in some countries, there is still controversy about whether calcitonin reduces fractures in osteoporosis, although recent studies suggest that this may be the case for selected doses of salmon calcitonin. With more powerful and less expensive options available for inhibiting bone resorption, the impetus behind developing new therapies based on calcitonin or its peptide analogues has diminished, although there is still some interest in improving drug delivery, e.g. by the oral route.

Other antiresorptive compounds

Other compounds currently under study include ipriflavone, a synthetic isoflavenoid similar to those derived from soya beans and with some oestrogen-like properties. The potential place for other phyto-oetrogens in the management of osteoporosis remains unclear, although differences in the dietary intake of such compounds may contribute to the geographical variations observed in the prevalence of osteoporosis.

The need for anabolic agents: parathyroid hormone, high-dose oestrogens, strontium salts, and other options

It would be a significant advance if anabolic agents could be developed that would enhance the formation of new bone and therefore produce bigger changes in bone mass and strength than can be achieved with current drugs, such as the BPs and oestrogens, which seem to only inhibit bone resorption. There is some evidence, both from experimental studies in animals and from clinical studies in women, that oestrogens may have a significant anabolic effect on bone, particularly at higher doses.

The only anabolic agent that has been used over many years in some countries is fluoride, which at appropriate doses produces an increase in bone mass. However, convincing effects on reducing fractures have not been seen in trials performed so far, and there is little prospect of substantial backing from industry for an agent that cannot be protected from generic use by patents. With regard to the future, the remarkable effect of fluoride may provide clues to biochemical mechanisms that might be exploited to produce anabolic effects on bone. At present, the mode of action of fluoride is only partially understood. Its effects on bone mineral to produce fluoroapatite may contribute, but there are probably additional effects on intracellular signalling pathways in bone cells.

The best known of the 'newer' potentially anabolic agents for bone is PTH, which has been studied on and off for about thirty years. There have been several recent initiatives by different companies to see through trials of a sufficient size not only to verify the large effects on bone mass but also to determine whether there is a significant effect on fractures. It now seems likely that PTH will become available as an anabolic agent in the near future.

A strontium salt (strontium ranelate) has been shown to be anabolic in animal and clinical studies, in that it produces increases in bone mass without corresponding increases in bone resorption. The gains in bone mass appear to be greater than can be accounted for by substitution of calcium by strontium in the hydroxyapatite crystal lattice. The mode of action of strontium remains unclear.

Among other known agents that show anabolic effects in experimental models there are several peptide hormones and growth factors. These include IGFs, BMPs, FGFs and amylin. Non-peptide factors such as prostaglandins (e.g. PGE_2), and statins that induce BMP2 also have anabolic effects.[24–26] It is perhaps unlikely that any of these can be utilized directly clinically, but identifying low molecular weight secretagogues for growth factors that might be selective for bone and could be given by mouth is an experimental approach worthy of exploration.

The challenges of developing new treatments

Within the next few years, several more new drugs are likely to be licensed for use in osteoporosis and other disorders of bone resorption. Unfortunately, the process of drug discovery and development is slow. It usually takes at least 10 years from the discovery of a new compound for it to be studied experimentally, for its safety to be established, and for clinical trials to be completed.

The clinical trial stage for drugs in osteoporosis requires 5 years or more from start to finish, in order to meet the current regulatory requirements for demonstrating a reduction in fractures.

Prospects for novel drugs in the future

Looking further ahead, there are several other ways in which novel agents might be developed based on the rapidly increasing knowledge of bone biology and the pathogenesis of osteoporosis. Greater knowledge of the agonists, antagonists and receptors for osteoblasts and osteoclasts and their cellular precursors, as well as defining the functional machinery within these cells, should offer many opportunities for rational drug design. Fig. 1.11 illustrates some of the practical and theoretical options available. Work towards defining the human genome will also contribute to this process of discovery, as will the use of informative transgenic animal models.

Other developments in biotechnology should make the process faster. For example, high-throughput screening methods and combinatorial chemistry have vastly increased the rate at which new pharmacological candidates may be identified.

In any case, we are moving rapidly from an era when only a few drugs were available to one in which more and better methods of treatment should soon exist.

METHODS OF INCREASING BONE MASS
Inhibition of Resorption

- Bisphosphonates
- Calcitonins & synthetic analogues
- Oestrogens & analogues
- SERMs (Post Raloxifene, eg Lasofoxifene)
- Other agents: testosterone, vitamin D and analogues, ipriflavone
- Anti-cytokines (eg IL-1 & TNFs with Abs, IL-1ra, TNF & IL-1 receptor constructs)
- Osteoprotegerin
- Proton (H^+) pump inhibitors
- Calcium receptor modulators
- Nitric oxide modulators
- Statins
- Enzyme inhibitors (eg metalloproteinases, cathepsin K, etc)
- Adhesion molecule inhibitors (RGD peptides)
- Intracellular signalling targets e.g., c-src, TRAFs, NF kB.

a

METHODS OF INCREASING BONE MASS
Stimulation of Bone Formation

- Parathyroid Hormone, PTH-RP & analogues
- Fluoride
- High dose oestrogens
- Strontium salts
- Prostaglandins and mimetics
- Growth Factors (GH, IGFs, FGFs etc)
- Endothelins and analogues
- Androgens
- Calcitriol & analogues
- Amylin & analogues
- Mechanoreceptor modulators eg Glutamate
- Purinergic modulators
- Statins
- Proteosome inhibitors
- Intracellular signalling targets eg SMADs

b

Fig. 1.11 This illustrates some of the many possibilities for drugs to be developed to increase bone mass either (a) by suppressing bone resorption, or (b) by augmenting bone formation.

REFERENCES

1. Favus MJ, ed., *Primer of the Metabolic Bone Diseases and Disorders of Mineral Metabolism*, 4th edn. Philadelphia: Lippincott, Williams and Wilkins, 1999.
2. Rodan GA, Harada S, The missing bone. *Cell* 1997; 89: 677–80.
3. Croucher PI, Russell RGG, Growth factors. In: *Dynamics of Bone and Cartilage Metabolism* (Seibel MJ, Robins SP, Bilezekian JP, eds). Academic Press, 1999: 83–95.
4. Wozney JM, Rosen V, Bone morphogenetic protein and bone morphogenetic protein gene family in bone formation and repair. *Clin Orthop* 1998; 346: 26–37.
5. Yamashita H, Ten Dijke P, Helden C-H, Miyazono K, Bone morphogenetic protein receptors. *Bone* 1996; 19: 569–74.
6. Kong Y-Y, Yoshida H, Sarosi I et al, OPGL is a key regulator of osteoclastogenesis, lymphocyte development and lymph-node organogenesis. *Nature* 1999; 397: 315–23.
7. Lanyon LE, Amplification of the osteogenic stimulus of load-bearing as a logical therapy for the treatment and prevention of osteoporosis. In: *Novel Approaches to Treatment of Osteoporosis* (Russell RGG, Skerry TM, Kollenkirchen U, eds). Springer, 1998.
8. Damien E, Price JS, Lanyon LE, The estrogen receptor's involvement in osteoblasts' adaptive response to mechanical strain. *J Bone Miner Res* 1998; 13: 1275–82.
9. Mason DJ, Suva LJ, Genever PG et al, Mechanically regulated expression of a neural glutamate transporter in bone. A role for excitatory amino acids as osteotropic agents? *Bone* 1997; 203: 199–205.
10. Plotkin LI, Weinstein RS, Parfitt AM, Roberson PK, Manolagas SC, Bellido T, Prevention of osteocyte and osteoblast apoptosis by bisphosphonates and calcitonin. *J Clin Invest* 1999; 104(10): 1363–74.
11. Felix R, Cecchini MG, Fleisch H, Macrophage colony stimulating factor restores in vivo bone resorption in the op/op osteopetrotic mouse. *Endocrinology* 1990; 127: 2592–4.
12. Hofbauer LC, Khosla S, Dunstan CR, Lacey DL, Boyle WJ, Riggs BL, The roles of osteoprotegerin and osteoprotegerin ligand in the paracrine regulation of bone resorption. *J Bone Miner Res* 2000; 15: 2–12.
13. Istova V, Caamano J, Loy J, Yang Y, Lewin A, Bravo R, Osteopetrosis in mice lacking NF-kappaB1 and NF-kappaB2. *Nature Med* 1997; 3: 1285–9.
14. Schwartzberg PL, Xing L, Hoffmann O et al, Rescue of osteoclast function by transgenic expression of kinase-deficient Src in *src*−/− mutant mice. *Genes Dev* 1997, 11: 2835–44.
15. Compston JE, Fogelman I, eds, *Osteoporosis. Key Advances in Clinical Management*, RSM Key Advances Symposia Series, Royal Society of Medicine, 1999.
16. Ralston SH, The genetics of osteoporosis. *Q J Med*, 1997; 90: 247–51.
17. Royal College of Physicians, *Osteoporosis: Clinical Guidelines for Prevention and Treatment*. Royal College of Physicians of London, 1999.
18. Eastell R, Treatment of postmenopausal osteoporosis. *N Eng J Med* 1998; 338: 736–46.
19. Macgregor JI, Jordan VC, Basic guide to the mechanisms of antiestrogen action. *Pharmacol Rev* 1998; 50: 151–96.
20. Bijvoet O, Fleisch H, Canfield RE, Russell RGG, eds, *Bisphosphonates on Bone*. Elsevier Science, 1995.
21. Russell RGG, The bisphosphonate odyssey. A journey from chemistry to the clinic. In: *Phosphorus, Sulfur, and Silicon and the Related Elements*, Proceedings or the XIV International Conference on Phosphorus Chemistry, Ohio, 1998 (Ebetino FH, McKenna CE, eds). 144–6, 1999: 793–820.
22. Fisher JE, Rogers MJ, Halasy JM et al, Mechanism of action of alendronate:geranylgeraniol, an intermediate of the mevalonate pathway, prevents inhibition of osteoclast formation, bone resorption and kinase activation in vitro. *Proc Natl Acad Sci USA* 1999; 96: 133–8.
23. van Beek E, Pieterman E, Cohen L, Lowik C, Papapoulos S, Farnesyl pyrophosphate synthase is the molecular target of nitrogen-containing bisphosphonates. *Biochem Biophys Res Commun* 1999; 264(1): 108–11.
24. Mundy G, Garrett R, Harris S et al, Stimulation of bone formation in vitro and in rodents by statins. *Science* 1999; 286(5446): 1946–9.
25. Rogers MJ, Statins: lower lipids and better bones? *Nature Med* 2000; 6(1): 21–3.
26. Sugiyama M, Kodama T, Konishi K, Abe K, Asami S, Oikawa S, Compactin and simvastatin, but not pravastatin, induce bone morphogenetic protein-2 in human osteosarcoma cells. *Biochem Biophys Res Commun* 2000; 271(3): 688–92.

27. Ducy P, Zhang R, Geoffroy V, Ridall AM, Karsenty K, Osf2/Cbfa1. A transcriptional activator of osteoblast differentiation. *Cell* 1997; 89: 747–54.
28. Gravallese EM, Manning C, Tsay A et al, Synovial tissue in rheumatoid arthritis is a source of osteoclast differentiation factor. *Arthritis Rheum* 2000; 43: 250–8.
29, Gravallese EM, Harada Y, Wang JT, Gorn AH, Thornhill TS, Goldring SR, Identification of cell types responsible for bone resorption in rheumatoid arthritis and juvenile rheumatoid arthritis. *Am J Pathol* 1998; 152: 943-51.
30. Klippel JH, ed., *Primer of the Rheumatic Diseases.* 11th edn. Atlanta, Georgia: Arthritis Foundation, 1997.
31. Romas E, Bakharevski O, Hards DK et al, Expression of osteoclast differentiation factor at sites of bone erosion in collagen-induced arthritis. *Arthritis Rheum* 2000; 43: 821–6.

2

Potential candidates for bone turnover markers—N-telopeptide cross-links of type I collagen (NTX)

Peter R Ebeling

Summary • Introduction • Metabolism of type I collagen in bone • Bone formation markers • Bone resorption markers • Measurement of bone resorption markers • Validation of markers of bone turnover • Clinical utility of biochemical bone turnover markers in osteoporosis • References

SUMMARY

Measurement of N-telopeptide (NTX) cross-links of bone type I collagen in urine or serum reflects the rate of bone resorption. The most important potential clinical use for these tests is in determining an individual's risk of subsequent bone loss and fracture independently of bone mineral density. However, these markers do not substitute for bone density measurement. Another important use is in monitoring the early efficacy of, and compliance with, antiresorptive therapy in osteoporosis. Measurements of urine and serum NTX are specific and clinically useful markers for bone resorption. The NTX epitope may be preferentially liberated from type I collagen in bone by osteoclastic hydrolysis with cathepsin K. It may be degraded further in the liver and kidney, in particular, where free deoxypyridinoline (DPD) may be generated. However, NTX acts as a dynamic marker for bone resorption, and urine NTX excretion is not affected by dietary collagen intake. Serum NTX may also be measured, and values decrease significantly, but less than for urine NTX, following antiresorptive therapy. The effect of dietary collagen intake on serum NTX levels has not yet been assessed. Urine NTX values agree extremely well with values measured by point-of-care devices. Point-of-care devices are able to precisely measure urine NTX in minutes, and allow for the rapid clinical assessment of the efficacy of, and patient compliance with, antiresorptive therapy.

INTRODUCTION

Throughout the skeleton during adult life, bone is continually being remodelled.[1] Old bone is resorbed by osteoclasts, and osteoblasts form new, mechanically stronger bone to replace it. This occurs at sites that are mechanically weakened by stress and where stress microfractures may develop. Bone remodelling thus allows the skeleton to respond to biomechanical forces. The actual stimulus for initiation of a new bone remodelling cycle is unknown. Normally, bone resorption is closely coupled with bone formation, so that there is no net bone loss during the remodelling cycle. When this balance is upset, bone loss occurs. In most bone diseases (osteoporosis, hyperthyroidism, hyperparathyroidism), bone resorption is increased more than bone formation,[2,3] although net bone loss may also occur when the bone turnover rate is reduced.

Changes in the bone turnover rate are important determinants of bone disease, and measurements correlating with this rate could theoretically be very useful in assessing patients with metabolic bone disease. Previously, the only way to accurately assess bone turnover was with bone histomorphometry following double-labelling with tetracycline.[1,4] However, this technique is invasive and expensive, and may not reflect bone turnover rates at other skeletal sites. Nevertheless, it does provide information regarding static and dynamic rates of bone formation and bone resorption, and the percentage of the bone undergoing active bone formation and bone resorption.

As well as liberating calcium and phosphorus during bone resorption, osteoclasts also digest bone type I collagen, which is released as a series of peptides that can be measured in the serum and urine to quantify the bone resorption rate of the entire skeleton. Similarly, new bone formation leads to the production of collagen cleavage products and other bone matrix proteins that are elaborated during different stages of osteoblast differentiation.

METABOLISM OF TYPE I COLLAGEN IN BONE

Type I collagen is formed in bone from the combination of two $\alpha 1$ and one $\alpha 2$ collagen polypeptides containing hydroxylated proline and lysine residues. This structure is known as procollagen. As procollagen is secreted from the osteoblast, the N-terminal and C-terminal regions are cleaved, and these propeptides are released into the extracellular fluid, although a proportion of the N-terminal propeptide is also incorporated into bone. Type I collagen is helical; the non-helical domains at the N- and C-termini are known as the N-telopeptide (NTX) and C-telopeptide (CTX) regions.

Side-chains of three hydroxylysine residues from different type I collagen molecules condense to form a pyridinium ring so that pyridinium cross-links are formed connecting three different collagen molecules, stabilizing the structure of type I collagen. Pyridinoline cross-links result from the combination of three hydroxylysine side-chains (hydroxylysylpyridinoline),[5] while deoxypyridinoline (DPD) cross-links result from the combination of two hydroxylysine side-chains and one lysine side-chain (lysylpyridinoline).[6] Pyridinoline cross-links occur in many types of collagen outside bone (including cartilage); however, DPD is relatively bone-specific, but can be found in other tissues such as dentine, skeletal muscle and the aorta (Table 2.1).[7,8] The pyridinium cross-link in the NTX region joins $\alpha 1$ type I collagen polypeptides to $\alpha 2$ type I collagen polypeptides;[9] by contrast, pyridinium cross-links in other tissues join $\alpha 1$ type I collagen polypeptides to $\alpha 1$ type I collagen polypeptides (e.g. C-telopeptide). This makes the NTX relatively bone-specific. In addition, two-thirds of deoxypyridinoline cross-links in bone type I collagen are NTX cross-links and only one-third are CTX cross-links.

BONE FORMATION MARKERS

The predominant product of osteoblasts is type I bone collagen, which comprises 95% of the extracellular non-mineral bone matrix. Other proteins such as osteopontin, osteonectin and osteocalcin are also secreted to form the osteoid or organic substrate in which mineralization occurs. Osteoblasts can be identified by staining for the enzyme alkaline phosphatase, attached to their cell membranes. This alkaline phosphatase is functionally similar but antigenically different to hepatic, intestinal or placental alkaline phosphatases.[10]

Table 2.1 Potential candidates—N-telopeptide cross-links of type I collagen (NTX).

Type I collagen	Bone, dentin
	Ligaments, fascia, tendon
	Aorta, vascular walls
	Skeletal muscle, intestine, skin
Deoxypyridoline cross-links	Bone, dentin (3.5 : 1)
	Vascular tissues and skeletal muscle (7 : 1)
N-telopeptide cross-links	67% of bone collagen DPD
	A specific product of osteoclastic bone resorption?

Bone-specific alkaline phosphatase

The serum concentration of bone-specific alkaline phosphatase (bone ALP) reflects the cellular activity of osteoblasts.[10–12] The ideal assay for bone ALP should have low cross-reactivity with other ALP isoenzymes, particularly the hepatic isoenzyme.

Osteocalcin

The serum concentration of osteocalcin reflects the rate of osteoblast synthesis of osteocalcin. However, only approximately 50% of newly synthesized osteocalcin is released into the circulation, while the remaining 50% is incorporated into hydroxyapatite. It has been proposed that another form of osteocalcin, undercarboxylated osteocalcin, may be an independent risk factor for hip fracture.[13]

Type I procollagen C-terminal propeptide

The serum concentration of the C-terminal propeptide of type I collagen (PICP) reflects changes in synthesis of new collagen, both by osteoblasts in bone and by fibroblasts in other connective tissues.[14] It is secreted into the circulation in its entirety.

Type I procollagen N-terminal propeptide

The serum concentration of the N-terminal propeptide of type I collagen (PINP) also reflects changes in synthesis of new collagen, both by osteoblasts in bone and by fibroblasts in other connective tissues.[15,16] However, unlike PICP, a proportion is also incorporated into bone as non-dialysable hydroxyproline and, thus, a component of the measured fragments might represent bone resorption. Nonetheless, PINP appears to be a more dynamic marker of changes in bone formation than PICP.

BONE RESORPTION MARKERS

This chapter will focus on the potential role of degradation products of type I bone collagen as biochemical markers of bone resorption. In particular, the utility of NTX cross-links as a bone resorption marker will be evaluated. Bone resorption is initiated by osteoclasts, which contain acid phosphatase.[1,17] Although acid phosphatase activity is present in other tissues, such as the prostate gland and blood cells, the type Vb osteoclast enzyme can be recognized by its insensitivity to inhibition by tartrate (tartrate-resistant acid phosphatase, TRACP).

Osteoclasts attach to the bone surface and secrete acid and hydrolytic enzymes that resorb

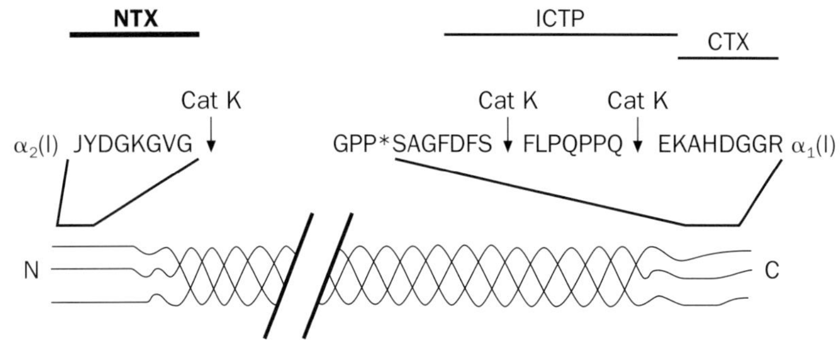

Fig. 2.1 Cleavage sites for cathepsin K in type I collagen and the peptides liberated by cleavage.[20]

bone, releasing bone minerals and collagen fragments. The NTX epitope is preferentially liberated from type I collagen in bone by osteoclastic hydrolysis with cathepsin K.[18] Cathepsin K is located in osteoclast intracellular vacuoles and in the subosteoclastic space. Within type I collagen telopeptide domains, there are three proteolytic sites for cathepsin K.[19] (Fig. 2.1). In addition, both serum NTX and CTX concentrations are reduced in pycnodyostosis, a disorder of reduced cathepsin K activity, while serum ICTP levels and urine free DPD excretion are elevated in this disorder.[20]

NTX cross-links may be degraded further in the liver and kidney in particular, where free DPD may be generated by the kidney. Serum NTX concentrations are elevated in chronic renal failure. Some of the collagen is completely digested by osteoclasts to its smallest units, free pyridinoline and deoxypyridinoline residues, which are excreted in the urine. The majority, however, appears to be incompletely digested, resulting in formation of pyridinium cross-links bound to fragments of the NTX α1 and α2 polypeptides; peptide-bound cross-links are also excreted in the urine.[9,21,22]

Table 2.2 N-telopeptide cross-links of type I collagen (NTX)—NTX assays.

Urine ELISA assay—hours to days
2nd morning void with no dietary restriction
Intra- and interassay CV ~7%
Point-of-care device—minutes
Urine assay
Serum NTX—hours to days
Avoids correction for urine creatinine

Urine calcium/creatinine ratio

The urinary excretion of calcium provides a measure of the rate of dissolution of bone, but also depends upon the renal threshold for calcium (determined in part by parathyroid hormone, PTH) and dietary calcium intake.

Urine hydroxyproline/creatinine

The urinary excretion of hydroxyproline reflects breakdown of collagen in bone, but is also influenced by the breakdown of collagen in other sites (cartilage, skin, components of the complement system) and, most importantly, by dietary collagen intake.[23]

Skeletal acid phosphatase

The serum concentration of skeletal acid phosphatase can be measured, both by immunoassay[17] and as tartrate-resistant type Vb acid phosphatase enzyme activity.[24]

Urine pyridinium cross-links of type I collagen

Urinary excretion of the hydroxypyridinium cross-links of type I collagen reflects bone resorption, and not dietary calcium or collagen intake. Thus, they are more precise indicators of bone resorption than urinary calcium or hydroxyproline excretion. Furthermore, because total DPD and the peptide-bound NTX and CTX pyridinium cross-links are almost exclusively derived from bone type I collagen, measurement of either acts as a specific marker of bone resorption. DPD is also found in relatively high concentrations in vascular tissue, such as the aorta, and in skeletal muscle; however, the metabolic turnover rate of these tissues is far lower than for bone. As a result, these tissues contribute little to the circulating pool of DPD.

MEASUREMENT OF BONE RESORPTION MARKERS

There are many different ways to measure these metabolites of collagen. Hydroxyproline is measured in a cumbersome colorimetric assay.[24] The original assays for the hydroxypyridinium cross-links required high-performance liquid chromatography (HPLC).[20] More recently, antibodies have been raised to various regions of the hydroxypyridinium cross-links, and they can now be measured in urine by a number of commercial assays. Both pyridinoline and DPD are excreted in the urine in both free (40%) and peptide-bound (60%) forms, and the ratio of free and bound forms is constant in normal individuals and in patients with metabolic bone disease.

High-performance liquid chromatography

Total (bound and free) pyridinoline and DPD is measured by HPLC following acid hydrolysis of urine samples and extraction by cellulose chromatography. Natural fluorescence of these amino acids is then compared with that of a known standard. Free cross-links can also be measured in unhydrolysed urine by HPLC. Two problems of this assay are the complexity and the lack of a suitable internal standard. External standards are derived from ovine or human bone and calibrated against purified pyridinoline.[25] Derivatized pyridinoline or isodesmosine, an elastin degradation product, are possible internal standards, but both are problematic with regard to their stability and fluorescent properties.

Enzyme-linked immunoabsorbent assays (ELISA)

Free DPD can be measured in the urine by ACS-DPD and Pyrilinks-D immunoassays,[26] and free DPD plus pyridinoline can be measured by the Pyrilinks immunoassay.[27] A coated-tube radioimmunoassay using the same antibody as that for Pyrilinks-D can be used to measure both free DPD in unhydrolysed urine and total DPD in hydrolysed urine. Free DPD has also been automated using Immulite technology.

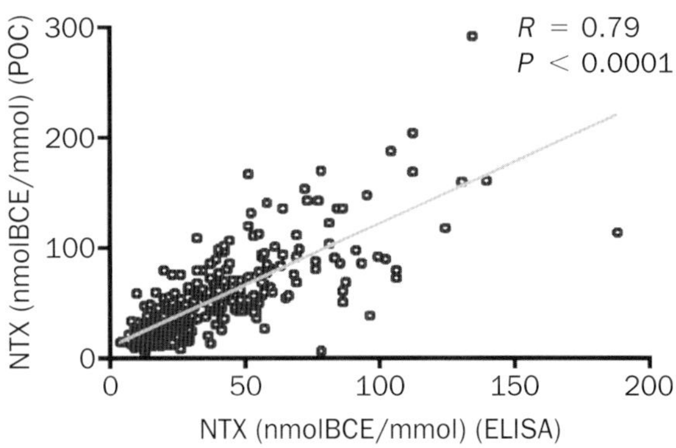

Fig. 2.2 Comparison of N-telopeptide levels measured by ELISA and point-of-care assays.[31]

Peptide-bound N-telopeptide and C-telopeptide cross-link assays

Peptide-bound NTX cross-links can be measured by an immunoassay called Osteomark,[9] and CTX cross-links can be measured in serum by an immunoassay called ICTP and Cross-Laps.[28,29] The monoclonal antibody used for the NTX assay is directed against a urinary pool of collagen cross-links derived from a patient with Paget's disease. Only the β-isomer of CTX is measured in the serum Cross-Laps assay, while both α- and β-isomers of CTX are measured in the urine Cross-Laps assay. The Cross-Laps immunoassay does not measure cross-links per se, but measures a sequence (octapeptide) of the CTX region of the α1 chain of type I collagen in the urine.[29] The NTX immunoassay has also been automated and modified to measure the NTX epitope in a serum ELISA assay. A recent clinical assessment and comparison of novel serum markers of bone resorption with established urinary indices in different disorders of bone and mineral metabolism showed that it measures deviations from normal as well, if not better, than the urine NTX assay.[30] A relatively inexpensive, single-use point-of-care device has also been recently devised for rapid measurement of urine NTX in a clinical outpatient setting. Measurements of urine NTX using this device can be made within minutes and correlate very well with the urine NTX ELISA assay.[31]

Urine for measurements of bone turnover markers can be collected either as random, non-fasting, 2-h post-voiding samples, or 24-h urine collections. It should be noted that the serum telopeptide assays may be affected by the non-fasting state, so that specimens for these assays should be collected during fasting conditions. The results in 2-h early-morning post-voiding samples correlate well with those in 24-h samples,[32] and rates of bone loss correlate with 2-h urinary DPD values better than with values from 24-h urine collections.[33] Short-term variability of 2-h post-voiding cross-link excretion (9–13%) is less than that of 24-h collections (26%), but differences in long-term variability may be less. Biological variability is also increased in women with postmenopausal osteoporosis. When ordering these tests through a local laboratory, it is important that samples are collected according to their instructions, so that results can be compared with the normal reference ranges established by that laboratory. Exposure of the sample to ultraviolet light should be avoided, because pyridinium cross-links are degraded by ultraviolet light.

VALIDATION OF MARKERS OF BONE TURNOVER

Three major criteria need to be satisfied before a biochemical test can be established as a biological marker of bone turnover:

1. The marker must change in parallel with changes in bone turnover measured by bone histomorphometry or calcium kinetics.[4]
2. The serum concentration or urinary excretion of the substance must be high in conditions characterized by high bone turnover, such as hyperparathyroidism,[1] hyperthyroidism[2] and Paget's disease.[1]
3. The serum concentration or urinary excretion of the substance must be low in conditions characterized by low bone turnover, e.g. after the administration of antiresorptive drugs.

Bone histomorphometry

Bone biopsies are expensive and invasive, and bone turnover measured at one skeletal site may not correlate with bone turnover at other sites. Correlations have been adequate in the studies that have been performed. The correlation coefficients between the results of histomorphometric measurements of bone resorption and urinary excretion of hydroxyproline, pyridinoline and DPD were 0.22, 0.77 and 0.80, respectively.[34–36]

Elevation of markers in conditions characterized by increased bone turnover

The values for most markers are elevated in patients with hyperthyroidism[15,37–39] and hyperparathyroidism,[15,17,40,41] and in postmenopausal women;[1,17,22] urinary DPD values tend to be more abnormal than those of pyridinoline.[32,42]

Most patients with Paget's disease[15,43,44] or osseous metastases[45–47] have high values for most tests, but their serum osteocalcin concentrations are often normal.[39,44,45] Urinary NTX excretion is a very sensitive indicator of bone resorption in patients with Paget's disease[43] and osseous metastases.[47]

Decline in markers after antiresorptive therapy

The serum concentrations or urinary excretion of most bone turnover markers decrease during antiresorptive therapy; however, the magnitude of the reduction depends upon both the therapy and the marker. There are differential effects of antiresorptive therapy on free and peptide-bound cross-link excretion.

First, when postmenopausal women are treated with oestrogen, values for most markers fall by 20–50%,[5,48,49] except for serum ICTP values which remain stable.[48] Second, when postmenopausal women are treated with bisphosphonates, serum ICTP and TRACP values tend to remain stable,[42] urinary total DPD excretion falls more than that of total pyridinoline, and urinary excretion of peptide-bound cross-links (NTX and CTX) falls to a much greater extent (60–80%) than that of free cross-links (30–40%).[19,38,42,43] The fall in serum NTX concentrations following antiresorptive therapy (24.4% and 28.4% for oestrogen and alendronate, respectively) is generally far less than the fall in urine NTX concentrations after the same therapy.[49,50]

The factors responsible for these differences are not known but the greater the fall in bone markers after initiation of antiresorptive therapy, the greater the benefit on bone mineral density (BMD). Peptide-bound cross-links may be further metabolized to free cross-links in the liver and kidney, particularly in the latter.[51] In addition, it is possible that peptide-bound NTX cross-links may be preferentially generated by osteoclasts from bone.[18] It is also likely that oestrogen and bisphosphonates affect osteoclastic bone degradation in different ways.[22]

CLINICAL UTILITY OF BIOCHEMICAL BONE TURNOVER MARKERS IN OSTEOPOROSIS

Biochemical bone turnover markers do not replace dual energy absorptometry (DXA) for the diagnosis of osteoporosis. However, the former may give some indication about the future risk for bone loss and fractures. They are also useful in monitoring the efficacy of antiresorptive therapy in patients with osteoporosis.

Diagnosis

The mean values for biochemical bone turnover markers are higher in patients with osteoporosis than in age- and sex-matched normal subjects. In recent studies, for example, the mean urinary excretion of DPD was 20–100% higher in patients with osteoporosis than in normal subjects.[52–55] Similar results of measurements of other markers are seen;[15,52–54] however, overlap with normal values occur.

In addition, BMD in patients with osteoporosis is inversely related to the levels of bone turnover markers. One study, for example, divided subjects into quartiles according to the excretion of cross-linked NTX of type 1 collagen, a marker of bone resorption.[56] There was an inverse relationship between the quartile of urinary NTX excretion and mean BMD. These findings are consistent with the concept that osteoporosis is characterized by an increase in both bone formation and resorption with resultant net bone loss, but it should be noted that this can occur at any level of bone turnover.

Biochemical markers are not useful in making the diagnosis of osteoporosis, because of overlap of values in normal subjects and patients with osteoporosis. The diagnostic utility of a single measurement is also limited because of its high biological variability. Measurements of markers of bone turnover should not be performed in order to make a diagnosis of osteoporosis.

Prediction of future risk of bone loss and fracture

A patient's current BMD is an important predictor of fracture risk.[57,58] However, a single measurement indicates only current BMD, not the anticipated rate of bone loss. Thus, patients in the middle tertile of BMD may still be at risk of osteoporosis and fracture if they are losing bone more rapidly.

Recent studies have demonstrated that markers of bone turnover may be useful in predicting rates of future bone loss, and may therefore provide independent information about fracture risk beyond that available from BMD measurements alone. In most studies there is a highly significant correlation between bone turnover markers and subsequent rates of bone loss.[42,59–63] The diagnostic utility of a single bone turnover measurement is limited because individuals with low levels of bone turnover have rates of bone loss that range from 0% to 10% per year;[60] the test has a low specificity. Nevertheless, a person with a high value of a bone turnover marker is generally at greater risk of bone loss than a person with a low value.

The relationship between bone markers and fracture risk has been demonstrated by prospective studies. A prospective case–control study found that for every 1 SD elevation of urinary free DPD excretion, the risk of hip fracture was elevated four-fold even after adjusting for age and bone density.[64] Women with urinary free DPD excretion above the normal limits for young women have twice the risk of fracture as compared with other women, even after adjusting for BMD.[65]

Women with the highest bone turnover appear to derive the greatest benefit from antiresorptive therapy with oestrogen,[62] calcium[62] or calcitonin,[66] while by comparison, women with the baseline urinary NTX values in the lowest quartile are no less likely to lose bone during administration of oestrogen than placebo.[62] Thus BMD can be measured at menopause if preventive therapy is being contemplated for women at risk for osteoporosis. If the BMD is in the middle tertile, measurement of a bone turnover marker such as the urinary excretion of NTX or CTX could be made. If the value is above the upper limit of normal for premenopausal women, treatment of the woman with an antiresorptive drug would be recommended.

Use in monitoring efficacy of antiresorptive therapy

The mean BMD in women receiving therapy with antiresorptive agents such as oestrogen, alendronate or calcitonin is stable or increases slightly. However, some women continue to lose bone. This has been estimated to occur in approximately one-third of women receiving oestrogen and one-sixth of those receiving alendronate.[33,61] Women taking antiresorptive therapy could be monitored to aid their compliance and to make sure the therapy is effective.

There are two possible approaches to monitoring antiresorptive therapy with bone densitometry and/or biochemical markers of bone turnover:

1. The BMD can be measured at baseline and again after 1 or 2 years of therapy. However, this approach would not detect early failure of the antiresorptive therapy or a lack of compliance.
2. The BMD and a bone turnover marker can be measured at baseline, followed by a repeat measurement of the marker in 3 months. If the marker falls appropriately, then there is evidence of compliance and drug efficacy, and therapy should be continued for 1 year, when BMD can be remeasured. This approach is supported by

the observation that the greater the fall in bone markers after initiation of antiresorptive therapy, the greater the benefit on BMD.[33,60,62,67] However, the correlation is relatively weak, accounting for only approximately 21% of the variance in change in BMD at 1 year, so its applicability to the individual patient with osteoporosis is uncertain. In addition, significant increases in BMD may be seen as early as 3 months with new, more potent bisphosphonates (alendronate).[68] The anticipated 3-month decline in urine NTX is 20–30% after treatment with calcitonin,[69] 35–50% after treatment with oestrogen,[22,48,60,70] more than 50% after treatment with alendronate[33] and even more with the combination of alendronate and oestrogen.[50]

Predictive value of bone markers

Controversy surrounds the question of which marker provides the most useful information about the subsequent risk of bone loss. The EPIDOS study found that among elderly women, urinary DPD excretion above the upper limit of normal for premenopausal women approximately doubled the fracture risk, while urinary NTX excretion was not predictive.[65] In the OFELY study, markers of bone turnover predicted postmenopausal forearm bone loss over 4 years.[71] Similarly, in a prospective study of early postmenopausal women, high baseline urinary NTX excretion predicted both rapid bone loss in women who were not treated and significant bone gain in those given oestrogen.[72] A low baseline urinary NTX excretion predicted little bone loss without and little benefit with oestrogen therapy; urinary DPD values were not predictive of subsequent bone loss.[73]

REFERENCES

1. Calvo MS, Eyre DR, Gundberg CM, Molecular basis and clinical application of biological markers of bone turnover. *Endocr Rev* 1996; 17: 333–68.
2. Benker G, Breuer N, Windeck R, Reinwein D, Calcium metabolism in thyroid disease. *J Endocrinol Invest* 1988; 11: 61–9.
3. Seeman E, Wahner HW, Offord KP et al, Differential effects of endocrine dysfunction on the axial and the appendicular skeleton. *J Clin Invest* 1982; 69: 1302–9.
4. Parfitt AM, Drezner MK, Glorieux FH et al, Bone histomorphometry: standardization of nomenclature, symbols, and units. *J Bone Miner Res* 1987; 2: 595–610.
5. Horgan DJ, King NL, Kurth LB, Kuypers R, Collagen crosslinks and their relationship to the thermal properties of calf tendons. *Arch Biochem Biophys* 1990; 281: 21–6.
6. Eyre DR, Koob TJ, Van Ness KP, Quantitation of hydroxypyridinium crosslinks in collagen by high-performance liquid chromatography. *Anal Biochem* 1984; 137: 380–8.
7. Robins SP, Collagen crosslinks in metabolic bone disease. *Acta Orthop Scand* 1995; 66: 171–5.
8. Eyre DR, Dickson IR, Van Ness K, Collagen cross-linking in human bone and articular cartilage. *Biochem J* 1988; 252: 495–500.
9. Hanson DA, Weis ME, Bollen A-M et al, A specific immunoassay for monitoring human bone resorption: quantitation of type I collagen cross-linked N-telopeptides in urine. *J Bone Miner Res* 1992; 7: 1251–8.
10. Gomez BJ, Ardakani S, Ju J et al, Monoclonal antibody assay for measuring bone-specific alkaline phosphatase activity in serum. *Clin Chem* 1995; 41: 1560–6.
11. Garnero P, Delmas PD, Assessment of the serum levels of bone alkaline phosphatase with a new immunoradiometric assay in patients with metabolic bone disease. *J Clin Endocrinol Metab* 1993; 77: 1046–53.
12. Hill CS, Wolfert RL, The preparation of monoclonal antibodies which react preferentially with human bone alkaline phosphatase and not liver alkaline phosphatase. *Clin Chim Acta* 1989; 186: 315–20.
13. Vergnaud P, Garnero P, Meunier PJ et al, Undercarboxylated osteocalcin measured with a specific immunoassay predicts hip fracture in elderly women: the EPIDOS study. *J Clin Endocrinol Metab* 1997; 82: 719–24.
14. Melkko J, Niemi S, Ristelli L, Risteli J, Radioimmunoassay of the carboxyterminal propeptide of human type I procollagen. *Clin Chem* 1990; 36: 1328–32.

15. Ebeling PR, Peterson JM, Riggs BL, Utility of type I procollagen propeptide assays for assessing abnormalities in metabolic bone diseases. *J Bone Miner Res* 1992; 7: 1243–50.
16. Melkko J, Kauppila S, Niemi S et al, Immunoassay for intact amino-terminal propeptide of human type I procollagen. *Clin Chem* 1996; 42: 947–54.
17. Cheung CK, Panesar NS, Haines C et al, Immunoassay of a tartrate-resistant acid phosphatase in serum. *Clin Chem* 1995; 41: 679–86.
18. Apone S, Lee MY, Eyre DR, Osteoclasts generate cross-linked collagen N-telopeptides (NTx) but no free pyridinolines when cultured on human bone. *Bone* 1997; 21(2): 129–36.
19. Atley LM, Mort JS, Lalumiere M, Eyre DR, Proteolysis of human bone collagen by cathepsin K: characterization of the cleavage sites generated by cross-linked N-telopeptide neoepitope. *Bone* 2000; 26(3): 241–7.
20. Nishi Y, Atley L, Eyre DR et al, Determination of bone markers in pycnodysostosis: effects of cathepsin K deficiency on bone matrix degradation. *J Bone Miner Res* 1999; 14(11): 1902–8.
21. Taylor AK, Lueken SA, Libanati C, Baylink DJ, Biochemical markers of bone turnover for the clinical assessment of bone metabolism. *Rheum Dis Clin North Am* 1994; 20: 589–607.
22. Garnero P, Gineyts E, Arbault P et al, Different effects of bisphosphonate and estrogen therapy on free and peptide-bound bone cross-links excretion. *J Bone Miner Res* 1995; 10: 641–9.
23. Eyre DR, New biomarkers of bone resorption. *J Clin Endocrinol Metab* 1992; 74: 470A.
24. Matsuki H, Kasuga H, Sugita M et al, An improved method for analysis of urinary hydroxyproline by an automated analyzer. *Tokai J Exp Clin Med* 1984; 9: 421–8.
25. Ebeling PR, Atley LM, Guthrie JR et al, Bone turnover markers and bone density across the menopausal transition. *J Clin Endocrinol Metab* 1996; 81: 3366–71.
26. Robins SP, Woitge H, Hesley R et al, Direct, enzyme-linked immunoassay for urinary deoxypyridinoline as a specific marker for measuring bone resorption. *J Bone Miner Res* 1994; 9: 1643–9.
27. Seyedin SM, Kung VT, Daniloff YN et al, Immunoassay for urinary pyridinoline: the new marker of bone resorption. *J Bone Miner Res* 1993; 8: 635–41.
28. Risteli J, Elomaa I, Niemi S et al, Radioimmunoassay for the pyridinoline cross-linked carboxy-terminal telopeptide of type I collagen: a new serum marker of bone collagen degradation. *Clin Chem* 1995; 39: 635–40.
29. Garnero P, Gineyts E, Riou JP, Delmas PD, Assessment of bone resorption with a new marker of collagen degradation in patients with metabolic bone disease. *J Clin Endocrinol Metab* 1994; 79: 780–5.
30. Woitge HW, Pecherstorfer M, Li Y et al, Novel serum markers of bone resorption: clinical assessment and comparison with established urinary indices. *J Bone Miner Res* 1999; 14: 792–801.
31. Hannon RA, Sacco-Gibson N, Mallinak N et al, Comparison of ELISA and Direct Response Device to measure urinary type I collagen N-telopeptide (NTX) in postmenopausal women. *Arth Rheum* 1999; 42: S290.
32. Uebelhart D, Schlemmer A, Johansen JS et al, Effect of menopause and hormone replacement therapy on the urinary excretion of pyridinium cross-links. *J Clin Endocrinol Metab* 1991; 72: 367–73.
33. Garnero, P, Shih, WJ, Gineyts E et al, Comparison of new biochemical markers of bone turnover in late postmenopausal osteoporotic women in response to alendronate treatment. *J Clin Endocrinol Metab* 1994; 79: 1693–700.
34. Delmas PD, Schlemmer A, Gineyts E et al, Urinary excretion of pyridinoline crosslinks correlates with bone turnover measured on iliac crest biopsy in patients with vertebral osteoporosis. *J Bone Miner Res* 1991; 6: 639–44.
35. Eriksen EF, Charles P, Melsen F et al, Serum markers of type I collagen formation and degradation in metabolic bone disease: correlation with bone histomorphometry. *J Bone Miner Res* 1993; 8: 127–32.
36. Parfitt AM, Simon LS, Villanueva AR, Krane SM, Procollagen type I carboxy-terminal extension peptide in serum as a marker of collagen biosynthesis in bone. Correlation with iliac bone formation rates and comparison with total alkaline phosphatase. *J Bone Miner Res* 1987; 2: 427–36.
37. Ohishi T, Takahashi M, Kushida K et al, Quantitative analyses of urinary pyridinoline and deoxypyridinoline excretion in patients with hyperthyroidism. *Endocr Res* 1992; 18: 281–90.
38. Rosen HN, Dresner-Pollak R, Moses AC et al, Specificity of urinary excretion of cross-linked N-telopeptides of type I collagen as a marker of bone turnover. *Calcif Tissue Int* 1994; 54: 26–9.
39. Duda RJ Jr, O'Brien JF, Katzmann JA et al,

Concurrent assays of circulating bone Gla-protein and bone alkaline phosphatase: effects of sex, age and metabolic bone disease. *J Clin Endocrinol Metab* 1988; 66: 951–7.
40. Seibel MK, Gartenberg F, Silverberg SJ et al, Urinary hydroxypyridinium cross-links of collagen in primary hyperparathyroidism. *J Clin Endocrinol Metab* 1992; 74: 481–6.
41. De La Piedra C, Diaz Martin MA, Diaz Diego EM et al, Serum concentrations of carboxyterminal cross-linked telopeptide of type I collagen (ICTP), serum tartrate resistant acid phosphatase, and serum levels of intact parathyroid hormone in parathyroid hyperfunction. *Scand J Clin Lab Invest* 1994; 54: 11–15.
42. Bonde M, Qvist P, Fledelius C et al, Applications of an enzyme immunoassay for a new marker of bone resorption (crosslaps): follow-up on hormone replacement therapy and osteoporosis risk assessment. *J Clin Endocrinol Metab* 1995; 80: 864–8.
43. Randall AG, Kent GN, Garcia-Webb P et al, Comparison of biochemical markers of bone turnover in paget disease treated with pamidronate and a proposed model for the relationships between measurements of the different forms of pyridinoline cross-links. *J Bone Miner Res* 1996; 11: 1176–84.
44. Alvarez L, Guanabens N, Peris P et al, Discriminative value of biochemical markers of bone turnover in assessing the activity of Paget's disease. *J Bone Miner Res* 1995; 10: 458–65.
45. Nakayama K, Fukumoto S, Takeda S et al, Differences in bone and vitamin D metabolism between primary hyperparathyroidism and malignancy-associated hypercalcemia. *J Clin Endocrinol Metab* 1996; 81: 607–11.
46. Abildgaard N, Nielsen JL, Heickendorff L, Connective tissue components in serum in multiple myeloma: analyses of propeptides of type I and type III procollagens, type I collagen telopeptide, and hyaluronan. *Am J Hematol* 1994; 46: 173–8.
47. Demers LM, Costa L, Chinchilli VM et al, Biochemical markers of bone turnover in patients with metastatic bone disease. *Clin Chem* 1995; 41: 1489–94.
48. Prestwood KM, Pilbeam CC, Burleson JA et al, The short term effects of conjugated estrogen on bone turnover in older women. *J Clin Endocrinol Metab* 1994; 79: 366–71.
49. Chestnut CH III, Bell NH, Clark GS et al, Hormone replacement therapy in postmenopausal women: urinary N-telopeptide of type I collagen monitors therapeutic effect and predicts response of bone mineral density. *Am J Med* 1997; 102: 29–37.
50. Bone HG, Greenspan SL, McKeever C et al, Alendronate and estrogen effects in postmenopausal women with low bone mineral density. Alendronate/Estrogen Study Group. *J Clin Endocrinol Metab* 2000; 85: 720–6.
51. Colwell A, Eastell R, The renal clearance of free and conjugated pyridinium cross-links of collagen. *J Bone Miner Res* 1996; 11: 197–8.
52. Bettica P, Taylor AK, Talbot J et al, Clinical performances of galactosyl hydroxylysine, pyridinoline, and deoxypyridinoline in postmenopausal osteoporosis. *J Clin Endocrinol Metab* 1996; 81: 542–6.
53. Seibel MJ, Woitge H, Scheidt-Nave C et al, Urinary hydroxypyridinium crosslinks of collagen in population-based screening for overt vertebral osteoporosis: results of a pilot study. *J Bone Miner Res* 1994; 9: 1443–40.
54. McLaren AM, Hordon LD, Bird HA, Robins SP, Urinary excretion of pyridinium crosslinks of collagen in patients with osteoporosis and the effects of bone fracture. *Ann Rheum Dis* 1992; 51: 648–51.
55. Seibel MJ, Cosman F, Shen V et al, Urinary hydroxypyridinium crosslinks of collagen as markers of bone resorption and estrogen efficacy in postmenopausal osteoporosis. *J Bone Miner Res* 1993; 8: 881–9.
56. Schneider DL, Barrett-Connor EL, Urinary N-telopeptide levels discriminate normal, osteopenic, and osteoporotic bone mineral density. *Arch Intern Med* 1997; 157: 1241–5.
57. Cummings SR, Black D, Bone mass measurements and risk of fracture in caucasian women: a review of findings from prospective studies. *Am J Med* 1995; 98(suppl 2A): 24S–28S.
58. Greenspan SL, Myers ER, Maitland LA et al, Fall severity and bone mineral density as risk factors for hip fracture in ambulatory elderly. *JAMA* 1994; 271: 128–33.
59. Slemenda C, Hui SL, Longcope C, Johnston CC, Sex steroids and bone mass: a study of changes about the time of menopause. *J Clin Invest* 1987; 80: 1261–9.
60. Johansen JS, Riis BJ, Delmas PD, Christiansen C, Plasma BGP: an indicator of spontaneous bone loss and of the effect of oestrogen treatment in postmenopausal women. *Eur J Clin Invest* 1988; 18: 191–5.

61. Cosman F, Nieves J, Wilkinson C et al, Bone density change and biochemical indices of skeletal turnover. *Calcif Tissue Int* 1996; 58: 236–43.
62. Chesnut III CH, Bell NH, Clark GS, Hormone replacement therapy in postmenopausal women: urinary N-telopeptide of type I collagen monitors therapeutic effect and predicts response of bone mineral density. *Am J Med* 1997; 102: 29–37.
63. Hansen MA, Overgaard K, Riis BJ, Christiansen C, Role of peak bone mass and bone loss in postmenopausal osteoporosis: 12 year study. *BMJ* 1991; 303: 961–4.
64. van Daele PLA, Seibel MJ, Burger H et al, Case–control analysis of bone resorption markers, disability, and hip fracture risk: the Rotterdam Study. *BMJ* 1996; 312: 482–3.
65. Garnero P, Hausherr E, Chapuy MC et al, Markers of bone resorption predict hip fracture in elderly women: the EPIDOS prospective study. *J Bone Miner Res* 1996; 11: 1531–8.
66. Civitelli R, Gonnelli S, Zacchei F et al, Bone turnover in postmenopausal osteoporosis: effect of calcitonin treatment. *J Clin Invest* 1988; 82: 1268–74.
67. Fuleihan GE-H, Brown EM, Curtis K et al, Effect of sequential and daily continuous hormone replacement therapy on indexes of mineral metabolism. *Arch Intern Med* 1992; 152: 1904–9.
68. Pols HA, Felsenberg D, Hanley DA et al, Multinational, placebo-controlled, randomized trial of the effects of alendronate on bone density and fracture risk in postmenopausal women with low bone mass: results of the FOSIT study. Foxamax International Trial Study Group. *Osteoporos Int* 1999; 9(5): 461–8.
69. Kraenzlin ME, Seibel MJ, Treschsel U et al, The effect of intranasal calcitonin on postmenopausal bone turnover as assessed by biochemical markers: evidence of maximal effect after 8 weeks of continuous treatment. *Calcif Tissue Int* 1996; 58: 216–20.
70. Rosen CJ, Chesnut III CH, Mallinak NJ, The predictive value of biochemical markers of bone turnover for bone mineral density in early postmenopausal women treated with hormone replacement or calcium supplementation. *J Clin Endocrinol Metab* 1997; 82: 1904–10.
71. Garnero P, Sornay-Rendu E, Duboeuf F, Delmas PD, Markers of bone turnover predict postmenopausal forearm bone loss over 4 years: the OFELY study. *J Bone Miner Res* 1999; 14: 1614–21.
72. Ravn P, Hosking D, Thompson D et al, Monitoring of alendronate treatment and prediction of effect on bone mass by biochemical markers in the early postmenopausal intervention cohort study. *J Clin Endocrinol Metab* 1999; 84: 2363–8.
73. Marcus R, Holloway L, Wells B et al, The relationship of biochemical markers of bone turnover to bone density changes in postmenopausal women: results from the postmenopausal Estrogen/Progestin Interventions (PEPI) Trial. *J Bone Miner Res* 1999; 14: 1583–95.

3

C-telopeptides

Elizabeth T Leary

Summary • Introduction • CTX • ICTP • Conclusion • References

SUMMARY

The degradation products of type I collagen have shown promise as biomarkers of bone resorption. At the present time, there are two categories of C-telopeptide methods, the CTX and the ICTP, both immunoassays. These recognize different domains of the C-terminal telopeptide region of the α1 chain of type I collagen and respond differently to bone metabolic processes. All of the current CTX assays are based on the CrossLaps antibodies which recognize the EKAHD-β-GGR octapeptide where the aspartate residue is β-isomerized. The degree of racemization and of isomerization of the aspartate have been shown to correlate with bone age. Urinary CTX is available as an ELISA assay (Osteometer BioTech A/S) and as a point-of-care device (Cortecs Diagnostics Ltd). Urine volume variation is normalized using urinary creatinine. To circumvent limitations associated with urine, serum CTX was first introduced as an ELISA assay (Osteometer BioTech A/S) and more recently has been automated (Elecsys, Roche Diagnostics GmbH). The two serum assays correlate well ($R = 0.87$–0.98). The comparison among the serum and the urine assays gave $R = 0.76$–0.88. Both urinary CTX and serum CTX have demonstrated marked response to antiresorptive therapies. They have also shown predictive value for fracture risk and reasonable correlation with bone density in population studies. The available CTX assays differ in performance features such as sensitivity, specificity, precision and assay design. The automated assays are more precise, less operator skill dependent and provide faster turnaround of results. Because of the large diurnal variation, samples for both urinary and serum CTX analyses must be carefully timed. Serum ICTP is insensitive to normal metabolic bone processes such as osteoporosis but may be a marker of bone degradation in pathological conditions. Further improvement of the C-telopeptide assays, including minimizing variability through more robust assays and carefully controlled sampling conditions, will facilitate their use as investigative tools in new areas.

INTRODUCTION

Type 1 collagen represents more than 90% of the organic matrix of bone. In recent years, the degradation products of mature type 1 collagen have been explored as biomarkers in the evaluation and management of bone metabolic diseases.[1,2] The new generation of bone resorption markers, including the C- and N-telopeptides

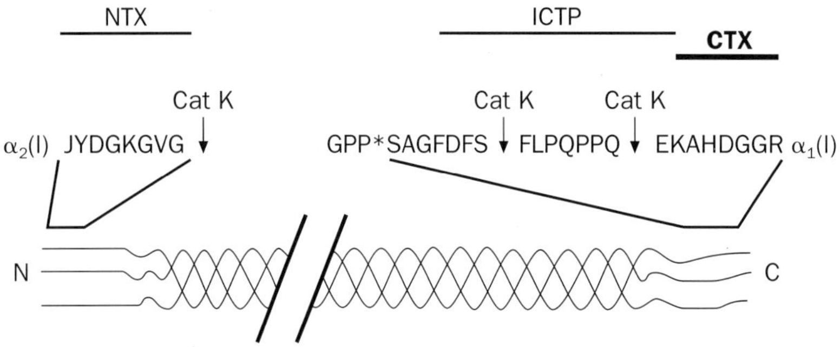

Fig. 3.1 The telopeptide assay epitopes in type I collagen. (Adapted from Nishi et al.[7])

and pyridinoline cross-links, represents a vast improvement over the older biomarkers in specificity and sensitivity for mature bone.[3–6] The currently available C-telopeptide methods are all immunoassays. They may be divided into two categories, CTX and ICTP. These recognize different domains of the C-terminus of type I collagen.[3,7,8] Significant understanding of these markers has been gained through continued improvement in analytical methods and from new clinical data. This chapter will focus on the available CTX assays and the clinical utility of CTX.

CTX

Analytical methods

All of the current CTX assay methods utilize the CrossLaps antibodies (Osteometer BioTech A/S, Denmark), which recognize an eight amino-acid (8AA) sequence (EKAHβDGGR) in the C-terminal telopeptide region of the α1 chain[3,4] (Fig. 3.1). Table 3.1 summarizes the features of the available C-telopeptide assays. The first CrossLaps-based assay employed a polyclonal competitive ELISA for assay of urinary CTX (CrossLaps ELISA, Osteometer BioTechA/S, Denmark). This assay recognizes the linear 8AA epitope with the aspartic acid in the β-isomerized configuration. During analysis, prediluted patient samples, as well as six levels of standard and control samples, are pipetted in duplicate into respective microtitre plate wells precoated with the synthetic octapeptide. Rabbit polyclonal antibody raised against the 8AA is then added to the wells. After 1 h of incubation at room temperature and the subsequent washing steps, a second peroxidase-conjugated antibody (goat anti-rabbit IgG) is added to the wells. At the completion of a second 1-h incubation and washing steps, the bound octapeptide is expressed by the addition of tetramethylbenzidine substrate. The peroxidase enzyme reaction is stopped and measured in a microtitre plate-reader at 450 nm. The assay can be completed in 3 h. Urine volume variation is normalized using urinary creatinine. Thus, the imprecisions of the marker and creatinine determinations both contribute to the imprecision of the final reported value. Because of the significant diurnal variation of CTX in urine, second morning void urine or 24-h collection are recommended.[9]

A semi-quantitative urine assay based on the CrossLaps antibodies is available in a point-of-care format (Osteosal, Cortecs Diagnostics Ltd). In this case, the colour development of a dipstick reaction is read in a small meter. Urine volume correction is by qualitative creatinine measurement using a second dipstick assay.[10]

The development of the CTX assay in serum has been prompted by the limitations associated with the urine assay. These include the large diurnal variation alluded to above, the need for urine volume correction by creatinine, the inability to quantitate CTX in very dilute urine,

Table 3.1 C-Telopeptide assays.

Product name	Manufacturer	Format	Method	Epitope	Sample	Reporting unit
CrossLaps ELISA	Osteometer BioTech	Microtitre plate/ELISA	Polyclonal competitive immunoassay Peroxidase TMB chromogen Visible signal (450 nm) Need urine creatinine	EKAHDβGGR β-Isomerized Asp	Second morning void urine	μg/mmol creatinine
Serum CrossLaps One Step ELISA	Osteometer BioTech	Microtitre plate/ELISA	Monoclonal sandwich immunoassay Biotinylated antibody with peroxidase conjugate Visible signal (450 nm)	EKAHDβGGR β-Isomerized ASP Need two copies cross-linked	Morning fasting serum or plasma	pmol/l
β-CrossLaps/serum	Roche Diagnostics	Automated analyser	Monoclonal sandwich assay Biotinylated antibody with ruthenium conjugate Electrochemiluminescent signal Automated on Elecsys	EKAHDβGGR β-Isomerized Asp Need two copies cross-linked	Morning fasting serum or plasma	ng/ml
ICTP	Orion Diagnostics	^{125}I RIA	Polyclonal radioimmunoassay	Cross-linked phenylalanine-rich domain near the C-terminus	Serum	μg/l
Osteosal	Cortecs Diagnostics	POC	Monoclonal competitive binding immunoassay	EKAHDβGGR	Urine	

the difficulty in obtaining carefully time-controlled specimens in certain patient populations (e.g. elderly or renal impaired), and the impracticality of 24-h urine collections.

The first CTX assay in serum was introduced as a two-site sandwich ELISA format (Serum CrossLaps One Step ELISA, Osteometer BioTech A/S).[4,11] The assay is based on two highly specific monoclonal antibodies recognizing an epitope that contains duplicate copies of the 8AA octapeptide in which the aspartic acid residue (D) is β-isomerized. Patient sera, six levels of standards and controls are pipetted, in duplicate, directly into the respective microtitre plate wells precoated with streptavidin. A mixture of the biotinylated monoclonal antibody and the peroxidase-conjugated monoclonal antibody prepared fresh within 30 min of use is then added to each well. After a 2-h incubation on a rotator at room temperature, the plate is washed five times. The enzyme substrate (tetramethylbenzidine) is then added, and the reaction is stopped after a 15 min incubation in a dark environment. The absorbance is measured at 450 nm with a 650-nm reference. The entire procedure can be completed in less than 2.5 h and involves less operator intervention than the urine assay.

Recently, serum CTX has been automated using the Elecsys analyzer (β-CrossLaps/serum, Roche Diagnostics GmbH).[12–14] As with the Osteometer assay, the Elecsys β-CrossLaps/serum is a double antibody sandwich assay, and is based on the same CrossLaps antibodies. It differs from the serum ELISA assay in that the first monoclonal antibody is biotinylated, but the second monoclonal antibody is conjugated with ruthenium rather than peroxidase. The bound immunocomplex is immobilized upon strepavidin-coated paramagnetic particles. The detection is based on the more sensitive electrochemiluminescent signal of the excitation and decay of ruthenium. A two-point calibration is performed with each reagent lot and repeated weekly or monthly (instrument model dependent). The reagents are supplied ready to use. Because of the wide dynamic range of the assay (linear range greater than 20× of premenopausal female mean), sample predilution is rarely necessary. The automated format significantly decreases the analysis time; 20 min to the first result, compared with 2–3 h with the ELISA format.

Tissue specificity

The β-CrossLaps antibody is specific for the EKAHβDGGR (8AA) octapeptide with β-isomerized aspartic acid in the C-terminus of the α1 chain of type I collagen, with the C-terminal arginine being essential.[11] There is <0.2% cross-reactivity with non-isomerized 8AA or octapeptides in other configurations. In the serum assay, two copies of β-isomerized 8AA must be present, leading to improved specificity. It has been reported that the Asp^{1211} residue of the ^{1209}AHDGGR^{1214} sequence undergoes post-translational non-enzymatic racemization and isomerization.[15,16] In addition to the native αL form, there are three additional potential isomers, βL, βD and αD. The αL form represents newly formed bone, while βL, βD and αD represent increasingly older bone. Therefore, the ratio between the αL form and each of the other isomers may be a potential index of bone age. This supposition is consistent with the notion that the double-isomerized epitope of the serum assay is more specific for mature bone than the urine assays, which require only a single octapeptide. Indeed, it has been reported that both serum CTX and bone-specific alkaline phosphatase discriminate between bone turnover rates of adolescent young women of different age groups, while the urine CTX assay does not.[17] In certain other population subsets, the serum CTX is found to be more specific as well.[18]

Correlation among assays

As expected, the two serum CTX assays are very highly correlated (Table 3.2, Fig. 3.2), with $R = 0.871–0.982$, depending on the population studied[12,14,19] (Roche Elecsys multicentre trial,

Table 3.2 CTX assay correlation.

		N	Correlation coefficient (*r*)
Elecsys β-CTX versus serum CTX ELISA			
	Lyon	120	0.904
	Prague	129	0.982
	Luxembourg	97	0.971
	Seattle	218	0.871
Elecsys β-CTX versus urinary CTX ELISA			
	Lyon	488	0.764
	Luxembourg	97	0.880
	Seattle	210	0.823
Serum CTX ELISA versus urinary CTX ELISA			
	Osteometer[2]	638	0.856
	Seattle	210	0.759

Elecsys β-CrossLaps/serum external evaluation data (Roche Diagnostics GmbH).

unpublished data). In our laboratory we have observed occasional discrepancies between the Elecsys assay results and the Serum CrossLaps One Step ELISA assay results. The latter appear to overestimate the CTX concentration as indicated by values of other bone markers analysed on the same samples. This overestimation is most likely as a result of interferents such as heterophilic antibodies. Fig. 3.2 depicts assay comparison data on paired serum and urine samples from 218 subjects, including 61 healthy males, 42 pre- and 41 postmenopausal females, 54 osteoporotics and 20 Paget's patients. The Elecsys assay has a dynamic range that is more than 10 times the premenopausal 95 percentile cut-off limit of 0.54 ng/ml. The wide range allows samples from the vast majority of patient populations to be analysed directly without additional sample dilution. On the other hand, the dynamic range of the serum ELISA is approximately four times the upper limit of the premenopausal range of ~5000 pmol/l.

The correlation between serum and urine CTX is not as strong ($R = 0.76$–0.88) as that observed for the two serum assays. This may be in part because the urine CTX assay is less specific for mature bone, and also because it requires creatinine-based volume correction, which contributes to analytical noise. Furthermore the assay requires only one epitope-containing peptide to be present for recognition, unlike the two sites required in the sandwich assay. One additional limitation of the urine ELISA assay is its low-end detection limit of 50 μg/l CTX. This limit prevents quantitation of CTX in very dilute urine, which represents ~2–5% of clinical study samples. Eight of the 218 samples shown in Fig. 3.2 could not be quantitated.[12]

Practical considerations

To assess clinical usefulness of a marker, the observed marker response must exceed the minimum significant change (MSC). The MSC

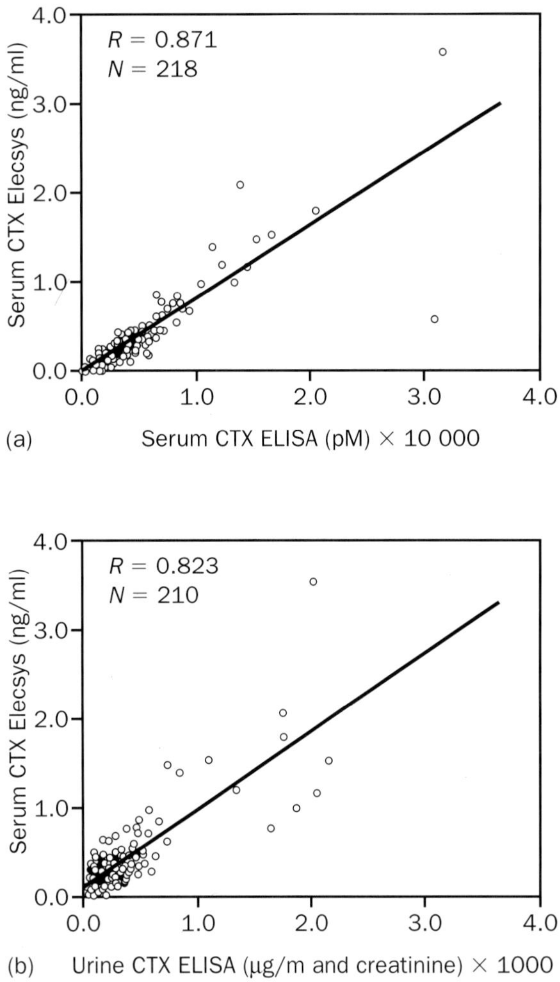

Fig. 3.2 CTX assay comparison. Paired serum and urine samples from 218 subjects (61 healthy males, 42 pre- and 41 postmenopausal females, 54 osteoporotics and 20 Paget's) were analysed by the Elecsys serum β-CrossLaps/serum assay (*y*-axis, a and b), the serum CrossLaps One Step ELISA assay (*x*-axis, a) and the urine CrossLaps ELISA assay (*x*-axis, b). In figure b, eight samples were excluded because the urine values were below the low-end detection limit of the assay.

represents the maximum potential change from the combined intra-individual variation of a control population and analytical variation.[9,20] Therefore, to minimize noise not related to clinical response, one must reduce both the analytical variability and the biological variability associated with sample collection. In addition to being scientifically and technically sound (accurate, precise and specific), a desirable analytical method must be developed with certain practical considerations in mind. These include pre-analytical issues such as patient preparation and sample handling (collection, processing and storage). Because of the significant diurnal variation of CTX in both urine and serum, well-defined specimen collection conditions are essential.[9,21,22] Peak serum CTX is observed during the early morning hours (2–6 a.m.), while a nadir occurs between noon and 3 p.m. The peak and nadir may be ±60% of the 24-h mean. This circadian rhythm is largely blunted by fasting.[23] Diurnal rhythm in urine lags behind serum somewhat, with a peak around 4–8 a.m. and a nadir in the afternoon between 3 and 6 p.m. While a 24-h urine collection and a 2 p.m. blood draw would provide least variation, patient compliance to these demands can present a problem. The current recommendations from the reagent manufacturers are second morning void urine and serum from morning blood draw after overnight fast. Specimen collection time should be consistent during monitoring visits. Sample stability claims for CTX on the Elecsys are 1 day at 4°C for serum and 7 days for EDTA plasma. For longer-term sample storage (>3 months) Roche recommend storage at −70°C. CTX recovery is equivalent among serum, EDTA and heparinized plasma. However, EDTA plasma presents the best storage stability. CTX is stable in urine for 6 days at refrigerated or room temperature. Samples not to be analysed fresh should be frozen in cryogenic vials as soon as possible within the acceptable window of stability.[12]

Analytical variability may be reduced by practical considerations such as convenient test format and robust test design. Issues which need to be addressed include assay dynamic range, calibration frequency, reagent and sample preparation, reagent stability and lot-to-lot consistency, operator skill level and hands-on time required, turnaround time, and overall throughput. The intra- and inter-assay coefficients of variation (CVs) for the ELISA assays (serum and urine) range from 5% to 13%, with increasing CV throughout the normal range.[12,14] The automated Elecsys serum assay has improved precision, with intra-assay CV in the 2–4% range and inter-assay CV of 3–10%. When evaluating performance data, it is important to ensure that the quality control material matrix reflects patient sample performance; the control samples must be handled similarly to patient samples (temperature, testing order, etc.) and the performance data must represent results from multiple analysts and reagent lots. In general, the ELISA assay format has an inherent imprecision limitation of 4–5% due to the complexity of the technique and the chemistries involved. With well-designed automation, where human factors are minimized, short-term and long-term precision may be reduced and overall efficiency improved.

Clinical utility

There are at least six categories where the bone markers may be of potential use: (1) monitoring effectiveness of therapy; (2) monitoring patient's therapy compliance; (3) prediction of bone loss; (4) predicting fracture risk; (5) prediction of bone mass; and (6) selection of patients for antiresorptive therapy.[1] Both the urine and serum CTX have shown decreases of 70% or more from baseline following bisphosphonate therapies, and approximately 50% in hormone replacement therapies.[18–20,24] The maximum effect may be seen after 3–6 months of therapy. Generally, urinary and serum CTX appear to provide similar information in studies following the same time course. However, in addition to circumventing some of the practical limitations of urine samples, it has been reported that serum CTX improves differentiation of control subjects and patients with primary vertebral osteoporosis, or primary hyperparathyroidism.[18] Unlike urinary CTX, serum CTX is also able to discriminate between bone status of adolescent girls of different age groups.[17] In a study of thyroid hormone-induced bone loss treated with intravenous pamidronate, fasting serum CTX was found to show low intra-individual variability and low minimum significant change. Compared to urinary N-telopeptide and deoxypyridinoline, serum CTX was found to have superior clinical response. Unfortunately, urine CTX was not measured in this study.[20]

The predictive value of bone markers for fracture risk is based on follow-up data from major prospective studies. In the EPIDOS study, elevated urinary CTX, deoxypyridinoline and serum CTX drawn at 1–2 p.m. were found to be associated with increased fracture risk in elderly women.[21,25] Interestingly, this association is lost when data from serum samples collected in the morning, without dietary restriction, are included in the analysis or analysed alone. Serum CTX from blood samples collected with fasting between 8 and 9:30 a.m. in the Ofely study is predictive of osteoporotic fracture.[26] CTX has been shown to be inversely correlated with bone mineral density ($R = 0.45$–0.65). Some investigators maintain that early marker values including CTX (at 3 or 6 months following therapy) may be predictive of bone density at 1 or 2 years of treatment.[23,27] The predictive value in individuals is complicated, and may depend on the skeletal site studied, as well as the variability in skeletal fragility of different populations. In addition to bone density, fracture risk is clearly related to other factors such as age, propensity to fall, and rate of bone turnover.[28] At this time, the marker predictive value of bone mass and fracture risk in individuals requires further investigation.[29]

Other potential applications of serum CTX include investigating its response in rheumatoid arthritis along with osteocalcin.[30] It has been observed that in patients with joint destruction there is an uncoupling of bone formation and resorption, with a decrease in serum osteocalcin and an increase in serum CTX. In patients without joint destruction, while osteocalcin is similarly decreased, serum CTX remains at the level of control groups. Serum CTX correlates positively with indices of disease activity, suggesting that it could be used in monitoring the treatment of joint destruction in rheumatoid arthritis.

Both CTX and ICTP are cleared by the kidneys and may accumulate in renal failure. However, this does not appear to prevent the markers from being clinically useful in these conditions.[31] Studies are being conducted in special populations such as renal dialysis patients with severely compromised bone metabolism.[32]

Table 3.3 C-Telopeptide assay improvements.

Standardize sample collection and handling procedure
Improve technical performance
Short-term and long-term precision/accuracy
Robust and user-friendly format
Standardize among assays (including reporting units)
Better, faster and cheaper assays (e.g. through automation)

ICTP

ICTP, available as a radioimmunoassay, measures a relatively large hydrophobic phenylalanine-rich, pyridinoline cross-link of the two α1 chains of the C-terminal telopeptides. This domain is located between the triple helix and the lysine-derived trivalent cross-link. It has been shown that ICTP and CTX are derived from different enzymatic processes. Hydrolysis by cathepsin K, a major osteoclastic proteinase, totally abolishes ICTP, whereas CTX is unaffected.[7,8,33] In addition, CTX and ICTP also respond differently in clinical situations. The CTX assays have demonstrated marked response to antiresorptive therapies such as bisphosphonates, hormone replacement therapy and SERMs, whereas ICTP has proven to be non-responsive. On the other hand, recent findings indicate that ICTP reflects increased type I collagen degradation in pathological conditions such as bone metastasis and rheumatoid arthritis.

CONCLUSION

There is now a large amount of data supporting the use of both urine and serum CTX for following antiresorptive therapies. Although there are some conflicting reports in the literature, preliminary data from prospective studies indicate that certain bone markers, including CTX, are predictive of fracture risk and bone mineral density. It is quite possible that conflicting reports may in part arise from study protocols which do not specify well-defined sample procurement procedures, such as fasting status, time of collection, sample processing and storage conditions. Furthermore, the analytical methods have been improved with time. The more specific and precise automated methods have only recently become available. In order to explore the clinical utility of the markers in treatments which may not produce as large a change as seen, for example, in the bisphosphonate therapy, it is essential to reduce the least significant change of the marker. With further improvements of the assays and of the markers themselves (Table 3.3), a more accurate clinical picture will emerge. The utility of bone markers will undoubtedly be investigated in new arenas, e.g. new disease states such as rheumatoid arthritis or bone metastasis. The potential of using serum CTX in differential diagnosis and assessment of bone age based on the degree of racemization and isomerization of aspartate in the CrossLaps octapeptide opens up exciting possibilities. Last, but not least, while much attention has been directed to generating new clinical data, serious efforts should be made to standardize the bone marker assays and the reporting of such results.

REFERENCES

1. Khosla S, Kleerekoper M, Biochemical markers of bone turnover. In: *Primer on the Metabolic Bone Diseases and Disorders of Mineral Metabolism*, 4th edn. Lippincott Williams & Wilkins: 128–34.
2. Miller P, Baran DT, Bilezikian JP et al, Practical clinical application of biochemical markers of bone turnover. *J Clin Densitometry* 1999; 2: 323–42.
3. Bonde M, Qvist P, Fledelius C et al, Immunoassay for quantifying type I collagen degradation products in urine evaluated. *Clin Chem* 1994; 40: 2022–5.
4. Rosenquist C, Fledelius C, Christgau S et al, Serum CrossLaps One Step ELISA. First application of monoclonal antibodies for measurement in serum of bone-related degradation products from C-terminal telopeptides of type I collagen. *Clin Chem* 1998; 44: 2281–9.
5. Hanson DA, Weis MAE, Bollen A, Maslan SL, Singer FR, Eyre DR, A specific immunoassay for monitoring human bone resorption: quantitation of type I collagen cross-linked N-telopeptides in urine. *J Bone Miner Res* 1992; 7: 1251–8.
6. Robins SP, Woitge H, Hesley R, Ju J, Seyedin S, Seibel M, Direct enyzme-linked immunoassay for urinary deoxypyridinoline as a specific marker for measuring bone resorption. *J Bone Miner Res* 1994; 9: 1643–9.
7. Nishi Y, Atley L, Eyre D et al, Determination of bone markers in pycnodysostosis: effects of cathepsin K deficiency on bone matrix degradation. *J Bone Miner Res* 1999; 14: 1902–8.
8. Sassi M, Eriksen H, Risteli L et al, Immunochemical characterization of assay for carboxyterminal telopeptide of human type I collagen: loss of antigenicity by treatment with cathepsin K. *Bone* 2000; 26: 367–73.
9. Hannon R, Blumsohn A, Naylor K, Eastell R, Response of biochemical markers of bone turnover to hormone replacement therapy: impact of biological variability. *J Bone Miner Res* 1998; 13: 1124–33.
10. Hannon RA, Branton R, Percival DA et al, Comparison of measurement of urinary CrossLaps™ by Osteosal™, a rapid point of care test, and by ELISA. *J Bone Miner Metab* 1998; 23: S630.
11. Bonde M, Garnero P, Fledelius C, QVist P, Delmas PD, Christiansen C, Measurement of bone degradation products in serum using antibodies reactive with an isomerized form of an 8 amino acid sequence of the C-telopeptide of type I collagen. *J Bone Miner Res* 1997; 12: 1028–34.
12. Leary ET, McLaughlin MK, Swezey D, Foster AP, Performance of β-Crosslaps/Serum on the Elecsys 2010 in comparison with six serum and urine bone resorption markers in microtiter assay format. *Clin Chem* 2000; 46: A145.
13. Hoyle NR, Kyriatsoulis A, The assessment of bone metabolic status with serum C-telopeptide, PTH and osteocalcin: three electrochemiluminescent immunoassays for the fully automated analysis of bone turnover. *Clin Chem* 1998; 44S6: 638.
14. Hoyle NR, Banauch D, Ebert C, An international technical and clinical evaluation of the Elecsys β-CrossLaps/serum assay. *J Bone Miner Metab* 1999; 14: S372.
15. Gineyts E, Cloos PAC, Borel O, Grimaud L, Delmas PD, Garnero P, Racemization and isomerization of type I collagen C-telopeptides in human bone and soft tissues: assessment of tissue turnover. *Biochem J* 2000; 345: 481–5.
16. Garnero P, Cloos PAC, Sornay-Rendu E, Qvist P, Delmas PD, Type I collagen racemization and izomerisation and the risk of fracture in postmenopausal women: the Ofely Prospective Study. *J Bone Miner Res* 2000; 15: S144.
17. Traba ML, Calero HA, Mendez-Davila C et al, Different behaviors of serum and urinary CrossLaps ELISA in the assessment of bone resorption in healthy girls. *Clin Chem* 1999; 45: 682–3.
18. Woitge HW, Pecherstorfer M, Li Y et al, Novel serum markers of bone resorption: clinical assessment and comparison with established urinary indices. *J Bone Miner Res* 1999; 14: 792–801.
19. Christgau S, Rosenquist C, Alexandersen P et al, Clinical evaluation of the serum CrossLaps One Step ELISA, a new assay measuring the serum concentration of bone-derived degradation products of type I collagen C-telopeptides. *Clin Chem* 1998; 44: 2290–300.
20. Rosen HN, Moses AC, Garber J et al, Serum CTX: a new marker of bone resorption that shows treatment effect more often than other markers because of low coefficient of variability and large changes with bisphosphonate therapy. *Calcif Tissue Int* 2000; 66: 100–3.
21. Chapurlat RD, Garnero P, Breart G, Meunier PJ, Delmas PD, Serum type I collagen breakdown product (serum CTX) predicts hip fracture risk in elderly women: The EPIDOS Study. *Bone* 2000; 27: 283–6.

22. Wilchers M, Schmidt E, Bidlingmaier F et al, Diurnal rhythm of CrossLaps in human serum. *Clin Chem* 1999; 45: 1858–60.
23. Christgau S, Bitsch Jensen O, Bjarnasson NH et al, Serum CrossLaps™ provides a rapid assessment of therapy response compared to BMD measurements. *J Bone Miner Res* 1999; 14: S200.
24. Garnero P, Shih W, Gineyts E et al, Comparison of new biochemical markers of bone turnover in late postmenopausal osteoporotic women in response to alendronate treatment. *J Clin Endocrinol Metab* 1994; 79: 1693–700.
25. Garnero P, Hausherr E, Chapuy MC et al, Markers of bone resorption predict hip fracture in elderly women: the EPIDOS prospective study. *J Bone Miner Res* 1996; 11: 1531–8.
26. Garnero P, Sornay-Rendu E, Claustrat B, Delmas P, Markers of bone turnover, endogenous hormone and the risk of fracture in postmenopausal women. *J Bone Miner Res* 1999; 14S: 171.
27. Štepán JJ, Vokrouhlická J, Comparison of biochemical markers of bone remodelling in the assessment of the effects of alendronate on bone in postmenopausal osteoporosis. *Clin Chim Acta* 1999; 288: 121–35.
28. Faulkner K, Bone matters: are density increases necessary to reduce fracture risk? *J Bone Miner Res* 2000; 15: 183–7.
29. Bauer D, Sklarin PM, Stone KL et al, Biochemical markers of bone turnover and prediction of hip bone loss in older women: the study of osteoporotic fractures. *J Bone Miner Res* 1999; 14: 1404–10.
30. Garnero P, Jouvenne P, Buchs N, Delmas PD, Miossec P, Uncoupling of bone metabolism in rheumatoid arthritis patents with or without joint destruction: assessment with serum type I collagen breakdown products. *Bone* 1999; 24: 381–5.
31. Urena P, Vernejoul MC, Circulating biochemical markers of bone remodeling in uremic patients. *Kidney Int* 1999; 55: 2141–56.
32. Leary ET, McLaughlin MK, Swezey D, Aggoune T, Carlson, TH, Foster AP, Evaluation of serum bone turnover markers in renal dialysis patients. *J Bone Miner Res* 2000; 15: S525.
33. Risteli J, Elomaa I, Niemi S, Novamo A, Risteli L, Radioimmunoassay for the pyridinoline cross-linked carboxy-terminal telopeptide of type I collagen: a new serum marker of bone collagen degradation. *Clin Chem* 1993; 39: 635–40.

4

Pyridinium cross-links as bone resorption markers

Simon P Robins

Summary • Development of cross-link assays • Specificity for bone • Comparison with telopeptide markers • Limitations of the cross-link assays • Standardization of cross-link assays • Concluding remarks • Acknowledgement • References

SUMMARY

In recent years, the increasing interest in osteoporosis and related disorders has provided a major drive for further research into the development of new markers for bone metabolism. Considerable advances have been made, and markers of bone turnover have increasing clinical applications in the detection and management of disorders involving the skeleton. For measurements of bone resorption, the newer markers have essentially replaced the archetypal assay, urinary hydroxyproline, and several assays now permit estimations of bone resorption in serum which provide additional utility for clinicians. The purpose of this brief review is to chart the evolution of the pyridinium cross-links, deoxypyridinoline and pyridinoline, as bone resorption indices and to consider their limitations and advantages in relation to the other group of newly developed assays, those involving collagen type I telopeptides.

DEVELOPMENT OF CROSS-LINK ASSAYS

Following the initial characterization of pyridinoline,[1] the potential value of this cross-link as a marker was suggested by the discovery that the compound could be detected and quantified in urine.[2,3] The development of a reversed-phase HPLC separation[4] facilitated analysis of tissue hydrolysates for both pyridinoline (PYD), also referred to as hydroxylysyl pyridinoline (HP), and deoxypyridinoline (DPD) (Fig. 4.1), an analogue derived from a lysyl residue in the collagen helix,[5] giving rise to the alternative nomenclature of lysyl pyridinoline (LP). Inclusion of a pre-fractionation step using cellulose partition chromatography facilitated analysis by HPLC of both cross-links in hydrolysed urine samples,[6] a technique which was later fully automated with the use of an internal standard.[7]

All of the initial analyses of cross-links were performed using acid hydrolysates to liberate all peptide-bound and other conjugated forms of the cross-link. Although preliminary immunodetection methods for PYD indicated that no cross-links were present in free form in

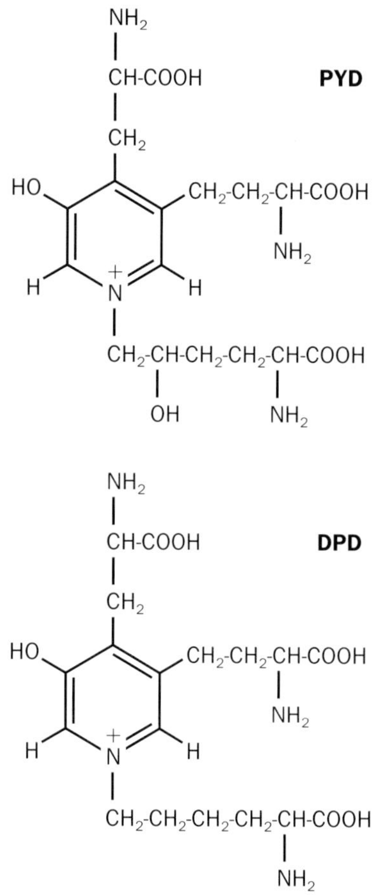

Fig. 4.1 Structure of the pyridinium cross-links, pyridinoline (PYD) and deoxypyridinoline (DPD).

urine, this was later shown to be due to the presence of specific diastereoisomers: the HPLC chromatography system combined with prefractionation on cellulose indicated that 40–50% of the cross-links were free and could be analysed directly without hydrolysis.[8,9] Further analysis of urine from patients with a series of different disorders associated with increased bone resorption revealed that a similar proportion of cross-link was free in each case, thus indicating that similar information about bone resorption could be obtained by direct analysis of urine without the time-consuming hydrolysis step. These results also suggested the possible application of direct immunoassays for the cross-links, and a number of ELISA systems for measuring free DPD[10] or free PYD and DPD combined[11,12] have been described.

SPECIFICITY FOR BONE

A primary impetus for developing the pyridinium cross-links as markers of bone turnover was to provide increased sensitivity and specificity in comparison with urinary hydroxyproline measurements. For the latter, metabolism in the liver, at a rate which is assumed to be constant, presents a major drawback, but the main problems with this analyte are the contributions from dietary sources and production of hydroxyproline from the degradation of newly synthesized collagen. In addition, a significant amount of hydroxyproline arises from the procollagen type I N-propeptides, which are removed during normal fibril formation. The pyridinium cross-links do not appear to be metabolized, and there is direct evidence for negligible dietary contribution to their excretion.[13] Because the pyridinium cross-links are formed only during the final stages of fibril formation, these cross-links indicate degradation only of mature, functional tissue and are unaffected by degradation of newly synthesized collagen.

Measurements of the pyridinium cross-links have advantages over hydroxyproline in conferring a high degree of tissue specificity. Cross-linking pathways are determined primarily by hydroxylation of the lysine residues in collagen telopeptides, a process which is now known to be accomplished by telopeptide lysyl hydroxylase, a different enzyme to that which hydroxylates lysines destined to be in the helix.[14] Consequently, pyridinium cross-links are essentially absent in skin, which contains no telopeptide hydroxylysine. The relative proportions of PYD and DPD are controlled by the activity of helical lysyl hydroxylase, and analyses have indicated that PYD has a much wider tissue distribution than DPD.[4,15] Initially, it appeared that DPD was present only in mineralized tissues, but it is now known to be present in cardiovascular

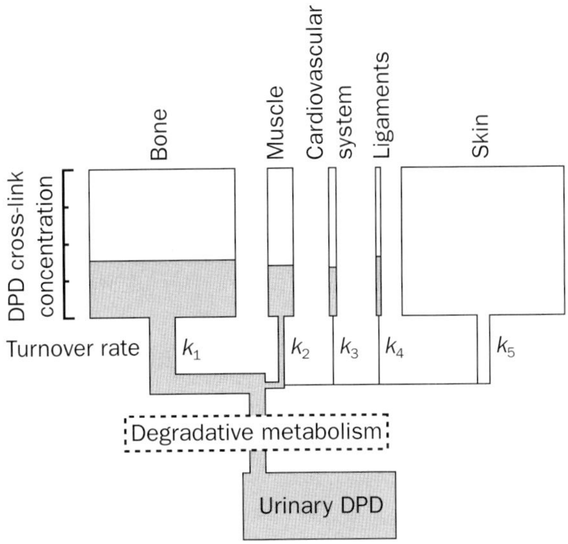

Fig. 4.2 Schematic of DPD excretion from body tissues. The area of the boxes represents the approximate body pool content of collagen type I, with the shaded area denoting DPD concentration: the thickness of the channels exiting the tissue boxes is roughly proportional to the turnover rates. Bone is the major source of urinary DPD, but measurements of peptide material may be complicated by changes in degradative metabolism.

tissue, intramuscular collagen and some ligaments. These findings have little bearing on the suitability of DPD as a bone-specific marker, because, when the pool size of the tissue and its turnover rate are considered (Fig. 4.2), it is clear that the low metabolic turnover of the main soft tissues in which DPD has been detected make their contribution to urinary DPD negligible.

COMPARISON WITH TELOPEPTIDE MARKERS

Several assays have been developed based on specific amino acid sequences close to the cross-linking sites rather than on the cross-links themselves. As the development of assays for the N-telopeptides (NTX) and C-telopeptides (CTX) are described in Chapters 2 and 3, those will not be considered in detail here. The peptide assays represent the main alternative to measuring pyridinium cross-links, and it is instructive to consider some of the factors which influence the specificity and clinical utility of the two types of assay. Assays for both NTX and CTX are in general specific for collagen type I, and there is little, if any, dependence on the type of cross-link present in the tissue. In practice, this presents no real difficulties in attaining a reasonable degree of specificity for bone, as its turnover rate is so high in comparison to other tissues (Fig. 4.2). The lack of dependence on the type of cross-link may be advantageous, as the assays will also detect peptides containing pyrroles and other as yet incompletely characterized cross-links in bone.[16,17] For the NTX assay, it was suggested that specificity for bone resulted from recognition by the antibody of a telopeptide sequence from the α2(I) chain of collagen, a feature that was much less common in other tissues.[18] However, the fact that a digest of skin collagen peptides showed the same molar reactivity with the NTX assay as a similar digest of bone[19] casts some doubt on whether the assay recognizes a neoepitope specific for bone.

A series of CTX assays has now emerged which measure cross-link-containing peptides with various isomeric forms of the Asp–Gly bond located in the α1(I) telopeptide.[17] The β-isomerization and racemization of this bond is an age-dependent process which potentially might allow the measurement of fragments derived from collagen with different turnover rates.[20,21] A similar modification to isoaspartyl residues has also been shown to occur in the α2(I) N-telopeptide,[22] but preliminary evidence suggests that the rate of equilibration in bone between the various aspartyl isomers is low relative to the average lifetime of collagen in bone. Consequently, the usefulness of assaying these different forms may be limited to diseases with very high turnover rates, such as Paget's disease.[23]

LIMITATIONS OF THE CROSS-LINK ASSAYS

For the convenience of a direct assay, it is necessary to measure free urinary DPD without hydrolysis, but for this to be a valid index of bone resorption, these values must represent a consistent proportion of the total DPD. Implicit in this statement is an understanding that measuring total DPD is the most accurate parameter available for determining the true rate of bone resorption, a fact supported experimentally by comparing this method with radioisotopic exchange.[24] The proportion of free cross-link was shown to be consistent in a series of studies including a range of patients with osteoporosis and other metabolic bone diseases.[8] Some disparity in urinary free/bound cross-link ratios has been noted, however, particularly in patients treated with amino-bisphosphonates, such that the proportion of free DPD was shown to increase during the 4-week treatment period.[25] In other studies of short-term intravenous administration of bisphosphonate, the change over 3 days in urinary free cross-link concentration was less than that for the total cross-links.[26] The same study showed much larger percentage changes during treatment for both the NTX and CTX assays. The most likely explanation for these observations is that some treatments for high bone resorption states may affect not only true bone resorption but also the patterns of free cross-links and peptides released into urine.[19] These effects may be at the osteoclast level in bone or at other sites of the body, such as liver or kidney. The potential effect of treatment on the degradative metabolism of bone collagen is an important factor influencing the interpretation of data for both free cross-link measurements and telopeptide assays, and there is still a need for more information on this topic. It should be borne in mind, however, that the problems referred to are largely overcome by measuring total DPD, which can be done either by HPLC or by immunoassay. This procedure currently requires a time-consuming chemical hydrolysis step, but all ambiguity in the interpretation of data is avoided.

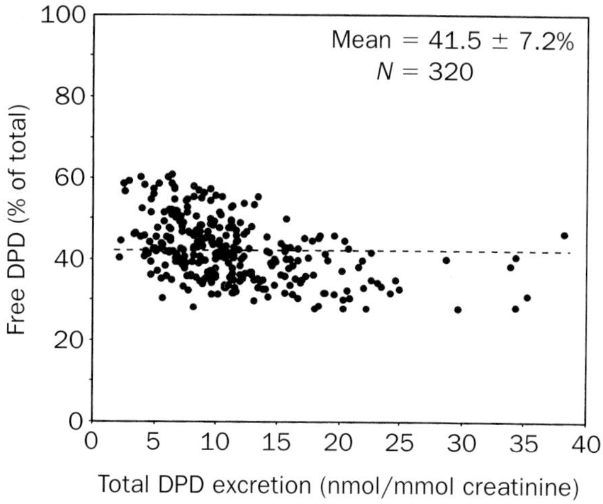

Fig. 4.3 Relationship between the proportion of free DPD in urine and the level of total excretion for a group comprising both healthy volunteers and patients with metabolic bone diseases. There was no marked decrease in the proportion of free DPD in those individuals with high total excretion.

It has been suggested that changes in free cross-link proportions are simply a function of the rate of bone collagen resorption and not related to any particular form of treatment. Some evidence for this was provided by an observed inverse relationship between the proportion of free DPD cross-link and the total amount of PYD excreted.[26] Our own experience in examining a wide range of data from both healthy volunteers and patients with metabolic bone disease not receiving treatment is that the relationship between the proportion of free DPD and the total DPD excreted is present (Fig. 4.3), but is unlikely to contribute significantly to the observed effects of treatment. Direct measurement of serum and urinary cross-links in children and adolescents has established that the clearance rate for DPD is greater than unity,[27] suggesting release of free DPD in the kidney, and similar conclusions have been drawn from other studies.[28] The results in Fig. 4.3 suggest, however, that this process is non-saturable and is unlikely to be the main determinant of free cross-link concentrations in urine.

Another limitation of the DPD assay is the current lack of a convenient, direct method for measurements in serum. Such an assay is desirable not only for increased clinical utility but also because of the lower biological variability in serum[29] and the lack of necessity for creatinine correction. The concentrations of free DPD in serum are too low for accurate quantification, but preliminary details of a method that allows measurement of total DPD in serum using an adapted immunoassay have been described.[30]

STANDARDIZATION OF CROSS-LINK ASSAYS

An important advantage of the DPD assay compared with the peptide measurements is the relative ease of standardization. The analyte is well defined, and criteria have been established for the quantification of standard solutions using primarily UV extinction coefficients.[31] Also, DPD with similar characteristics to the isolated material has now been chemically synthesized in several centres.[32–34] The monoclonal antibody used in the assay was shown to react only with DPD in urine.[10] As the concentrations of DPD in bone change little with age,[35] the results of total urinary output can be related directly to an equivalent mass of bone being resorbed. By contrast, for the peptide assays, the exact nature of the analytes is unclear, with the consequent difficulties in setting up appropriate calibrators. For the NTX assay, standards are prepared from a collagenase digest of bone,[18] and the results are calculated in terms of 'bone collagen equivalents', even though the patterns of peptides being analysed may be distinctly different from those of the prepared calibrators. The situation for the various forms of CTX assay is increasingly complex, and calibrators may be based on incompletely characterized peptides isolated from urine. These difficulties in standardization appear to have led to major discrepancies in the reported reference values for β-CTX.[36,37]

CONCLUDING REMARKS

Free DPD represents a simple, convenient urinary analyte for measuring bone resorption rates. These assays are now widely available as, in addition to automated HPLC methods, the reagents developed for immunoassay kits have been incorporated into other platforms for use in multiple clinical analysers. In addition, a point-of-care assay for measuring urinary DPD is being developed, which may have applications in future risk assessment screening for osteoporosis, perhaps in conjunction with simple bone quality measures such as ultrasound.

ACKNOWLEDGEMENT

I am indebted to the Scottish Environment and Rural Affairs Department for support.

REFERENCES

1. Fujimoto D, Moriguchi T, Ishida T et al, The structure of pyridinoline, a collagen crosslink. *Biochem Biophys Res Commun* 1978; 84: 52–7.
2. Gunja-Smith Z, Boucek RJ, Collagen crosslink components in human urine. *Biochem J* 1981; 197: 759–62.
3. Robins SP, An enzyme-linked immunoassay for the collagen crosslink, pyridinoline. *Biochem J* 1982; 207: 617–20.
4. Eyre DR, Koob TJ, Van Ness KP, Quantitation of hydroxypyridinium crosslinks in collagen by high-performance liquid chromatography. *Anal Biochem* 1984; 137: 380–8.
5. Ogawa T, Ono T, Tsuda M et al, A novel fluor in insoluble collagen: a crosslinking molecule in collagen molecule. *Biochem Biophys Res Commun* 1982; 107: 1252–7.
6. Black D, Duncan A, Robins SP, Quantitative analysis of the pyridinium crosslinks of collagen in urine using ion-paired reversed-phase high-performance liquid chromatography. *Anal Biochem* 1988; 169: 197–203.
7. Pratt DA, Daniloff Y, Duncan A et al, Automated analysis of the pyridinium crosslinks of collagen in tissue and urine using solid-phase extraction and reversed-phase high-performance liquid chromatography. *Anal Biochem* 1992; 207: 168–75.

8. Robins SP, Duncan A, Riggs BL, Direct measurement of free hydroxy-pyridinium crosslinks of collagen in urine as new markers of bone resorption in osteoporosis. In: Christiansen C, Overgaard K, eds. *Osteoporosis 1990*. Copenhagen: Osteopress ApS, 1990: 465–8.
9. Abbiati G, Bartucci F, Longoni A et al, Monitoring of free and total urinary pyridinoline and deoxypyridinoline in healthy volunteers: sample relationships between 24-h and fasting early morning urine concentrations. *Bone Miner* 1993; 21: 9–19.
10. Robins SP, Woitge H, Hesley R et al, Direct, enzyme-linked immunoassay for urinary deoxypyridinoline as a specific marker for measuring bone resorption. *J Bone Miner Res* 1994; 9: 1643–9.
11. Seyedin SM, Kung VT, Daniloff YN et al, Immunoassay for urinary pyridinoline: the new marker of bone resorption. *J Bone Miner Res* 1993; 8: 635–41.
12. Gomez B, Ardakani S, Evans B et al, Monoclonal antibody assay for free urinary pyridinium crosslinks. *Clin Chem* 1996; 42: 1168–75.
13. Colwell A, Russell RG, Eastell R, Factors affecting the assay of urinary 3-hydroxy pyridinium crosslinks of collagen as markers of bone resorption. *Eur J Clin Invest* 1993; 23: 341–9.
14. Bank RA, Robins SP, Wijmenga C et al, Defective collagen crosslinking in bone, but not in ligament or cartilage, in Bruck syndrome: indications for a bone-specific telopeptide lysyl hydroxylase on chromosome 17. *Proc Natl Acad Sci USA* 1999; 96: 1054–8.
15. Seibel MJ, Robins SP, Bilezikian JP, Urinary pyridinium crosslinks of collagen: specific markers of bone resorption in metabolic bone disease. *Trends Endocrinol Metab* 1992; 3: 263–70.
16. Hanson D, Eyre D, Molecular site specificity of pyridinoline and pyrrole cross-links in type I collagen of human bone. *J Biol Chem* 1996; 271: 26508–16.
17. Fledelius C, Johnsen AH, Cloos PAC et al, Characterization of urinary degradation products derived from type I collagen. Identification of a beta-isomerized Asp–Gly sequence within the C-terminal telopeptide (α1) region. *J Biol Chem* 1997; 272: 9755–63.
18. Hanson DA, Weis MA, Bollen AM et al, A specific immunoassay for monitoring human bone resorption: quantitation of type I collagen cross-linked N-telopeptides in urine. *J Bone Miner Res* 1992; 7: 1251–8.
19. Robins SP, Collagen crosslinks in metabolic bone disease. *Acta Orthop Scand Suppl* 1995; 266: 171–5.
20. Cloos PAC, Fledelius C, Collagen fragments in urine derived from bone resorption are highly racemized and isomerized: a biological clock of protein aging with clinical potential. *Biochem J* 2000; 345: 473–80.
21. Gineyts E, Cloos PAC, Borel O et al, Racemization and isomerization of type I collagen C-telopeptides in human bone and soft tissues: assessment of tissue turnover. *Biochem J* 2000; 345: 481–5.
22. Brady JD, Ju J, Robins SP, Isoaspartyl bond formation within n-terminal sequences of collagen type I: implications for their use as markers of collagen degradation. *Clin Sci* 1999; 96: 209–15.
23. Garnero P, Gineyts E, Schaffer AV et al, Measurement of urinary excretion of nonisomerized and β-isomerized forms of type I collagen breakdown products to monitor the effects of the bisphosphonate zoledronate in Paget's disease. *Arthritis Rheum* 1998; 41: 354–60.
24. Eastell R, Colwell A, Hampton L et al, Biochemical markers of bone resorption compared with estimates of bone resorption from radiotracer kinetic studies in osteoporosis. *J Bone Miner Res* 1997; 12: 59–65.
25. Tobias J, Laversuch C, Wilson N et al, Neridronate preferentially suppresses the urinary excretion of peptide-bound deoxypyridinoline in postmenopausal women. *Calcif Tissue Int* 1996; 59: 407–9.
26. Garnero P, Gineyts E, Arbault P et al, Different effects of bisphosphonate and estrogen therapy on free and peptide-bound bone cross-links excretion. *J Bone Miner Res* 1995; 10: 641–9.
27. Colwell A, Eastell R, The renal clearance of free and conjugated pyridinium cross-links of collagen. *J Bone Miner Res* 1996; 11: 1976–80.
28. Randall A, Kent G, Garcia-Webb P et al, Comparison of biochemical markers of bone turnover in Paget disease treated with pamidronate and a proposed model for the relationships between measurements of the different forms of pyridinoline cross-links. *J Bone Miner Res* 1996; 11: 1176–84.
29. Hannon R, Blumsohn A, Naylor K et al, Response of biochemical markers of bone turnover to hormone replacement therapy: impact of biological variability. *J Bone Miner Res* 1998; 13: 1124–33.

30. Weitz S, Benham P, Leung S, Total deoxypyridinoline in serum and urine as measured by a novel adaptation of the Pyrilinks-D enzyme immunoassay. *J Bone Miner Res* 1999; 14(suppl 1): S371.
31. Robins SP, Duncan A, Wilson N et al, Standardization of pyridinium crosslinks, pyridinoline and deoxypyridinoline, for use as biochemical markers of collagen degradation. *Clin Chem* 1996; 42: 1621–6.
32. Adamczyk M, Johnson DD, Reddy RE, Collagen cross-links: synthesis of pyridinoline, deoxypyridinoline and their analogues. *Tetrahedron* 1999; 55: 63–88.
33. Allevi P, Longo A, Anastasia M, Total synthesis of deoxypyridinoline, a biochemical marker of collagen turnover. *Chem Commun* 1999; 559–60.
34. Adamczyk M, Johnson DD, Reddy RE, An efficient one-pot synthesis of (+)-deoxypyridinoline. *Tetrahedron Lett* 1999; 40: 8993–4.
35. Eyre DR, Dickson IR, VanNess KP, Collagen crosslinking in human bone and cartilage: age-related changes in the content of mature hydroxypyridinium residues. *Biochem J* 1988; 252: 495–500.
36. Bonde M, Qvist P, Fledelius C et al, Applications of an enzyme immunoassay for a new marker of bone resorption (CrossLaps): follow-up on hormone replacement therapy and osteoporosis risk assessment. *J Clin Endocrinol Metab* 1995; 80: 864–8.
37. Garnero P, Fledelius C, Gineyts E et al, Decreased β-isomerization of the C-terminal telopeptide of type I collagen α1 chain in Paget's disease of bone. *J Bone Miner Res* 1997; 12: 1407–15.

5

Bone-specific alkaline phosphatase

Laurence M Demers

Summary • Introduction • Bone-specific alkaline phosphatase • Methods for the determination of bone-specific alkaline phosphatase • Cross-reactivity of bone alkaline phosphatase with other tissue non-specific alkaline phosphatase isoforms • Clinical use of the bone alkaline phosphatase assay • Future developments • References

SUMMARY

Alkaline phosphatase has been used for many years as a marker of bone metabolism and for monitoring the treatment of patients with metabolic bone disease. The recent development of immunoassay-based markers with monoclonal antibodies directed to the bone-specific isoform of alkaline phosphatase has improved both the specificity and sensitivity of this enzyme for the clinical management of patients with Paget's disease, osteoporosis, metastatic bone disease and other metabolic bone diseases. Although current monoclonal antibodies used to measure bone-specific alkaline phosphatase show some cross-reactivity with the liver isoform of this enzyme, the specificity of the immunoassay reagents for assessing bone formation and osteoblast cell function is still much improved over the previous use of total alkaline phosphatase for the clinical management of patients with metabolic bone disease.

INTRODUCTION

Alkaline phosphatase (ALP) (EC 3.1 3.1: orthophosphoric-monoester phosphohydrolase) is a ubiquitous enzyme located in most tissues and generally found at or near the membrane of cells.[1] Although the exact metabolic function of ALP is unknown, it is an enzyme that catalyses the alkaline hydrolysis of monophosphate ester groups and is present in high concentration in the intestinal epithelium, kidney tubules, bone, liver and placenta.[2] ALP is present in several isoenzyme forms that are encoded by four human genes.[3] Three genes are located on chromosome 2q34-37 and encode for intestinal, germ cell and placental ALP isoenzymes. The germ cell isoenzyme is present in testis, cervix, thymus and lung. The fourth gene is located on chromosome 1p36.1-34 and encodes for a single-tissue non-specific isoenzyme that is found in liver, bone and kidney.[4] Although bone, liver and kidney ALP share the same primary amino acid sequence, different isoforms exist in different tissue sites due to post-translational modification of the enzyme. Both sialylation and glycosylation of the enzyme produce the different isoforms.[5] Of the total ALP found in the circulation, 95% of the enzyme in blood originates from either liver or bone. In health, the ratio of the bone to liver isoforms is approximately 1 : 1. ALP is also expressed in different tumours, and the placenta is a rich source of the enzyme during pregnancy.

BONE-SPECIFIC ALKALINE PHOSPHATASE

The bone isoform of ALP is produced by osteoblasts as a tetramer and is initially anchored to the outer surface of the osteoblast cell membrane to inositol via a glycan ester (glycosylphosphatidylinositol).[3] Bone ALP is subsequently released from the osteoblast cell surface via a glycosylphosphatidylinositol-specific phospholipase D enzyme. The phospholipase enzyme cleaves ALP from the outer surface of the membrane, releasing it into the circulation in two forms: (1) an anchorless, soluble dimeric form; and (2) an anchor-intact, insoluble form.[6] The anchorless, soluble form predominates and constitutes 35–40% of the total ALP found in the circulation in health. With either hepatobiliary or bone disease, these different isoforms differ in their concentration in blood. Although the exact function of ALP in cells is presently unknown, its primary physiological role in bone is associated with calcification of the skeleton and bone formation. Bone ALP catalyses the hydrolysis of phosphate esters at the osteoblast cell surface to provide a high phosphate concentration for the bone mineralization process as part of the osteoblast cell role in bone remodelling.[7] As a result, bone ALP levels are raised in the circulation during periods of active bone formation and bone growth. During life, there are two age-dependent physiological peaks of high bone ALP activity, during infancy and at the time of puberty when bone growth is accelerated by the effects of sex steroids.[8] Bone ALP activity has been shown to correlate better physiologically with intact parathyroid hormone (PTH) than total ALP and shows a better correlation with height velocity than total enzyme concentration, lending credence to the importance of the bone isoform of ALP.[9]

METHODS FOR THE DETERMINATION OF BONE-SPECIFIC ALKALINE PHOSPHATASE

Over the years, a variety of methods have been developed that specifically isolate and measure the circulating bone-specific isoform of ALP distinct from the liver and kidney isoforms (Table 5.1). With liver disease, for example, such as hepatitis or cirrhosis, marked elevations in total ALP occur which alter the normal ratio of liver to bone isoforms in the circulation, making it difficult to discern the source of the enzyme elevation. With metabolic bone disease, increases in the bone isoforms occur with conditions such as Paget's disease, metastatic bone disease and osteoporosis that also alter the liver/bone isoform ratio of ALP in the circulation. Thus it is important in certain clinical situations to identify the source of the elevation and to differentiate the contribution from bone or liver to the elevated total ALP level found in patients with co-existing liver and bone disease.

Table 5.1 Methods used to measure bone specific alkaline phosphatase.

Heat inactivation
Amino acid inhibition
Urea inhibition
Zonal electrophoresis
Wheat germ lectin precipitation
Concanavalin A precipitation
HPLC
Immunoassay
EIA (activity-based assays)
IRMA (mass-based assay)

Earlier methods for bone-specific ALP measurements relied on the physicochemical properties of the enzyme and on the instability of the enzyme in the present of elevated temperatures.[10] The heat-inactivation methods take advantage of the fact that the bone isoform is more susceptible to higher temperatures than the liver isoform. Patient samples are incubated at 56°C for 10–20 min, and the difference in enzyme activity before and after incubation at 56°C typically reflects the bone isoform contribution to total enzyme activity. Electrophoresis on agarose gels following precipitation with neuraminidase or high-performance liquid chromatography has also been used to separate the liver and bone isoforms; however, these methods are tedious, time-consuming and not practical in the setting of routine clinical laboratory testing.[11,12] The precipitation of bone ALP with wheat germ lectin in combination with electrophoresis has also been used to distinguish bone from liver ALP isoforms, but the time-consuming nature of these methods and the lack of sufficient sensitivity has also precluded the routine use of this method.[13]

The development of monoclonal antibodies which preferentially react with the bone-specific isoform of ALP in the late 1980s was hailed as a breakthrough for the routine assessment of the bone-specific form of this enzyme by immunoassay.[14,15] The first commercial reagents using monoclonal antibodies targeted to bone-specific ALP soon became available, and the first assay to receive widespread attention measured the mass amount of enzyme in a two-site immunoradiometric assay (Hybritech Tandem-R Ostase).[16,17] This assay required an overnight incubation and appeared to provide reasonable sensitivity and specificity for the routine determination of bone-specific ALP (Table 5.2). A low but significant degree of cross-reactivity with liver ALP, however, was noted with these reagents. Several laboratories subsequently reported on the clinical application of this assay in patients with Paget's disease, primary hyperparathyroidism, osteoporosis and other metabolic bone diseases.[18,19] Most laboratories reported 14–16% cross-reactivity with the liver isoform, and although this was acceptable, it did pose a problem in patients with co-existing liver disease. Shortly thereafter, a second commercial immunoassay based on enzyme activity measurements for bone-specific ALP (Metra Biosystems, Alkphase-B) became available that employed a single monoclonal antibody to capture bone ALP from serum followed by reactivity with the substrate *p*-nitrophenyl phosphate to assess enzyme activity.[20] This assay required a 3-h incubation and improved the assay turn-around time over the Tandem-R Ostase assay. Initial reports with this activity-based immunocapture assay for bone-specific ALP suggested that the cross-reactivity with the liver isoform was less than that observed with the IRMA mass assay.[20] Other studies, however, put the liver cross-reactivity with this bone ALP activity assay at about the same (8–15%) level as the mass-based immunoradiometric assay, depending on how the cross-reactivity studies were done.[19,21,22] To improve on assay time for bone-specific ALP, Hybritech subsequently developed an activity-based assay (Tandem-MP Ostase) that uses a single biotinylated monoclonal antibody to bind the serum bone ALP, followed by the addition of the *p*-nitrophenyl phosphate substrate to produce an activity-based result.[23] This activity assay uses a 60-min incubation time. The performance characteristics of the mass and activity Ostase assays (Tandem-R and Tandem-MP) are virtually identical, with a regression line of $y = 1.03x + 0.22$ μg/l and a correlation of 0.97

Table 5.2 Immunoassay methods for bone-specific alkaline phosphatase.

Tandem-R Ostase
Two-site IRMA; Hybritech, detection limit, 1.0 μg/l
Standard: dimeric bone ALP from SAOS-2 cells, overnight incubation
Reporting units: μg/l
Tandem-MP Ostase
EIA; Hybritech, detection limit, 0.6 μg/l
Standard: dimeric bone ALP from SAOS-2 cells, 60-min incubation
Reporting units: μg/l
Alkphase-B
EIA, Metra Biosystems, detection limit, 0.7 U/l
Standard; dimeric bone ALP from SAOS-2 cells, 90-min incubation
Reporting units: U/l

from patient result comparison studies.[23] Reactivity with the liver ALP isoform was shown to be similar between the two methods.

CROSS-REACTIVITY OF BONE ALKALINE PHOSPHATASE WITH OTHER TISSUE NON-SPECIFIC ALKALINE PHOSPHATASE ISOFORMS

The issue of liver isoform cross-reactivity with the antibodies used to measure bone ALP has been of concern to investigators evaluating the clinical use of this bone formation marker in the setting of co-existing liver and renal disease. Although the extent of cross-reactivity that has been reported in the literature is relatively modest, nevertheless most studies have shown that when the patient's liver ALP activity level exceeds 110 U/l, cross-reactivity can influence the measured bone ALP result and interpretation. A number of different studies have addressed the liver isoform cross-reactivity issue with the commercially available bone ALP assays. Two approaches have been used to calculate the percentage cross-reactivity with the different assays: (1) heat-inactivation studies; and (2) the use of a slope comparison approach, where the ratio of liver to bone ALP is calculated from bone ALP assay results over an increasing concentration range of liver isoform spiked samples. Table 5.3 compares the liver cross-reactivity results from two such studies that used somewhat different sample populations. The study by Price et al[22] used samples from children and patients with Paget's disease, while the study by Broyles et al.[23] used samples obtained primarily from Paget's disease patients. Overall, the liver cross-reactivity results reported for all three immunoassays were quite similar and in the range of 8–15%. The differences in cross-reactivity observed between the heat-inactivation and slope ratio methods has been ascribed to the fact that the slope liver to bone ratio method does not account for the presence of the bone isoform in the liver-enriched samples. Price et al[22] also noted a difference between the Tandem-R and Alkphase-B assay using the slope ratio method with their samples. They suggested that this may be due to differences in the bone isoform present in children compared to adults and/or the presence of different isoforms of bone origin. In support of the multiple bone isoform possibility, Magnusson et al,[12] using an HPLC method to fractionate the tissue non-specific isoforms, noted at least two bone and three liver isoforms with their chromatography method. In terms of multiple bone isoforms, Magnusson's observations are consistent with the reported presence in the circulation of the glycosylphosphatidylinositol anchor and anchorless forms of bone ALP released via phospholipase D cleavage from the osteoblast cell surface.[6]

Table 5.3 The cross-reactivity of liver alkaline phosphatase in three immunoassays for bone-specific alkaline phosphatase.

Study 1[22]	Study 2[23]
Heat inactivation	Heat inactivation
Tandem-R, 7.1%	Tandem-R, 8.3%
Alkphase-B, 7.9%	Alkphase-B, 10%
	Tandem-MP, 8.1%
Slope ratio	Slope ratio
Tandem-R, 12.7%	Tandem-R, 16.7%
Alkphase-B, 8.7%	Alkphase-B, 14.9%
	Tandem-MP, 16.2%

Given the similarity in amino acid homology between the liver and bone isoforms, it is not surprising to find some cross-reactivity of the liver isoforms in the bone ALP monoclonal antibody-based assay. Investigators, however, generally agree that understanding the potential for a false elevation in the bone ALP result in patients with liver disease should allow for the rational and appropriate use of the bone-specific immunoassay for the clinical purposes intended, namely for monitoring the effects of therapy in patients with established metabolic bone disease.

CLINICAL USE OF THE BONE ALKALINE PHOSPHATASE ASSAY

Biochemical markers of bone turnover have been used to complement radiographic assessment of patients with metabolic bone disease with increasing frequency in the past 10 years, as a result of more specific markers of bone remodelling. Historically, total ALP has been used by clinicians for many years to determine response to medical therapy in patients with bone disease. With the availability of more specific markers of bone resorption and bone formation, physicians treating metabolic bone disease now have more biochemical test options to specifically assess either the bone resorption or bone formation phase of the bone turnover process. In the management of patients with metabolic bone diseases such as Paget's disease, osteoporosis, hyperparathyroidism and metastatic bone disease, the bone markers can help guide therapeutic decisions for selecting antiresorptive therapies, can help confirm patient compliance in taking their medications, and can provide an earlier indication of response to therapy than bone density measurements by radiographic methods (Table 5.4).

Table 5.4 Clinical use of bone alkaline phosphatase measurements.

To select and adjust appropriate antiresorptive therapies for metabolic bone disease
To evaluate the efficacy of new forms of antiresorptive therapy, such as the newer bisphosphonates
To document patient compliance in taking their antiresorptive medications
To confirm and predict a patient's response to antiresorptive medications
To identify the source of elevated total ALP in patients with little evidence of liver disease

There are several advantages to the measurement of bone-specific ALP. Bone ALP determination by immunoassay can and does provide clinicians with a more selective means of determining increases in the rate of bone formation than by simply measuring total ALP enzyme activity, in spite of the low but significant liver isoform cross-reactivity found with the current immunoassay methods. Measurement of bone ALP along with a specific bone resorption marker allows the bone specialist to determine whether coupling of bone resorption to bone formation is normal and whether medical therapy corrects an abnormal resorption to formation coupling event. With certain metabolic bone diseases such as metastatic bone disease, this coupling process is significantly altered in favour of increased bone resorption.

The disadvantages of measuring bone-specific ALP as a marker of bone turnover include the following. Most bone diseases effect a demineralization process, and thus most therapies are directed at inhibiting bone resorption. Measurement of a bone resorption marker is thus more pertinent to the assessment of increased bone loss with disease. In addition, the earliest biochemical changes in bone markers with antiresorptive therapy are seen with biochemical markers of bone resorption. Changes in bone ALP can lag by several weeks the suppression observed with the resorption markers following the start of antiresorptive therapy, as the coupling process is normalized. Finally, the use of bone ALP is compromised somewhat in patients with liver disease with the immunoassay methods currently available.

FUTURE DEVELOPMENTS

The current immunoassays for bone ALP represent a major improvement in the methods developed to date that determine the activity level of this enzyme in blood in health and disease. However, several issues need to be addressed to further improve the selectivity and clinical use of these assays in clinical practice. One of the major issues is standardization between assays. For example, there is no international standard preparation available for bone ALP. From the studies reported to date, it is clear that the immunoassays currently available for bone-specific ALP are different and perhaps may be seeing different epitopes or different isoforms of the bone enzyme. Both manufacturers used the same osteosarcoma cell line for monoclonal antibody development, but differences between the two major commercial activity assays remain. Development of a monoclonal antibody with lower cross-reactivity with the liver enzyme (i.e. less than 5%) should be an additional goal for improving the current assays for bone ALP. Although the current 8–15% cross-reactivity is acceptable, it still complicates the intended use of this assay in discerning increases or decreases in the bone-specific isoform with disease.

REFERENCES

1. Moss DW, Alkaline phosphatase isoenzymes. *Clin Chem* 1982; 28: 2007–16.
2. Harris H, The human alkaline phosphatases: what we know and what we don't know. *Clin Chim Acta* 1989; 180: 177–88.
3. Seargeant LE, Stinson RA, Evidence that three structural genes code for human alkaline phosphatase. *Nature* 1979; 281: 152–4.
4. Weiss MJ, Henthorn PS, Lafferty MA et al, Isolation and characterization of a CBNA encoding a human liver/bone/kidney-type alkaline phosphatase. *Proc Natl Acad Sci USA* 1986; 83: 7182–6.
5. Weiss MJ, Ray K, Henthorn PS et al, Structure of the human liver/bone/kidney-type alkaline phosphatase gene. *J Biol Chem* 1988; 263: 12002–10.
6. Anh D, Dimai H, Hall S, Farley J, Skeletal alkaline phosphatase activity is primarily released from human osteoblasts in an insoluble form and the net released is inhibited by calcium and skeletal growth factors. *Calcif Tissue Int* 1998; 62: 332–40.
7. Fishman WH, Alkaline phosphatase isoenzymes: recent progress. *Clin Biochem* 1990; 23: 99–104.
8. Behnke B, Kemper M, Kruse HP, Muller-Weifel D, Bone alkaline phosphatase in children with chronic renal failure. *Nephrol Dial Transplant* 1998; 13: 662–7.
9. Nailk RB, Gosling P, Price CP, Comparative study of alkaline phosphatase isoenzymes, bone histology and skeletal radiography in dialysis bone disease. *BMJ* 1977; 1: 1307–10.
10. Moss DW, Witby LG, A simplified heat inactivation method for investigating alkaline phosphatase isoenzymes in serum. *Clin Chim Acta* 1975; 61: 63–71.
11. Moss DW, Edwards RK, Improved electrophoretic resolution of bone and liver alkaline phosphatases resulting from partial digestion with neuraminidase. *Clin Chim Acta* 1984; 143: 177–82.
12. Magnusson P, Lofman O, Larsson L, Determination of alkaline phosphatase isoenzymes in serum by HPLC with post-column reaction detection. *J Chromatogr* 1992; 576: 79–86.
13. Rosalki SB, Foo AY, Two new methods for separating and quantifying bone and liver alkaline phosphatase isoenzymes in plasma. *Clin Chem* 1984; 30: 1182–6.
14. Bailyes EM, Seabrook RN, Calvin J et al, The preparation of monoclonal antibodies to human bone and liver alkaline phosphatase and their use in immuno-affinity purification and in studying these enzymes when present in serum. *Biochem J* 1987; 244: 725–33.
15. Hill CS, Wolfert RL, The preparation of monoclonal antibodies which react preferentially with human bone alkaline phosphatase and not liver alkaline phosphatase. *Clin Chim Acta* 1989; 186: 315–20.
16. Garnero P, Delmas PD, Assessment of the serum levels of bone alkaline phosphatase with a new IRMA in patients with metabolic bone disease. *J Clin Endocrinol Metab* 1993; 77: 1046–53.
17. Panigrahi K, Delmas PD, Singer F et al, Characteristics of a two site IRMA for human skeletal alkaline phosphatase in serum. *Clin Chem* 1994; 40: 822–8.
18. Kyd PA, DeVooght K, Kerkhoff F et al, Clinical usefulness of bone alkaline phosphatase in osteoporosis. *Ann Clin Biochem* 1998; 35: 717–25.
19. Withold W, Schulte U, Reinauer H, Method for determination of bone alkaline phosphatase activity; analytical performance and clinical usefulness in patients with metabolic and malignant bone diseases. *Clin Chem* 1996; 42: 210–17.
20. Gomez B, Ardakani S, Ju J et al, Monoclonal antibody assay for measuring bone specific alkaline phosphatase activity in serum. *Clin Chem* 1995; 41: 1560–6.

21. Price CP, Mitchell CA, Moriaty J et al, Mass versus activity: validation of an immunometric assay for bone alkaline phosphatase in serum. *Ann Clin Biochem* 1995; 32: 405–12.
22. Price CP, Milligan TP, Darte C, Direct comparison of performance characteristics of two immunoassays for bone isoform of alkaline phosphatase in serum. *Clin Chem* 1997; 43: 2052–7.
23. Broyles DL, Nielsen RG, Bussett EM et al, Analytical and clinical performance characteristics of Tandem-MP ostase, a new immunoassay for serum bone alkaline phosphatase. *Clin Chem* 1998; 44: 2139–47.

6

Osteocalcin

Caren M Gundberg

Summary • Overview • Circulating osteocalcin • Biological considerations • Analytical considerations • N-terminal mid-molecule fragment • Conclusions • References

SUMMARY

Osteocalcin, a small protein synthesized by mature osteoblasts, odontoblasts and hypertrophic chondrocytes, is a highly specific bone marker which has been used to assess relative degrees of bone turnover. This chapter provides an overview of current ideas on the sources of osteocalcin and its role in bone turnover. The application of osteocalcin assays to routine clinical practice has been limited by the multiple forms of osteocalcin present in the circulation and differences in the ability of available assays to detect these forms. In the immediate future, assays with known specificity and those that are not affected by sample handling and storage should be used. In the longer term, the forms and sources of osteocalcin need to be precisely identified, and all the factors that influence their synthesis, secretion and catabolism need to be established.

OVERVIEW

Osteocalcin is a small protein (49 amino acids) synthesized by mature osteoblasts, odontoblasts, and hypertrophic chondrocytes. The function of osteocalcin has not been precisely defined, but gene targeting has produced a mouse which has the osteocalcin gene ablated. Studies in these animals suggest that osteocalcin may play a role in the regulation of bone turnover. The protein is characterized by the presence of three residues of γ-carboxyglutamic acid (Gla). Vitamin K is required for the addition of a second γ-carboxyl group to the side-chain of specific glutamate residues, resulting in the production of Gla residues in the newly synthesized protein. This reaction, which is inhibited by warfarin, is identical to that involved in the activation of the vitamin K-dependent blood coagulation factors. In osteocalcin, these Gla residues are responsible for facilitating the binding of the protein to hydroxyapatite and maintaining its secondary structure. In human bone, the first potential Gla residue at position 17 is only partially carboxylated (55–89%), while residues 21 and 24 are greater than 90% γ-carboxylated. This is due to partial carboxylation of the protein during synthesis.[1] Presumably, the circulating form has the same carboxylation features, but this has never been directly tested.

CIRCULATING OSTEOCALCIN

Serum osteocalcin is considered a marker of bone formation. This is based on several observations: (1) serum osteocalcin is correlated with bone formation as determined by histomorphometry or calcium kinetics;[2] (2) osteocalcin synthesis increases with mineralization and with progressive osteoblastic differentiation;[3] and (3) early animal studies suggested that circulating osteocalcin originated from new bone synthesis and not from breakdown of bone matrix.[4] Although osteocalcin mRNA has also been detected in bone marrow megakaryocytes and peripheral blood platelets, only very low levels of the protein itself are detected, and it is unlikely that platelet osteocalcin contributes to serum levels of the protein.[5]

The presence of Gla-containing proteins in bone suggests that vitamin K may play a role in skeletal health. Recent epidemiological studies indicate that vitamin K intake is inversely related to fracture incidence.[6] Assessment of undercarboxylated osteocalcin (ucOC) is potentially useful for determining subtle changes in vitamin K status. Methods have been developed with the recognition that the affinity of osteocalcin for hydroxyapatite depends on the number of Gla residues in the protein, and the portion of immunoassayable osteocalcin that does not bind to hydroxyapatite has been taken to be undercarboxylated. In elderly institutionalized women, undercarboxylated osteocalcin was found to be elevated compared to young, healthy premenopausal women.[7] In a subset of subjects from the EPIDOS study, serum ucOC, but not total osteocalcin, was predictive of future hip fractures,[8] and serum levels of ucOC showed a significant negative correlation with bone mineral density (BMD). When bone mineral density and serum ucOC were considered together, the ability to predict fractures was further increased. Whether vitamin K nutrition per se, total nutritional status, or some other underlying biochemical mechanism is responsible for the association between ucOC and risk of fracture has not been defined.

Table 6.1 Osteocalcin considerations.

1. Tissue specificity
   - Bone, dentin, calcified cartilage
   - Low levels of mRNA found in megakaryocytes (does not contribute to circulating levels)
2. Clearance
   - Renal (serum half-life of 20 min)
3. Assays
   - Serum (heparin, EDTA or citrated plasma, dependent upon assay)
   - RIA, ELISA, IRMA, analyser

BIOLOGICAL CONSIDERATIONS

Major deviations from normal concentrations of circulating osteocalcin are a consequence of changes in the synthesis or degradation of the protein. Such changes may result from physiological alterations in skeletal homeostasis that accompany normal development or may be associated with specific disease states (Table 6.1).

Osteocalcin levels follow a circadian rhythm characterized by a decline during the morning to a low around noon, followed by a gradual rise which peaks after midnight. The difference between the peak and nadir in a 24-h period can range from 10% to 20%, depending upon the assay used. Serum osteocalcin levels have also been reported to vary significantly during the menstrual cycle, with the highest levels observed during the luteal phase.[9,10]

The rate of glomerular filtration or renal catabolism of osteocalcin influences circulating osteocalcin levels. The chief route of circulating osteocalcin catabolism is renal glomerular filtration and degradation. Administration of radiolabelled osteocalcin to rats shows primarily renal uptake, with minor amounts of radioactivity being concentrated in bone, liver, and other soft tissues.[11] The plasma half-life is about 20 min in humans. When renal glomerular function is impaired, circulating osteocalcin increases. This occurs when the glomerular filtration rate (GFR) is below 20–30 ml/min per

Table 6.2 Advantages and limitations of using osteocalcin as a clinical index of bone turnover.

Advantages
Bone specific
Correlates with bone formation as assessed by histomorphometry or calcium kinetics
Correlates with increases or decreases in bone turnover
Widely available
Limitations
Poor sample stability
Wide variation in reported results with different kits
Universal standards unavailable
Both intact protein and fragments circulate

1.73 m^2 body surface area, or serum creatinine is greater than 160 μmol/l.[12] In children, increases occur at GFR below 40 ml/min per 1.73 m^2.[13]

The major advantages in using osteocalcin as a clinical index of bone turnover is its tissue specificity, its wide availability, and relatively low within-person variation (between 12% and 22%)[14] (Table 6.2). In general, serum levels of osteocalcin are elevated in patients with diseases characterized by high bone turnover rate, and reflect the expected changes in bone formation following surgical or therapeutic intervention. An exception is found in Paget's disease, in which serum alkaline phosphatase (either total or bone specific) is a better predictor of severity of disease than osteocalcin.[15]

ANALYTICAL CONSIDERATIONS

Several commercial assay kits are available for serum osteocalcin measurements. These include bovine or human, monoclonal or polyclonal antibody-based immunoassays. These are in radioimmunoassay or ELISA format. Automated methods are available on the Nichols Advantage and Roche ELECSYS systems (Table 6.3).

The major difficulty when choosing an assay for measuring osteocalcin is related to the fact that fragments of the protein circulate. It is thought that approximately one-third of circulating osteocalcin is composed of the intact molecule, one-third is a large N-terminal mid-molecule fragment, and one-third consists of smaller fragments.[16,17] Although the large N-terminal mid-molecule fragment has not been directly sequenced, it is thought to encompass residues 1–43, based on the fact that osteocalcin contains a tryptic site at residue 43. Other circulating fragments have also been detected but have not been fully characterized. Recent data indicate that there are smaller N-terminal immunoreactive species of osteocalcin in serum in addition to the larger N-terminal mid-molecule fragment.[18–20] They are in greatest abundance in normal children, in patients with osteoporosis and in hyperparathyroidism, suggesting that these immunoreactive species are produced in a high-turnover state (Table 6.2).

Such circulating heterogeneity has resulted in inconsistent results when various kits are compared and a wide range of values has been reported in both control and patient populations. Several studies have compared both research assays and commercial kits and found that results could not be compared among assays.[21,22] Some of the differences between assays may be due to standards and methods, but the most likely source of variation is the specificity of the antibodies and their ability to recognize the circulating fragments of osteocalcin. In some assays, epitope specificity and the degree of reactivity with multiple circulating forms of the protein are unknown (Table 6.3).

Several laboratories have developed two-site immunoassays with the intention of measuring only the intact molecule. In some cases, however, these have not improved the specificity of the measurement, and nor have they provided better discrimination between normal and pathological samples. The failure to improve specificity with a two-site assay can be

Table 6.3 Commercially available osteocalcin kits.

Name	Company	Type	Sample	Sensitivity
Intact kits[a]				
ELSA-OST-NAT	CIS-bio International, Gif-sur-Yvette, France	IRMA	50 μl serum	0.3 μg/l
Intact Osteocalcin	Biomedical Technologies, Stoughton, MA, USA	ELISA	20 μl serum	0.5 μg/l
Human Osteocalcin	Quest, San Juan Capistrano, CA, USA	IRMA	25 μl serum	0.1 μg/l
N-tact Osteo-SP	Incstar Corporation, Stillwater, MN, USA	IRMA	20 μl serum	0.2 μg/l
Two-site kits[b]				
Elecsys N-Mid Osteocalcin	Roche Diagnostics GmbH, Mannheim, Germany	ECLIA*	20 μl serum or hep or EDTA plasma	0.5 μg/l
ELSA-OSTEO	CIS-bio International Gif-sur-Yvette, France	IRMA Capture: 25–37 Detection: 5–13	50 μl serum	0.4 μg/l
Active Human Osteocalcin	Diagnostic Systems Laboratories, Webster, TX, USA	IRMA Capture: ? Detection: 30–49	25 μl serum	0.3 μg/l
N-Mid Osteocalcin	Osteometer Biotech, Rodovre, Denmark	IRMA Capture: 20–43 Detection: 7–19	25 μl serum	2.0 μg/l
Mid-tact OC	Biomedical Technologies, Stoughton, MA, USA	ELISA Capture: 5–19 Detection: 21–43	25 μl serum	0.5 μg/l
BGP	Mitsubishi Yuka IDS, Boldon, Tyne and Wear, UK	IRMA Capture: 12–33 Detection: 30–49	50 μl serum	0.4 μg/l
Single-site kits[c]				
Ostk-PR	CIS-bio International, Gif-sur-Yvette, France	RIA; Polyclonal antibody to bovine OC	100 μl serum or hep plasma	1 μg/l
Osteocalcin	Diagnostic Systems Laboratories, Webster, TX, USA	RIA; Polyclonal antibody to human OC	25 μl serum or hep plasma	0.7 μg/l
OSCAtest	Brahms-Henning, Berlin, Germany	RIA; Polyclonal antibody to human OC	50 μl serum	0.5 μg/l
NovoCalcin	METRA Biosystems, Mountain View, CA, USA	ELISA; Single monoclonal antibody to bovine OC	25 μl serum	0.45 μg/l

[a]Measures only intact osteocalcin. [b]Variable reactivity dependent upon antibodies used. [c]Precise epitope unknown. *ECLIA: Electro chemiluminescense immunoassay.

attributed to the fact that these new assays may measure large circulating fragments as well as the intact protein that they were specifically designed to assay. As an example, Deftos et al developed two separate assays for intact osteocalcin in two-site formats. The same capture antibody was used in both assays, but two different monoclonal antibodies, both directed to the same 12 amino acid N-terminal sequence, were employed for read-out. These assays gave different absolute values with the same serum samples and using the same standards, and suggest circulating heterogeneity in the N-terminus of the molecule.[23]

Most two-site assays have been developed using polyclonal or monoclonal antibodies derived from animals injected with the intact osteocalcin molecule, and region-specific antibodies, subsequently affinity purified or cloned. However, in many cases the exact epitopes which the antibodies recognize are unknown. Only those assays which require the absolute extremes of the molecule can be regarded as truly intact. Two recent studies showed that when intact and C-terminal fragments of osteocalcin were removed from patient sera, 8 of 10 commercial kits detect residual immunoreactivity ranging from 13% to 75% of their original value, suggesting that most kits will detect N-terminal osteocalcin fragments, potentially of variable size.[24,25]

Calcium dependence of the antibodies may also account for some of the differences between kits. Some research assays have been reported to be sensitive to calcium-chelating agents and are thought to have conformational epitopes.[26,27] With some assays, differences have been observed between serum and plasma and are influenced by the type of anticoagulant. Osteocalcin levels are often higher in serum compared to plasma prepared with sodium citrate or EDTA.[28–30] This may relate to sensitivity of the antibodies to conformational changes in osteocalcin resulting from calcium chelation. Alternatively, loss of secondary structure may make the protein more susceptible to proteolysis. Haemolysis and lipaemia will also affect assay results. Osteocalcin may bind to lipid, rendering it non-immunoreactive. Peptide hydrolases released by erythrocytes degrade the protein, resulting in reduced values.[26]

N-TERMINAL MID-MOLECULE FRAGMENT

The large N-terminal mid-molecule fragment can be generated by proteolysis during sample processing and storage. Osteocalcin levels decrease with incubation at room temperature when measured by conventional RIAs or by intact assays, but values are stable with assays that recognize both the intact and large N-terminal mid-molecule fragment. With these assays, the apparent instability of osteocalcin in the circulation and during sample handling is eliminated, and measurement of intact plus the large fragments provides a more comprehensive picture of osteocalcin biosynthesis.[16,17,31] Furthermore, these assays are superior to others in their ability to assess the influence of age, renal function and menopause on BMD, and, in general, correlate better with BMD than do single-site assays.[32,33]

It should be noted that not all such assays measure intact osteocalcin and the 1–43 fragment equivalently but can overestimate the 1–43 species.[34] If degradation occurs during processing and storage, increased reactivity of these assays to a catabolic fragment will result in falsely elevated levels of total osteocalcin, resulting in misclassification of the patient. In this setting, the use of an intact assay combined with proper sample handling and storage may provide the most reliable results.

The N-terminal mid-molecule fragment is found in fresh serum samples when all precautions are taken to limit proteolysis, suggesting that some production occurs in vivo. Based on the observation that antiresorptive therapy has no effect on circulating levels, this fragment appears not to be produced by osteoclastic dissolution of bone. However, the N-terminal mid-molecule fragment was detected in conditioned media from human osteoblast-like cells, but whether serum in the culture medium provided the proteolytic enzymes was not tested.[16] Degradation of intact osteocalcin in the circulation or by the osteoblast in situ are also potential co-existing sources. The source of this fragment, therefore, has never been exactly determined.

In order to investigate this further, we have developed a number of two-site assays for osteocalcin based on monoclonal antibodies to synthetic peptides of osteocalcin. An intact assay utilizes antibodies made to residues 1–9 and 30–49 and requires the extremes of the protein. An N-terminal assay is derived from antibodies specific for 11–19 and 1–9, while an N-terminal mid-molecule is derived from antibodies specific for residues 1–9 and 17–29. These assays can detect multiple forms of osteocalcin in the circulation if they exist. Thus, the N-terminal assay recognizes potential peptides encompassing 1–19 and greater, while the N-terminal mid-molecule assay sees peptides 1–29 and greater. All assays are indifferent to the presence or absence of Gla residues. Importantly, these assays display equal affinity for the intact molecule and potential fragments.[20] With these assays we observe that in children only 20% of the circulating immunoreactivity is intact osteocalcin, whereas in adults we find that about 50% is intact. These data suggest that the appearance of smaller osteocalcin fragments during skeletal growth may have some important regulatory function.

In order to investigate the source of osteocalcin fragments more thoroughly, we have used a transgenic mouse model that carries the full-length human osteocalcin cDNA. These animals provide us with the ability to culture primary osteoblasts which produce high levels of human osteocalcin for study and characterization (300–400 ng/ml of media). The trHOC mice express both human and mouse osteocalcin in bone developmental patterns indistinguishable from each other and are regulated consistent with species differences.[35] We prepared primary osteoblasts from these animals and maintained them in culture for 4 weeks, during which time they mineralized and produced matrix proteins in a maturation-dependent fashion, as demonstrated in other species.[36] Intact osteocalcin and smaller fragments were found in very

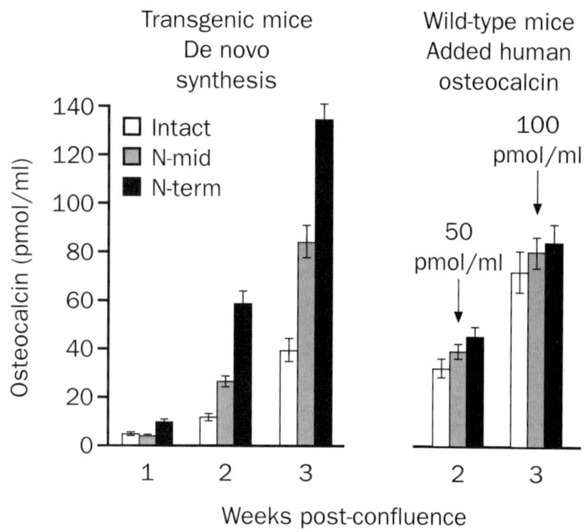

Fig. 6.1 Osteocalcin in primary mouse osteoblasts. Data represent accumulation or incubation for >2 h.

low concentrations at confluence and 1 week post-confluence. At 2 and 3 weeks post-confluence, when the cultures were maximally depositing mineral, there is an abrupt and progressive increase in the appearance of both intact osteocalcin and smaller fragments ≥1–19 or ≥1–29. The ratio of fragment to intact osteocalcin ranged from 3 : 1 for the peptides larger than 29 residues to 5 : 1 for peptides 19 residues and longer. To control for degradation by serum in the media, purified human osteocalcin was added to primary osteoblasts derived from wild-type mice that did not possess the human osteocalcin cDNA. Approximately 30% of the intact osteocalcin was degraded to smaller peptides (Fig. 6.1). Taken together, the data suggest that osteocalcin peptides are not derived only from degradation by serum proteases but that the osteoblast itself may be the origin of some of these peptides. We speculate that the catabolism of osteocalcin may play a regulatory role in limiting its deposition onto the mineral phase or that the fragment itself may play a role in osteoblast function.

CONCLUSIONS

Circulating osteocalcin is a highly specific osteoblastic marker which has been used for assessing relative degrees of bone turnover in many clinical studies. However, application of osteocalcin assays to routine clinical practice has been limited by the existence of multiple forms of osteocalcin in the circulation and by discrepancies in the ability of available assays to detect these forms. To achieve more reliable and uniform information from osteocalcin assays in the immediate future, assays with known specificity and those that are not affected by sample handling and storage should be recommended. Uniform standards and/or controls for use with these assays would be advantageous. For the long term, the forms and sources of osteocalcin must be precisely identified and all the factors which influence their synthesis, secretion and catabolism must be established. Since circulating levels of osteocalcin may be related to its biological function, it is of obvious importance to ultimately gain a better understanding of the role of osteocalcin in bone biology.

REFERENCES

1. Cairns JR, Price PA, Direct demonstration that the vitamin K-dependent bone Gla protein is incompletely g-carboxylated in humans. *J Bone Miner Res* 1994; 9: 1989–97.
2. Eastell R, Delmas PD, Hodgson SF et al, Bone formation rate in older normal women: concurrent assessment with bone histomorphometry, calcium kinetics, and biochemical markers. *J Clin Endocrinol Metab* 1988; 67: 741–8.
3. Owen TA, Aronow M, Shalhoub V et al, Progressive development of the rat osteoblast phenotype in vitro: reciprocal relationships in expression of genes associated with osteoblast proliferation and differentiation during formation of the bone extracellular matrix. *J Cell Physiol* 1990; 143: 420–30.
4. Price PA, Williamson MK, Lothringer JW, Origin of the vitamin K-dependent bone protein found in plasma and its clearance by kidney and bone. *J Biol Chem* 1981; 256: 12760–6.
5. Thiede MA, Smock SL, Petersen DN et al, Presence of messenger ribonucleic acid encoding osteocalcin, a marker of bone turnover, in bone marrow megakaryocytes and peripheral blood platelets. *Endocrinology* 1994; 135: 929–37.
6. Booth SL, Tucker KL, Chen H et al, Dietary vitamin K intakes are associated with hip fracture but not with bone mineral density in elderly men and women. *Am J Clin Nutr* 2000; 71: 1201–8.
7. Plantalech L, Guillaumont M, Vergnaud P et al, Impairment of gamma carboxylation of circulating osteocalcin (bone Gla protein) in elderly women. *J Bone Miner Res* 1991; 6: 1211–16.
8. Vergnaud P, Garnero P, Meunier PJ et al, Undercarboxylated osteocalcin measured with a specific immunoassay predicts hip fracture in elderly women: the EPIDOS study. *J Clin Endocrinol* 1997; 82: 719–24.
9. Gundberg CM, Markowitz ME, Mizruchi M, Rosen JF, Osteocalcin in human serum: a circadian rhythm. *J Clin Endocrinol Metab* 1985; 60: 736–9.
10. Nielsen H, Brixen K, Kassem M et al, Diurnal rhythm in serum osteocalcin: relation to sleep, growth hormone, and PTH (1–84). *Calcif Tissue Int* 1991; 49: 373–7.
11. Farrugia W, Melick RA, Metabolism of osteocalcin. *Calcif Tissue Int* 1986; 39: 234–8.
12. Delmas PD, Wilson DM, Mann KG, Riggs BL, Effect of renal function on plasma levels of bone Gla-protein. *J Clin Endocrinol Metab* 1983; 57: 1028–30.
13. Friedman A, Heiliczer J, Gundberg C et al, Serum osteocalcin concentrations in children with chronic renal insufficiency who are not undergoing dialysis. *J Pediatr* 1990; 116: S55–9.
14. Hannon R, Blumsohn A, Naylor K, Eastell R, Response of biochemical markers of bone turnover to hormone replacement therapy: impact of biological variability. *J Bone Miner Res* 1998; 13: 1124–33.
15. Delmas PD, Demiaux B, Malaval L et al, Serum bone GLA-protein is not a sensitive marker of bone turnover in Paget's disease of bone. *Calcif Tissue Int* 1986; 38: 60–1.
16. Garnero P, Grimaux M, Seguin P, Delmas PD, Characterization of immunoreactive forms of human osteocalcin generated in vivo and in vitro. *J Bone Miner Res* 1994; 9: 255–64.
17. Rosenquist C, Quist P, Bjarnason N, Christiansen C, Measurement of a more stable region of osteocalcin in serum by ELISA with two monoclonal antibodies. *Clin Chem* 1995; 41: 1439–45.

18. Chen JT, Hosoda K, Hasumi K et al, Serum N-terminal osteocalcin is a good indicator for estimating responders to hormone replacement therapy in postmenopausal women. *J Bone Miner Res* 1996; 11: 1784–92.
19. Gorai L, Hosoda K, Taguchi Y et al, A heterogeneity in serum osteocalcin N-terminal fragments in Paget's disease: a comparison with other biochemical indices in pre- and post-menopause. *J Bone Miner Res* 12(S1): T678.
20. Gundberg CM, Nieman SD, Abrams S, Rosen H, Vitamin K status and bone health: an analysis of methods for the determination of 'undercarboxylated' osteocalcin. *J Clin Endocrinol Metab* 1998; 83, 3258–66.
21. Delmas PD, Christiansen C, Mann KG, Price PA, Bone gla protein (osteocalcin) assay standardization report. *J Bone Miner Res* 1990; 5: 5–10.
22. Masters PW, Jones RG, Purves DA et al, Commercial assays for serum osteocalcin give clinically discordant results. *Clin Chem* 1994; 40: 358–63.
23. Deftos LJ, Wolfert RL, Hill CS et al, Two-site assays of bone Gla protein (osteocalcin) demonstrate immunochemical heterogeneity of the intact molecule. *Clin Chem* 1992; 38: 2318–21.
24. Diego EMD, Guerrero R, Piedra C, Six osteocalcin assays compared. *Clin Chem* 1994; 40: 2071–7.
25. Souberbielle JC, Marque D, Bonnet P et al, Simple method to evaluate specificity of osteocalcin immunoassays. *Clin Chem* 1997; 43,1663–4.
26. Tracy RP, Andrianorivo A, Riggs BL, Mann K, Comparison of monoclonal and polyclonal antibody-based immunoassays for osteocalcin: a study of sources of variation in assay results. *J Bone Miner Res* 1990; 5: 451–61.
27. Kuronen I, Kokko H, Parviainen M, Production of monoclonal and polyclonal antibodies against human osteocalcin sequences and development of a two-site ELISA for intact human osteocalcin. *J Immunol Methods* 1993; 163: 223–40.
28. Power MJ, O'Dwyer B, Breen E, Fottrell PF, Osteocalcin concentrations in plasma prepared with different anticoagulants. *Clin Chem* 1991; 37: 281–5.
29. Durham BH, Robinson J, Fraser WD, Differences in the stability of intact osteocalcin in serum, lithium heparin plasma and EDTA plasma. *Ann Clin Biochem* 1995; 32: 422–3.
30. Colford JW, Lueddecke BA, Salvati M et al, Immunoradiometric assay for intact human osteocalcin (1–49) without cross-reactivity to breakdown products. 1999; *Clin Chem* 45: 526–31.
31. Blumsohn A, Hannon RA, Eastell R, Apparent instability of osteocalcin in serum as measured with different commercially available immunoassays. *Clin Chem* 1995; 41: 318–20.
32. Dumon J, Wantier C, Mathieu H, Body JJ, Technical and clinical validation of a new immunoradiometric assay for human osteocalcin. *Eur J Endocrinol* 1996; 135: 231–7.
33. Minisola S, Rosso R, Romangnoli E et al, Serum osteocalcin and bone mineral density at various skeletal sites: a study performed with three different assays *J Lab Clin Med* 1997; 129: 422–9.
34. Hellman J, Käkönen S-M, Matikainen M-T et al, Epitope mapping of nine monoclonal antibodies against osteocalcin; combinations into two-site assays affect both assay specificity and sample stability. *J Bone Miner Res* 1996; 11: 1165–75.
35. Clemens TL, Tang H, Maeda S et al, Analysis of human osteocalcin expression in transgenic mice reveals a species difference in vitamin D regulation of mouse and human osteocalcin genes. *J Bone Miner Res* 1997; 12: 1570–6.
36. Lian JB, Shalhoub V, Aslam F et al, Species-specific glucocorticoid and 1,25 dihydroxyvitamin D responsiveness in mouse MC3T3-E1 osteoblasts: dexamethasone inhibits osteoblast differentiation and vitamin D down-regulates osteocalcin gene expression. *Endocrinol* 1997; 138: 2117–27.

7

The N- and C-terminal propeptides of human procollagen type I (PINP and PICP): molecular heterogeneity and assay technology

Jette Brandt, Jette Klose Frederiksen, Charlotte Harken Jensen and Børge Teisner

Summary • Introduction • Materials and methods • SDS–PAGE • Results of PINP analysis • Results of PICP analysis • Discussion • References

SUMMARY

The N- and C-terminal propeptides of human procollagen type I (PINP and PICP) were analysed in relation to available assay technologies. Circulating PINP (α1) exists in a trimeric and monomeric form, and transition takes place at 37°C. The anti-PINP antibodies applied in the PINP-RIA kit and those utilized in the PINP-ELISA recognize the PINP (α1) in both molecular forms. However, only the ELISA technique measured the PINP (α1) monomer efficiently, whereas results of the RIA kit are highly sensitive to the size distribution of PINP and to the thermal transition which is an ongoing in vivo process. The PICP-ELISA technique developed showed significant correlation with the results obtained with the PICP-RIA kit. Comparison of serum concentrations of PINP and PICP in children, and in patients with primary hyperparathyroidism before and after surgical treatment, suggests PINP to be a more sensitive biomarker than PICP in relation to growth and bone metabolism. The heterogeneity of PINP and the labile nature of the trimeric structure have to be taken into consideration in the choice of assay technology and during development of new generations of PINP assays.

INTRODUCTION

Collagen type I is present in soft connective tissues and bone, where it constitutes more than 90% of the organic matrix. Collagen type I is derived from procollagen type I, which is secreted from fibroblasts and osteoblasts. Procollagen type I contains N- and C-terminal extensions, which are removed by specific proteases during the conversion of procollagen to collagen.[1] The extensions are referred to as the C- and N-terminal propeptides of procollagen type I (PICP and PINP).

Human PINP was originally isolated from

amniotic fluid and, prior to its identification as a homomer of α1 chains of PINP, was referred to as fetal antigen 2.[2] Polyclonal antibodies against the α1 chain have formed the basis for immunoassays for quantification of PINP, i.e. electroimmunoassay and ELISA[3,4] as well as RIA.[5,6]

The anti-PINP antibodies used in electroimmunoassay and ELISA recognize two molecular forms of PINP which appear as distinct peaks following size chromatography,[3,4,7,8] but when the same antibodies were applied in a competitive RIA using ^{125}I-labelled PINP, only the high molecular weight form of PINP was detected.[5] In relation to assay technology, it is important to realize that the trimeric structure of PINP is held together only by non-covalent forces within the helical region (51 amino acid residues), and the thermal stability of soluble collagen is low.[9]

Analyses of PINP in patients with hypovitaminosis D-induced hyperparathyroidism,[4] Paget's disease[10] and postmenopausal osteoporosis[11] suggest PINP as a useful marker of bone metabolism.

The trimeric structure of PICP is stabilized by interchain disulphide bonds, and a well-established RIA kit for quantification of serum PICP has been commercially available for several years.[12]

The aims of the present review were, in relation to molecular heterogeneity, to summarize our observations on: (1) PINP antigen in different body fluids; (2) the lability of the trimeric structure of PINP; (3) the specificity of the available antibodies; (4) the extraction of PINP from circulation; (5) the development of a PICP ELISA technique; and (6) PINP and PICP as biomarkers in relation to growth and bone metabolism.

MATERIALS AND METHODS

Biological samples

Amniotic fluid, cord serum and sera from children were made available from the Department of Clinical Genetics, Vejle County Hospital, Vejle, and the Departments of Obstetrics and Gynaecology and Paediatrics at Odense University Hospital. Corresponding sera from femoral arterial blood and the hepatic and renal veins were collected as described previously.[13] Sera used to establish the normal reference intervals were obtained from the Blood Bank at Odense University Hospital.

Immunoassays

Electroimmunoassays

Rocket immunoelectrophoresis and fused rocket immunoelectrophoresis were performed in agarose containing monospecific rabbit anti-PINP as described previously.[3]

ELISA techniques

The ELISAs for quantification of PINP and PICP were developed as sandwich techniques using immunospecifically purified rabbit anti-PINP or anti-PICP antibodies as catcher antibody in Maxisorp microtitre plates, and biotinylated anti-PINP or anti-PICP antibodies as indicator.[4,14] The intra- and interassay CVs for both ELISA techniques were below 5%.

Direct ELISA applied to compare the specificity of the anti-PINP antibodies supplied with the PINP RIA kit (Orion Diagnostica, Espoo, Finland) and the anti-PINP used in the development of the ELISA was performed on Maxisorp microtitre plates coated with dilution series of size chromatography fractions of purified PINP. Following washing, the plates were incubated with primary antibodies for 18 h at 4°C. Biotin-labelled goat anti-rabbit Ig (DAKO Ltd, Denmark) was added and the plate processed as described previously.[7]

Radioimmunoassay techniques (RIA)

The Procollagen, Intact ^{125}I-PINP and PICP RIA kits were purchased from Orion Diagnostica, and used in accordance with the manufacturer's recommendations.

Size chromatography

Two hundred microlitres of purified PINP or the biological fluid to be analysed were applied to a Superose 12 HR 10/30 column connected to an Äcta FPLC System (Pharmacia). The chromatography was performed in phosphate-buffered saline (PBS), pH 7.3, at a flow rate of 0.5 ml/min, and 0.5-ml fractions were collected.

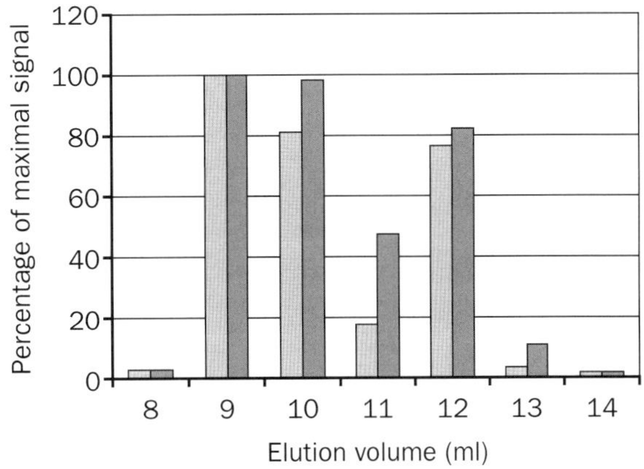

Fig. 7.2 ELISA analysis of the anti-PINP antibodies supplied with the PINP-RIA kit (grey columns) and the antibodies used for PINP-ELISA development (black columns). The microtitre plates were coated with fractions (elution volume) of purified PINP after size chromatography.

SDS-PAGE

SDS-PAGE analysis was performed under reducing and non-reducing conditions on 4–20% gradient gels with discontinuous buffers and silver stained.

RESULTS OF PINP ANALYSIS

Fig. 7.1 shows a PINP antigen profile analysed by rocket immunoelectrophoresis following size chromatography of purified PINP.

Diluted fractions were coated onto microtitre plates, and after washing, the antibodies to be analysed, i.e. the rabbit anti-PINP supplied with the RIA kit or the IgG fraction of the rabbit anti-PINP used for ELISA development, were added. The results (Fig. 7.2) clearly demonstrate that both antibodies reacted with both molecular forms of PINP.

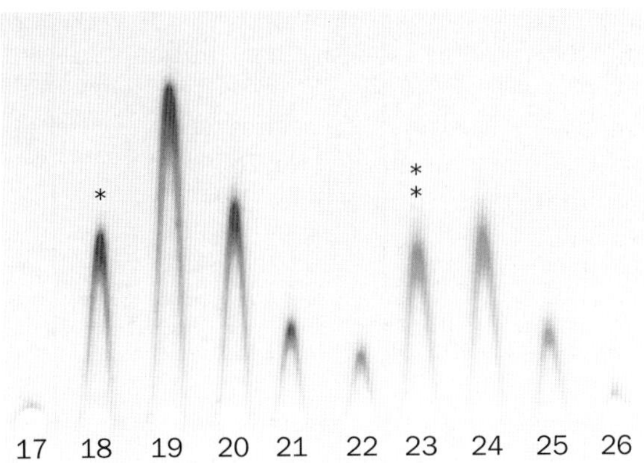

Fig. 7.1 The antigen profile of purified PINP following size chromatography analysed by rocket immunoelectrophoresis.

Characterization of the two molecular forms of PINP

The fractions marked with an asterisk in Fig. 7.1 were compared in fused rocket immunoelectrophoresis, and the interaction between the precipitates was that of immunological identity (Fig. 7.3a). Analysis in SDS-PAGE (Fig. 7.3b) revealed migration of both forms corresponding to a molecular mass of 27 kDa. Mass spectrometry of the two forms of purified PINP verified intact PINP $\alpha 1$ chains with a molecular mass of 14 329 Da.[8]

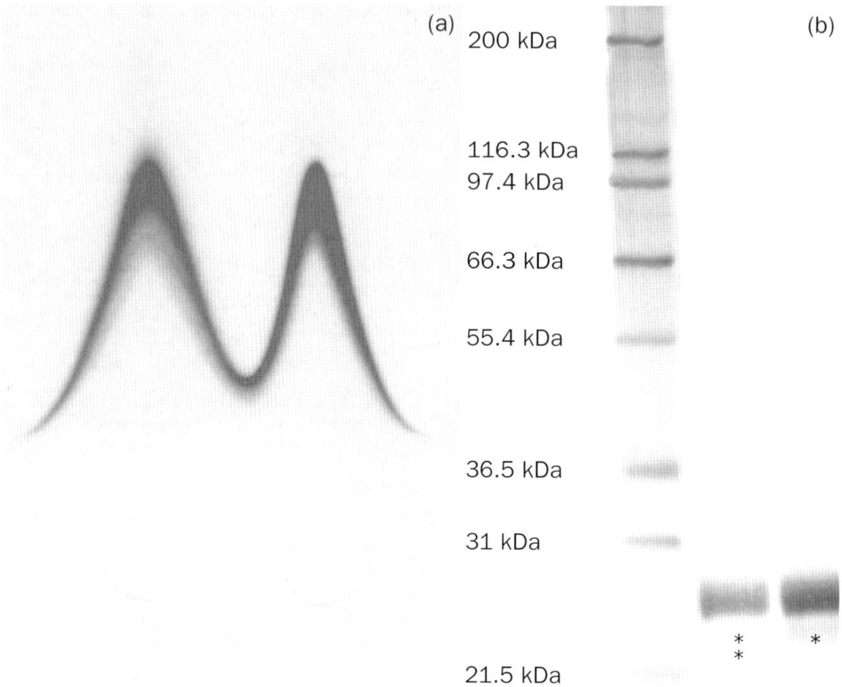

Fig. 7.3 Demonstration of immunological identity and identical apparent mass (27 kDa) of the two molecular forms of PINP analysed by fused rocket immunoelectrophoresis (a) and SDS-PAGE (b).

Heterogeneity in biological fluids and thermal transition of PINP

In order to exclude the possibility that the molecular heterogeneity was due to an artefact introduced during purification of PINP pools of cord sera, amniotic fluids and normal human sera were analysed. Because of the pronounced differences in PINP concentrations (cord serum 2 μg/ml, amniotic fluid 25 μg/ml, human serum 0.16 μg/ml), the results are expressed as percentages of the maximal signal at 492 nm of fractions diluted to fit the dynamic range of the ELISA. Fig. 7.4 shows heterogeneity in all three biological fluids.

The high molecular weight form of PINP represents intact α1 chains in a trimeric structure, and the low molecular weight form the monomer.[8] However, the trimeric structure is labile, and transition takes place at 37°C (Fig. 7.5). This thermal transition has also been reported in normal human serum, and in this context it is important to consider that the RIA only measures α1 chains in the trimeric form efficiently.[7,8]

Measurement of PINP in cord serum by ELISA and RIA revealed concentrations of 2.2 and 0.75 μg/ml, respectively. This difference seems to be due to the size distribution in cord

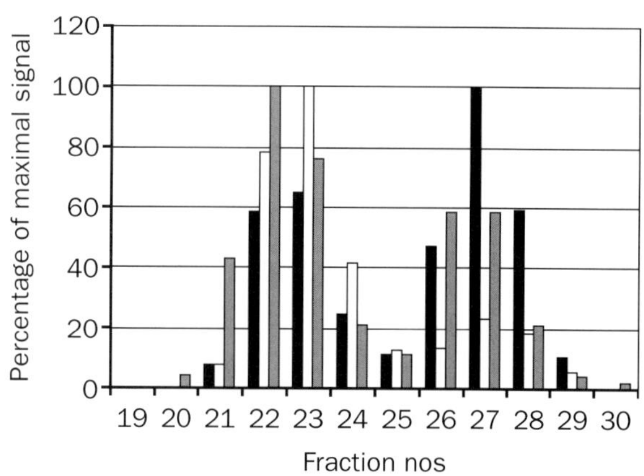

Fig. 7.4 Size distribution of PINP (α1) antigen of cord serum (black), normal human serum (white) and amniotic fluid (grey) as analysed by ELISA following size chromatography.

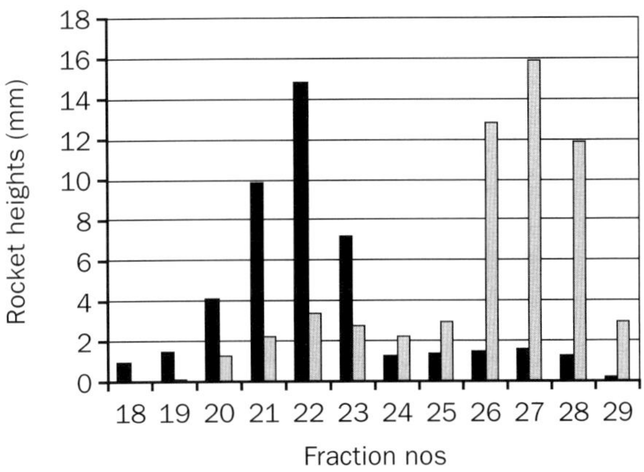

Fig. 7.5 Size distribution analysed by rocket immunoelectrophoresis of purified PINP before (black columns) and after (grey columns) incubation at 37°C for 72 h.

serum (Fig. 7.4). Following incubation of cord serum at 37°C for 24 h, the apparent concentration of PINP measured by ELISA remained unchanged (2.2 versus 2.1 μg/ml), whereas the concentration obtained by RIA decreased from 0.75 to 0.15 μg/ml.

Extraction of PINP from the circulation in relation to molecular heterogeneity

PINP concentrations were measured in samples obtained during catheterization of the renal and hepatic veins, and the results compared to simultaneously collected femoral arterial blood samples.

The PINP concentration in the hepatic vein was 102 ng/ml, and in the renal vein it was 117 ng/ml, whereas the corresponding arterial PINP concentrations were 123 ng/ml and 130 ng/ml respectively. Fig. 7.6 shows the size distribution of PINP in the femoral artery and renal as well as hepatic veins, indicating that both molecular forms are cleared from the circulation by both renal and hepatic extraction.

RESULTS OF PICP ANALYSIS

SDS-PAGE analysis of PICP purified from amniotic fluid revealed a band at 97 kDa (unreduced) which, under reducing conditions, migrated corresponding to a molecular mass of 31–33 kDa. The N-terminal amino acid

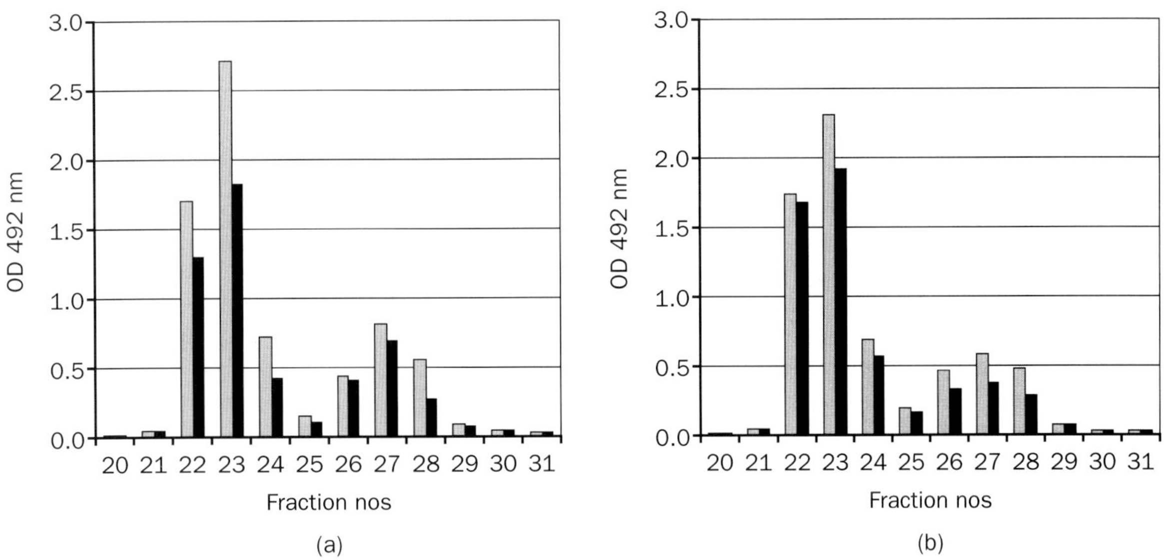

Fig. 7.6 Size distribution of PINP in hepatic vein (a) and renal vein (b) compared to simultaneously collected samples from the femoral artery (grey columns).

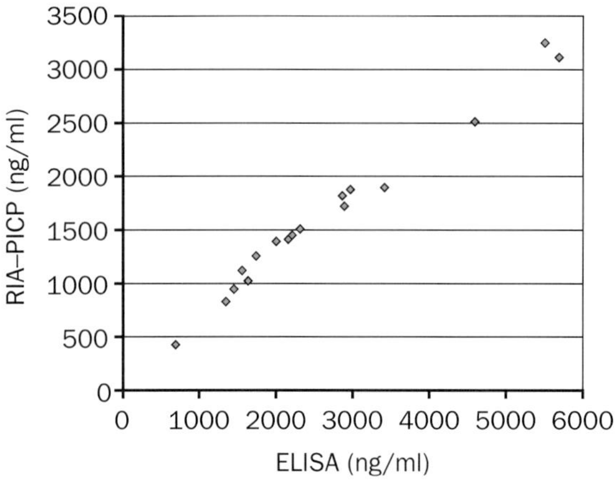

Fig. 7.7 Comparison of PICP concentrations in serum samples from 17 children analysed by the PICP-RIA kit and the developed PICP-ELISA technique.

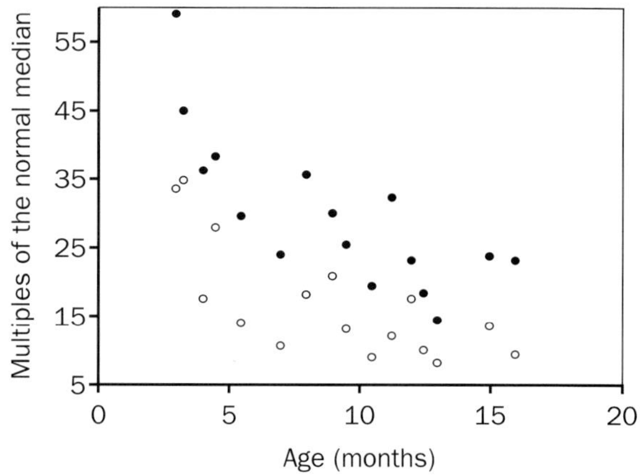

Fig. 7.8 Serum PINP (●) and PICP (○) concentrations expressed as multiples of the normal median during the first 18 months of life ($N = 16$).

sequence of purified PICP verified its identity.[14]

The rabbit anti-PICP antibody specifically reacted with the ^{125}I-labelled tracer supplied with the PICP radioimmunoassay kit.[14]

Fig. 7.7 shows the results of PICP quantification by ELISA and RIA in sera from 17 children ($r_s = 0.993$, $P < 0.001$). In spite of the significant correlation, the absolute concentration in the ELISA was higher than that obtained by the RIA. This may be due to different methods used for calibration of the assays where the ELISA results were based on quantitative amino acid analysis.[14]

Comparison of PINP and PICP in individuals with increased procollagen type I metabolism

Concentrations of PINP and PICP were analysed in children and patients with primary hyperparathyroidism in relation to the median of normal adults ($N = 170$; age 20–60 years). The normal median for PINP and PICP were 58 and 164 ng/ml, respectively.

In order to analyse the deviations in relation to the normal reference interval for adults, the results are presented as multiples of the normal median (MOM), and the upper limits (95th centile) of the normal MOM were 1.5 and 1.8 for PINP and PICP, respectively. Fig. 7.8 shows the MOM values from 16 children during the first 1.5 years of life. It can be seen that the MOM values for PINP were significantly higher than those for PICP, and that both values decreased during the observation period. Eight patients with primary hyperparathyroidism were analysed in the same way prior to and after surgical treatment. As can be seen from Fig. 7.9a, seven out of eight pretreatment samples exceeded the 95th centile of the PINP MOM, whereas only one of the pretreatment samples exceeded the corresponding value for PICP. It can be seen from the figure that the response to surgical treatment was more pronounced for PINP than for PICP.

DISCUSSION

PINP and PICP are released into the extracellular fluid during production of procollagen type I. Since collagen type I constitutes more than 90% of the organic matrix, and since PINP and

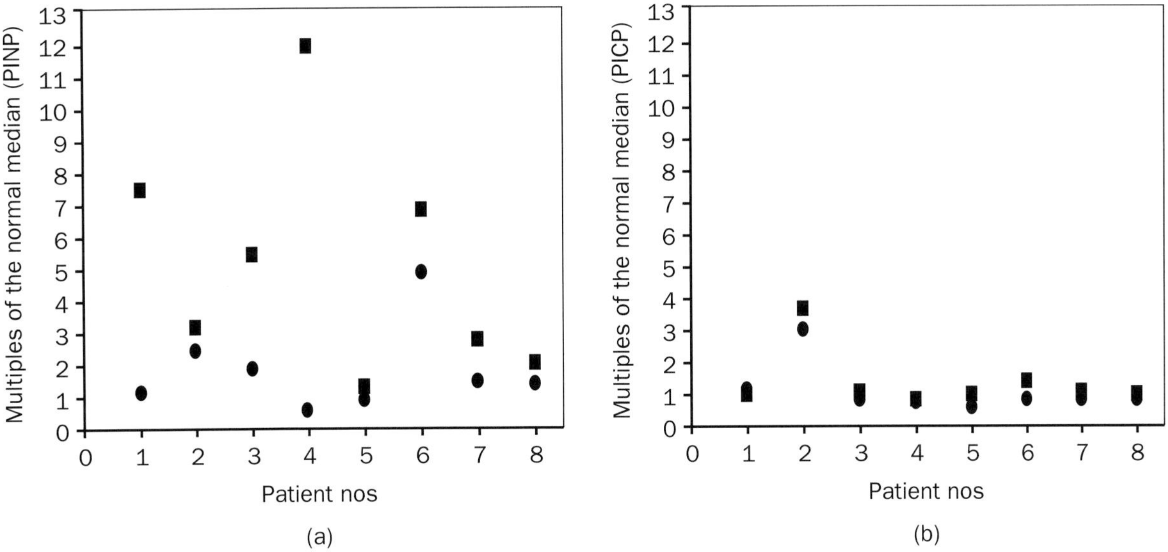

Fig. 7.9 Multiples of the normal median in serum samples from patients with primary hyperparathyroidism ($N = 8$) before (■) and after (●) surgical treatment: (a) PINP; (b) PICP.

PICP enter the circulation from the extracellular fluids, these molecules are obvious candidates as serum markers for changes in bone and connective tissue metabolism. Although neither PINP nor PICP are bone-specific biomarkers, studies on pigs indicate that the contribution of PICP from lymph draining the skin and other soft connective tissues to the serum pool is negligible. Thus the major part of PICP in normal serum (and this may also be the case for PINP) probably originates from procollagen metabolism in relation to formation of the organic bone matrix.[15]

With respect to PICP assay technology, we found no differences between the ELISA and RIA techniques applied in the present report. This is not surprising, since the trimeric structure of PICP is stable due to the interchain disulphide bonds. Analyses of the clinical usefulness of PICP have so far mainly been carried out using the RIA kit.[12] However, in several diseases the serum concentrations of PICP do not correlate with, or reflect changes less pronounced than, other markers of bone formation, among these PINP.[4,16–18]

The molecular heterogeneity of PINP (α1) makes the considerations in relation to assay technology more complicated. The α1 chains of PINP are found in all biological fluids in a trimeric and monomeric form; thermal transition takes place at 37°C and is an ongoing process in vivo.[8]

The molecular heterogeneity, the thermal transition and the differences in size distribution between biological fluids as well as between individual samples of the same biological fluid are events which have to be taken into consideration in PINP quantification.[7] All immunoassays developed for quantification of PINP have been based on polyclonal rabbit anti-PINP antibodies with specificity for the α1 chain, and these antibodies react equally well with α1 chains in trimeric and monomeric form. Price et al[5] demonstrated that a PINP-RIA using the same antibodies which were used in our ELISA[4] failed to detect the low molecular weight form of PINP. Likewise, the RIA kit fails to detect the monomeric α1 chains, and in the booklet supplied with the kit the manufacturer claims that the 'assay is not sensitive to small molecular weight degradation products of the propeptide ... cleaved from degradation of tissue type I pN-collagen'. Here and previously[7,8] we have demonstrated that these undocumented suggestions are wrong, and suboptimal assay technology should not be used to investigate the biology of PINP. As the specificities of the antibodies applied in ELISA and RIA are identical, the differences are most likely due to different functional affinities of the antibodies for the trimeric and monomeric structures, resulting in diminished reactivity in competitive (antibody-limiting) RIA techniques, but not in ELISA (reagent excess).

Since the data presented here suggest PINP to be a more sensitive biomarker than PICP in relation to growth and altered bone metabolism, it is important that the biology of PINP is taken into account during the development of assays for PINP quantification.

REFERENCES

1. Goldberg B, Taubman MB, Radin A, Procollagen peptidase: its mode of action on the native substrate. *Cell* 1975; 4: 45–50.
2. Teisner B, Rasmussen HB, Højrup P et al, Fetal antigen 2 (FA2): an amniotic protein identified as the aminopropeptide of the α1 chain of procollagen type I. *APMIS* 1992; 100: 1106–14.
3. Fay TN, Jacobs I, Teisner B et al, Two fetal antigens (FA1 and FA2) and endometrial proteins (PP12 and PP14) isolated from amniotic fluid: characterization and the distribution in fetal and maternal tissues. *Eur J Obstet Gynecol Reprod Biol* 1988; 29: 73–85.
4. Ørum O, Hansen M, Jensen CH et al, Procollagen type I N-terminal propeptide (PINP) as an indicator of type I collagen metabolism: ELISA development, reference interval, and hypovitaminosis D induced hyperparathyroidism. *Bone* 1996; 19: 157–63.
5. Price KM, Silman R, Armstrong P et al, Development of a radioimmunoassay for fetal antigen 2. *Clin Chim Acta* 1994; 224: 95–102.
6. Melkko J, Kauppila S, Niemi S et al, Immunoassay for intact amino-terminal propeptide of human type I procollagen. *Clin Chem* 1996; 42: 947–54.
7. Jensen CH, Hansen M, Brandt J et al, Quantification of the N-terminal propeptide of human procollagen type I (PINP): comparison of ELISA and RIA with respect to different molecular forms. *Clin Chim Acta* 1998; 269: 20–41.
8. Brandt J, Krogh NT, Jensen CH et al, Thermal instability of the trimeric structure of the N-terminal propeptide of human procollagen type I in relation to assay technology. *Clin Chem* 1999; 45(1): 47–53.
9. Berg RA, Prockop DJ, The thermal transition of non-hydroxylated form of collagen. Evidence for a hydroxyproline in stabilizing the helix of collagen. *Biochem Biophys Res Commun* 1973; 52:115–20.
10. Alvarez L, Peris P, Pons et al, Relationship between biochemical markers of bone turnover and bone scintigraphic indices in assessment of Paget's disease activity. *Arthritis Rheum* 1997; 40(3): 461–8.
11. Dominguez Caberra C, Sosa Henriques M, Traba M et al, Biochemical markers of bone formation in the study of postmenopausal osteoporosis. *Osteoporos Int* 1998; 8(2): 147–51.
12. Melkko J, Niemi S, Risteli L et al, Radioimmunoassay of the carboxyterminal propeptide of human type I procollagen. *Clin Chem* 1990; 36(7): 1328–32
13. Schytte S, Hansen M, Møller S et al, Hepatic and renal extraction of circulating type I procollagen aminopropeptide in patients with normal liver functions and in patients with alcoholic cirrhosis. *Scand J Clin Lab Invest* 1999; 59: 627–34.
14. Frederiksen JK, The C-terminal propeptide of human procollagen type I (PICP): preparation of immunoreactants, ELISA development, and comparison with PINP in relation to bone metabolism. MSc Thesis, Odense University, 1998.
15. Jensen LT, Olesen HP, Risteli J et al, External thoracic duct-venous shunt in conscious pigs for long term studies of connective tissue metabolism in lymph. *Lab Anim Sci* 1990; 40(6): 620–4.
16. Joffe H, Heaf JG, Jensen LT, Type I procollagen propeptide in patients on CAPD: its relationships with bone histology, osteocalcin, and parathyroid hormone. *Nephrol Dial Transplant* 1995; 10(10): 1912–17.
17. Katagiri M, Ohtawa T, Fukunaga M et al,

Evaluation of bone loss and the serum markers of bone metabolism in patients with hyperparathyroidism. *Surg Today* 1995; 25(7): 598–604.
18. Saggese G, Bertelloni S, Baroncelli GI et al, Serum levels of carboxyterminal propeptide of type I procollagen in healthy children from 1st year of life to adulthood and in metabolic bone diseases. *Eur J Pediatr* 1992; 151(10): 764–8.

8

Bone sialoprotein

Berthold Fohr, Henning W Woitge and Markus J Seibel

Summary • Biochemistry • Functions • Clinical data • References

SUMMARY

Bone sialoprotein is one of the most abundant proteins in bone. This chapter summarizes the biochemistry of bone sialoprotein, its possible functions, and cliwnical data on its levels in biological fluids. Bone sialoprotein appears to be a sensitive marker of bone turnover, and clinical data obtained from benign osteopathies suggest that its serum levels mainly reflect processes related to bone resorption. Furthermore, cell-bound as well as soluble bone sialoprotein may play an important role in the homing, growth and propagation of malignant tumour cells. These biological mechanisms may explain the prognostic value of tissue and serum levels in malignant diseases such as breast cancer and multiple myeloma.

BIOCHEMISTRY

Bone sialoprotein (BSP) was one of the first non-collagenous proteins isolated and purified from the mineral compartment of bone. Early reports came from Herring and Kent,[1,2] who purified a ~24-kDa breakdown product of today's known BSP and described the properties of its extensive carbohydrate modifications. Similar fragments to those described by Herring et al, can be generated by trypsin digestion of the intact molecule, leading to the assumption that BSP becomes degraded subsequent to its synthesis or after deposition in the skeletal matrix. This phenomenon has also been observed for osteonectin and bone proteoglycans.[3]

BSP is a 70–80-kDa phosphorylated and sulphated, sialic acid-rich glycoprotein. It is one of the most abundant proteins in bone and comprises 10–15% of the non-collagenous proteins in the mineral compartment. BSP undergoes extensive post-translational modifications; approximately 50% of its mass can be attributed to *N*- and *O*-linked oligosaccharides and sialic acid residues.[3–5] Further known post-translational modifications are tyrosine sulphation and serine/threonine phosphorylation. BSP has four Asn–X–Ser/Thr consensus repeats for Asn-linked glycosylation, and three tyrosine-rich repeats, two of which flank a C-terminal Arg–Gly–Asp (RGD) tripeptide. Three regions of BSP are rich in acidic amino acids ('polyglutamic acid regions'). Because of extensive post-translational carbohydrate modifications, accounting for ~50% of the mass of the molecule, X-ray diffraction analysis should not be used on BSP. A carbohydrate-free, C-terminal stretch of ~60 amino acids enriched with ^{15}N was analysed by NMR and found to be flexible.[6] Proton NMR spectra of the full-length molecule revealed that BSP is a random coil and completely flexible in solution.[7]

BSP can be purified from developing bone with standard biochemical techniques.[3,4] However, this approach might affect BSP's structure and functional properties, as it is subjected to denaturants. The rat osteosarcoma cell line UMR-106-BSP has been reported to be an in vitro source of a highly sulphated form of BSP.[8] Recently, the in vitro synthesis of pure, full-length BSP has been described, employing human bone marrow stromal cells and an adenoviral expression system.[7] Several polyclonal antisera raised against different epitopes of the human, bovine and rat BSP are available.[9,10] Furthermore, a monoclonal antibody against rat BSP is available through ATCC's Hybridoma Data Bank (ATCC, Manassas, VA, USA).

Although BSP is the original name, it was known as BSP II for a short time, because of confusion with osteopontin, which was then known as BSP I.[11] The gene name of BSP is *IBSP* (integrin-binding-sialoprotein), since the name for the BSP locus in the mouse genome had been used for another gene (brain serine protease). The complete coding cDNA of BSP was sequenced and published by Fisher et al, and can be accessed with the GenBank number J05213.[5] Oldberg et al were the first to fully clone the rat BSP cDNA species in 1988,[12] followed by the human,[5] mouse,[13,14] bovine,[15] chicken[16] and hamster sequence.[17] In humans, the 317 amino acid coding gene is organized in six small exons, ranging in size from 51 to 159 bp, and a last exon 7 spanning 2.6 kb (containing a large 3′-UTR and ALU repeats). Exon 1 is entirely non-coding, and the 5′-flanking region contains TATAA and CCAAT boxes as well as an AP-1 site. These sites have been shown to be involved in transcriptional regulation of several genes, including two other bone-related genes, osteopontin and osteocalcin.[18] BSP is a member of a family of small integrin-binding, secreted phosphoproteins that also include osteopontin (OPN), dentin matrix protein (DMP) and the dentin sialophosphoprotein (DSPP). All four related genes are clustered on the long arm of chromosome 4.[18]

BSP expression, as detected by immunohistochemistry and in situ hybridization, can be found in mature osteoblasts, odontoblasts, cementoblasts, osteoclasts, hypertrophic chondrocytes and cement lines of developing bone and dentin.[19,20] BSP expression in bone cells is considered to be a marker of the mature osteoblastic phenotype.[21,22] To date, only a few extraskeletal tissues have been reported to express BSP in detectable quantities; among them placenta trophoblasts[20] and certain cancers. Interestingly, BSP expression has been found in carcinomas that have a propensity to metastasize to bone, as in breast,[23] prostate[24] or thyroid cancer.[25] Because of its restricted expression pattern in trophoblasts and bone-forming cells, BSP expression in tumours has been proposed to play a role in either microcalcification[26,27] or in metastasis homing to bone.[28–31] In patients with breast cancer or multiple myeloma, the degree of BSP expression correlates positively with disease severity.[30,31]

FUNCTIONS

One of the major functions that BSP was thought to be involved in is the mineralization of newly deposited bone matrix and/or the calcification of extraskeletal tissues. BSP is a highly acidic protein with strong affinity for hydroxyapatite crystals.[6] In vitro studies have shown that BSP nucleates such crystals, and electron-microscopic localization studies have confirmed that secreted BSP is associated with the earliest mineral crystals.[32] While it seems that BSP (over)expression causes mineral to form in various cell and tissue types,[26,27] the primary function of BSP still remains elusive. Like other members of the family (osteopontin, DMP1), BSP has an Arg–Gly–Asp (RGD) domain, which has been discovered to serve as a ligand for the cell surface integrins $\alpha_v\beta_3$ and $\alpha_v\beta_5$. These integrin receptors allow cell attachment and cell signalling through matrix proteins and other molecules.[33] Two other non-RGD cell attachment domains have been shown in two of the tyrosine-rich domains.[34] Thus, BSP could be involved in cell attachment, migration and signalling. In bone cells, BSP and OPN are both upregulated by PGE_2.[35] Furthermore, BSP synthesis has been shown to be induced by dexamethasone and suppressed by $1,25\text{-}(OH)_2D_3$.[36] In serum, BSP binds with high affinity to complement factor H. Furthermore, in vitro experiments have shown that BSP-mediated sequestration of factor H to the cell surface protects cancer cells from complement-mediated lysis by inhibition of the C3 cascade.[37,38] A recent study has demonstrated BSP expression in human endothelial cells (HUVEC). With these cells, BSP was shown to have an effect on cell attachment and migration and to promote vascular growth in vitro.[39]

To date, no known disease has been attributed to changes in the *IBSP* gene. The most commonly reported abnormalities on the long arm of chromosome 4 include terminal deletions (4q31-qter), which result in clinical manifestations with multisystem defects.[40] A BSP-knockout mouse has been reported; the mice are smaller, have smaller bone marrow spaces, smaller secondary ossification sites and wider articular cartilage.[41]

CLINICAL DATA

In the last few years, several immunoassays for the determination of BSP in biological fluids such as serum, plasma or synovial fluid serum have been developed. All of these assays are based upon polyclonal antisera against BSP.[42–44] So far, most clinical experience has been gained with a novel homologous radioimmunoassay[10,42,45] using native BSP isolated from human bone. The anti-BSP antibodies were raised in chicken. In this assay, the coefficients of variation are 6.1–7.0% for intra-assay variability, and 9.2–9.4% for inter-assay variability. The lower detection limit is 0.7 ng/ml, and recovery after spiking of human samples with purified BSP ranged between 92% and 108%. Importantly, no cross-reactivity was observed with a number of non-collageneous proteins, such as osteocalcin, osteopontin, and osteonectin, or with bone alkaline phosphatase.[10] Using this assay, a mean serum concentration of 12.1 ± 5 ng immunoreactive BSP/ml was established in healthy adults (5–95% interval 5.0–21.6 ng/ml, median 10.5 ng/ml). In healthy females, a linear correlation was found between serum BSP levels and age ($R = 0.51$, $P < 0.01$), with significantly higher levels in healthy postmenopausal than in premenopausal women (13.3 ± 4.8 ng/ml versus 9.0 ± 3.8 ng/ml, $P < 0.01$). In healthy individuals, serum immunoreactive BSP levels correlate weakly with the urinary excretion of the pyridinium cross-links pyridinoline (PYD) and deoxypyridinoline (DPD), markers of bone resorption. In contrast, no co-variation was seen with serum levels of total alkaline phosphatase or osteocalcin, or other variables such as the body mass index, bone mineral density, serum calcium or serum creatinine.[45] Serum levels of immunoreactive BSP exhibit intra-individual diurnal variation as seen in most bone markers, with higher values in the early morning and lower values in the afternoon. In contrast, day-to-day variability is negligible.[46]

Elevated levels of serum and synovial fluid BSP have been described in patients with active rheumatoid arthritis (RA),[43,47] in patients with ankylosing spondylitis[48] and in normal pregnancy.[44] In patients with early-stage RA, high serum and synovial fluid levels of BSP seem to predict an erosive course of the disease.[43] Compared to healthy controls, significantly elevated serum BSP levels were further found in patients with active metabolic bone disease such as primary hyperparathyroidism (pHPT), Paget's disease of bone (PD), osteoporosis (OP) or osteomalacia.[45] In most osteopathies, close correlations were observed between serum BSP and bone resorption markers such as the pyridinium cross-links ($R < 0.8$, $P < 0.001$), but not with markers of bone formation.[45] In patients with active PD, intravenous ibandronate induced a rapid and lasting reduction in serum immunoreactive BSP levels. The post-treatment changes in serum BSP were similar to the bisphosphonate-induced decline in urinary DPD, and occurred much earlier than the reduction in total or bone-specific alkaline phosphatase. A relapse in disease activity was associated with a significant increase in serum immunoreactive BSP.[49]

In cross-sectional studies, patients with untreated vertebral osteoporosis show significantly elevated serum BSP levels over healthy age-matched controls.[50,51] Although ROC analyses suggest sufficient diagnostic power when calculated against healthy women, serum BSP, like any single biochemical marker, is certainly not a diagnostic tool for primary osteoporosis.[50] Treatment of older women with low-dose oestrogen, calcium and vitamin D induced a significant fall in serum BSP values.[51]

Serum BSP concentrations are usually elevated in patients with breast cancer metastasized to bone.[50] In contrast, mixed normal and abnormal values are found in non-metastasized disease. In patients with metastatic breast cancer, intravenous bisphosphonates induce a rapid reduction of serum BSP levels to approximately 40% of baseline within 4 days of treatment.[45]

As mentioned above, tissue studies in primary breast cancers have shown that expression of BSP by the tumour is associated with a high incidence of bone metastases in the course of the disease.[23–27] In a 30-month prospective study on 388 women with initially non-metastatic primary breast cancer, serum BSP concentrations were shown to be highly predictive of future bone metastases. While preoperative serum BSP levels correlated with the size of the primary tumour (but not with other prognostic factors), multivariate regression analysis proved serum BSP to be an independent prognostic factor for the development of skeletal metastases (relative risk = 94, $P < 0.001$). Thus, patients with preoperatively elevated serum BSP levels had a significantly increased risk of subsequent bone metastases within the first years after primary surgery. Although the precise mechanism and the role of BSP in the pathogenesis of skeletal metastases are unclear, it is conceivable that BSP, via its integrin recognition sequence, may facilitate adhesion of BSP-expressing tumour cells to the bone matrix. Therefore, breast cancer patients with elevated serum BSP levels may benefit from adjuvant bisphosphonate therapy even if no bone metastases are present.[52,53]

In patients with untreated multiple myeloma, serum BSP values are usually elevated. Interestingly, serum BSP concentrations increase continuously with disease progression, and most patients in stage III have serum BSP levels far above the upper limit of normal (Woitge et al, unpublished data). Within the same disease stage, patients with osteolytic lesions have higher serum BSP levels than individuals diagnosed with non-lytic bone disease. Furthermore, serum BSP levels correlate with the bone marrow plasma cell content and serum β_2-microglobulin in patients with multiple myeloma. ROC analyses of different disease populations have shown that serum BSP concentrations discriminate between patients with multiple myeloma and patients with (1) monoclonal gammopathy of undetermined significance (MGUS), (2) benign vertebral osteoporosis (OPO), or (3) healthy controls. This was not the case for the urinary pyridinium crosslinks or for serum osteocalcin.[54]

The interaction between tumour cells and osteoclasts seems to be pivotal for the growth and propagation of cancer cells in the bone marrow.[55] Fig. 8.1 shows a proposed model for the role of BSP in the relationship between osteotropic cancer cells and osteoclasts. In vitro, BSP increases bone resorption in a dose-dependent manner.[56] We therefore hypothesize that, following the homing of cancer cells to the bone marrow, their secretion of BSP will increase bone resorption. This process will in turn lead to the release of several matrix proteins from the bone matrix. In this context, it should also be noted that certain osteotropic cancer cells, i.e. myeloma cells, express integrin receptors that bind to the RGD region of BSP.[57] Thus, BSP and other adhesion molecules released from the bone matrix during this process may stimulate the proliferation and homing of an increased number of tumour cells, thereby leading to a vicious cycle resulting in tumour-induced osteolysis.

In summary, bone sialoprotein appears to be a sensitive marker of bone turnover, and clinical data obtained from benign osteopathies suggest

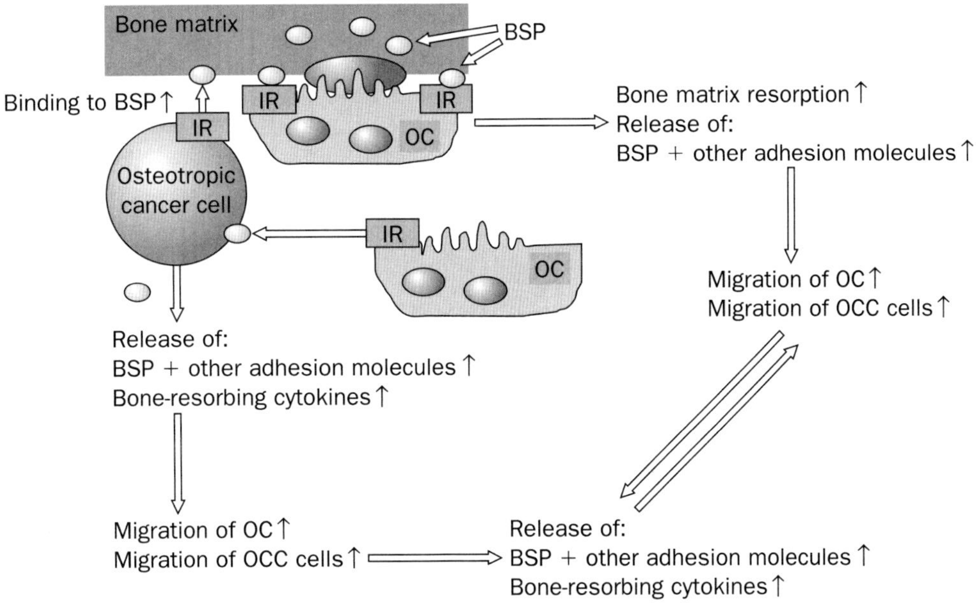

Fig. 8.1 Hypothetical role of bone sialoprotein (BSP) in the interaction between bone, bone cells and osteotropic cancer cells (OCC). IR, integrin receptor; OC, osteoclasts.

that its serum levels predominantly reflect processes related to bone resorption. Furthermore, and perhaps more importantly, cell-bound as well as soluble BSP may play an important role in the homing, growth and propagation of malignant tumour cells. These biological mechanisms may explain the prognostic value of tissue and serum BSP in malignant diseases such as breast cancer or multiple myeloma.

REFERENCES

1. Herring GM, Kent PW, *Biochem J* 1961; 81: 5P.
2. Herring GM. In: Bourne GH, ed. *The Biochemistry and Physiology of Bone*, Vol. I. Academic Press, 1972; 127–89.
3. Fisher LW. In: Slavkin H, Price P, eds. *Chemistry and Biology of Mineralized Tissues*. Elsevier Science Publishers, 1992: 177–86.
4. Fisher LW, Whitson SW, Avioli LV, Termine JD, Matrix sialoprotein of developing bone. *J Biol Chem* 1983; 258(20): 12723–7.
5. Fisher LW, McBride OW, Termine JD, Young MF, Human bone sialoprotein. Deduced protein sequence and chromosomal localization. *J Biol Chem* 1990; 265(4): 2347–51.
6. Stubbs JT 3rd, Mintz KP, Eanes ED et al, Characterization of native and recombinant bone sialoprotein: delineation of the mineral-binding and cell adhesion domains and structural analysis of the RGD domain. *J Bone Miner Res* 1997; 12(8): 1210–22.
7. Fohr B, Torchia DA, Fedarko NS et al, Recombinant human bone sialoprotein and osteopontin: structure–function studies including solving their structures in solution by NMR. *J Bone Miner Res* 1999; 14(suppl 1): 311.
8. Midura RJ, McQuillan DJ, Benham KJ et al, A rat osteogenic cell line (UMR 106–01) synthesizes a highly sulfated form of bone sialoprotein. *J Biol Chem* 1990; 265(9): 5285–91.
9. Fisher LW, Stubbs JT III, Young MF, Antisera and cDNA probes to human and certain animal model bone matrix noncollagenous proteins. *Acta Orthop Scand* 1995; 66(suppl 266): 61–5.

10. Karmatschek M, Woitge HW, Armbruster FP et al, Improved purification of human bone sialoprotein and development of a homologous radioimmunoassay. *Clin Chem* 1997; 43/11: 2076–82.
11. Franzen A, Heinegard D, Isolation and characterization of two sialoproteins present only in bone calcified matrix. *Biochem J* 1985; 232(3): 715–24.
12. Oldberg A, Franzen A, Heinegard D, The primary structure of a cell-binding bone sialoprotein. *J Biol Chem* 1988; 263: 19430–2.
13. Wuyts W, Tylzanowski P, Merregaert J, Sequence of mouse bone sialoprotein II (BSP). GenBank Accession number: L23801.
14. Young MF, Ibaraki K, Kerr JM et al, Murine bone sialoprotein (BSP): cDNA cloning, mRNA expression, and genetic mapping. *Mamm Genome* 1994; 5(2): 108–11.
15. Chenu C, Ibaraki K, Gehron Robey P et al, Cloning and sequence analysis of bovine bone sialoprotein cDNA: conservation of acidic domains, tyrosine sulfation consensus repeats, and RGD cell attachment domain. *J Bone Miner Res* 1994; 9(3): 417–21.
16. Yang R, Gotoh Y, Moore MA et al, Characterization of an avian bone sialoprotein (BSP) cDNA: comparisons to mammalian BSP and identification of conserved structural domains. *J Bone Miner Res* 1995; 10(4): 632–40.
17. Sasaguri K, Chen J, Cloning and characterization of hamster bone sialoprotein (BSP). GenBank Accession number: U65889.
18. Kerr JM, Fisher LW, Termine JD et al, The human bone sialoprotein gene (IBSP): genomic localization and characterization. *Genomics* 1993; 17(2): 408–15.
19. Chen J, Shapiro HS, Sodek J, Development expression of bone sialoprotein mRNA in rat mineralized connective tissues. *J Bone Miner Res* 1992; 7(8): 987–97.
20. Bianco P, Fisher LW, Young MF et al, Expression of bone sialoprotein (BSP) in developing human tissues. *Calcif Tissue Int* 1991; 49(6): 421–6.
21. Bianco P, Riminucci M, Bonucci E et al, Bone sialoprotein (BSP) secretion and osteoblast differentiation: relationship to bromodeoxyuridine incorporation, alkaline phosphatase, and matrix deposition. *J Histochem Cytochem* 1993; 41(2): 183–91.
22. Pinero GJ, Farach-Carson MC, Devoll RE et al, Bone matrix proteins in osteogenesis and remodelling in the neonatal rat mandible as studied by immunolocalization of osteopontin, bone sialoprotein, alpha 2HS-glycoprotein and alkaline phosphatase. *Arch Oral Biol* 1995; 40(2): 145–55.
23. Bellahcène A, Merville MP, Castronovo V, Expression of bone sialoprotein, a bone matrix protein, in human breast cancer. *Cancer Res* 1994; 54(11): 2823–6.
24. Waltregny D, Bellahcène A, Van Riet I et al, Prognostic value of bone sialoprotein expression in clinically localized human prostate cancer. *J Natl Cancer Inst* 1998; 90(13): 1000–8.
25. Bellahcène A, Albert V, Pollina L et al, Ectopic expression of bone sialoprotein in human thyroid cancer. *Thyroid* 1998; 8(8): 637–41.
26. Castronovo V, Bellahcène A, Evidence that breast cancer associated microcalcifications are mineralized malignant cells. *Int J Oncol* 1998; 12(2): 305–8.
27. Bellahcène A, Castronovo V, Expression of bone matrix proteins in human breast cancer: potential roles in microcalcification formation and in the genesis of bone metastases. *Bull Cancer* 1997; 84(1): 17–24.
28. Sung V, Stubbs JT 3rd, Fisher L et al, Bone sialoprotein supports breast cancer cell adhesion proliferation and migration through differential usage of the alpha(v)beta3 and alpha(v)beta5 integrins. J Cell Physiol 1998; 176(3): 482–94.
29. Craig AM, Bowden GT, Chambers AF et al, Secreted phosphoprotein mRNA is induced during multi-stage carcinogenesis in mouse skin and correlates with the metastatic potential of murine fibroblasts. Int J Cancer 1990; 46(1): 133–7.
30. Bellahcène A, Kroll M, Liebens F, Castronovo V, Bone sialoprotein expression in primary human breast cancer is associated with bone metastases development. *J Bone Miner Res* 1996; 11(5): 665–70.
31. Bellahcène A, Menard S, Bufalino R et al, Expression of bone sialoprotein in primary human breast cancer is associated with poor survival. *Int J Cancer* 1996; 69(4): 350–3.
32. Bianco P, Riminucci M, Silvestrini G et al, Localization of bone sialoprotein (BSP) to Golgi and post-Golgi secretory structures in osteoblasts and to discrete sites in early bone matrix. *J Histochem Cytochem* 1993; 41(2): 193–203.
33. Oldberg A, Franzen A, Heinegard D et al, Identification of a bone sialoprotein receptor in osteosarcoma cells. *J Biol Chem* 1988; 263: 19433–6.
34. Mintz KP, Grzesik WJ, Midura RJ et al, Purification and fragmentation of nondenatured

bone sialoprotein: evidence for a cryptic, RGD-resistant cell attachment domain. *J Bone Miner Res* 1993; 8(8): 985–95.
35. Kaji H, Sugimoto T, Kanatani M et al, Prostaglandin E2 stimulates osteoclast-like cell formation and bone-resorbing activity via osteoblasts: role of cAMP-dependent protein kinase. *J Bone Miner Res* 1996; 11(1): 62–71.
36. Oldberg A, Jirskog-Hed B, Axelsson S et al, Regulation of bone sialoprotein mRNA by steroid hormones. *J Cell Biol* 1989; 109: 3183–6
37. Fedarko NS, Fisher LW, Bone sialoprotein in serum is predominantly masked by a serum protein: unmasking shows BSP is elevated in osteotropic cancer patient sera. *J Bone Miner Res* 1999; 14(suppl 1): S322.
38. Fedarko NS, Fohr B, Robey PG et al, Factor H binding to bone sialoprotein and osteopontin enables tumor cell evasion from complement-mediated attack. *J Biol Chem* 2000; 275: 16666–72.
39. Bellahcène A, Bonjean K, Fohr B et al, Bone sialoprotein mediates human endothelial cell attachment and migration and promotes angiogenesis. *Circ Res* 2000; 86: 885–91.
40. Mitchell JA, Packmann S, Loughman WD et al, Deletions of different segments of the long arm of chromosome 4. *Am J Med Genet* 1981; 8: 73–89.
41. Aubin JE, Gupta AK, Zirngbl R, Rossant J, Knockout mice lacking bone sialoprotein expression have bone abnormalities. *J Bone Miner Res* 1996; 11(suppl 1): S102.
42. Seibel MJ, Woitge HW, Pecherstorfer M, Serum immunoreactive bone sialoprotein as a new marker of bone turnover in metabolic and malignant bone disease. *Bone Depeche* 1997; 3: 49–52.
43. Saxne T, Zunino L, Heinegard D, Increased release of bone sialoprotein into synovial fluid reflects tissue destruction in rheumatoid arthritis. *Arthritis Rheum* 1995; 38: 82–90.
44. Ohno U, Matsuyama T, Ishii S et al, Measurement of human serum immunoreactive bone sialoprotein (BSP) in normal adults and in pregnant women. *J Bone Miner Res* 1995; 10(suppl 1): S476.
45. Seibel MJ, Woitge HW, Pecherstorfer M et al, Serum immunoreactive bone sialoprotein as a new marker of bone turnover in metabolic and malignant bone disease. *J Clin Endocrinol Metab* 1996; 81(9): 3289–94.
46. Li Y, Woitge W, Kissling C et al, Biological variability of serum immunoreactive bone sialoprotein. *Clin Lab* 1998; 44: 553–5.
47. Mansson B, Carey D, Alini M et al, Cartilage and bone metabolism in rheumatoid arthritis: differences between rapid and slow progression of disease identified by serum markers of cartilage metabolism. *J Clin Invest* 1995; 95: 1071–7.
48. Acebes A, De la Piedra C, Traba ML et al, Biochemical markers of bone remodeling and bone sialoprotein in ankylosing spondylitis. *Clin Chim Acta* 1999; 289: 99–110.
49. Woitge HW, Oberwittler H, Heichel S et al, Short- and long-term effects of ibandronate treatment on bone turnover in Paget's disease of bone. *Clin Chem* 2000; 46: 684–90.
50. Woitge HW, Pecherstorfer M, Li Y et al, Novel serum markers of bone resorption: clinical assessment and comparison with established urinary indices. *J Bone Miner Res* 1999: 14: 792–801.
51. Prestwood KM, Thompson DL, Kenny AM et al, Low dose estrogen and calcium have an additive effect on bone resorption in older women. *J Clin Endocrinol Metab* 1999; 84: 179–83.
52. Diel I, Solomayer EF, Seibel MJ et al, Serum bone sialoprotein in patients with primary breast cancer as a prognostic marker for subsequent bone metastasis. *Clin Cancer Res* 1999; 5(12): 3914–19.
53. Seibel MJ, Diel IJ, Use of bone sialoprotein in body fluids as a marker of early skeletal metastases in cancer patients and of survival in patients with multiple myeloma. 1999; Patent No. 19813633.
54. Pecherstorfer M, Seibel MJ, Woitge H et al, Urinary pyridinium crosslinks in multiple myeloma, MGUS and osteoporosis. *Blood* 1997; 90: 3743–50.
55. Mundy GR, Bisphosphonates as anticancer drugs. *N Engl J Med* 1998; 339: 398.
56. Raynal C, Delmas PD, Chenu C, Bone sialoprotein stimulates in vitro bone resorption. *Endocrinology* 1996; 137: 2347.
57. Van Riet I, Van Camp B, The involvement of adhesion molecules in the biology of multiple myeloma. *Leuk Lymphoma* 1993; 9: 441.

Discussion paper A

Panel: *Laurence M Demers, Peter R Ebeling, Solomon Epstein, Caren M Gundberg, Elizabeth T Leary, Christopher P Price, Juha Risteli, Graham Russell, Simon P Robins, Markus J Seibel and Børge Teisner*

Gundberg Dr Robins, can you clarify the reasons for the apparent inability of DPD compared to the NTX or CTX to assess alendronate therapy?

Robins Yes, I think the simple answer is that alendronate not only decreases bone resorption per se, but it affects the way that free DPD is produced; alendronate treatment increases the proportion of free cross-links, and so there is a decrease in the apparent responsiveness. Now the obverse of this finding is that measurements of telopeptides overestimate the effect of bisphosphonate, and the point I was trying to make is that if total DPD is measured, the problems caused by changes in degradative metabolism are overcome and you get the right answer in my opinion.

Risteli I would like to ask, with respect to DPD, what is the proportion of it from the total cross-links in bone collagen?

Robins Of all cross-links?

Risteli Yes, known and unknown.

Robins The total pyridinium cross-links represent something between 30% and 40% of the total.

Risteli I have seen a figure as low as 5% only for DPD.

Robins Well for DPD, yes, it would be, because DPD is in a 4 : 1 ratio as a minor partner; I was talking of total pyridinium cross-links.

Risteli One problem is that if there is only 5% of DPD and that proportion increases to 6% as a result of post-translational modifications, there is a 1% change in the total amount of cross-links. However, this means a 20% change in the degradation of DPD. Thus DPD does not tell us exactly how much collagen was actually degraded.

Robins That's true, but there have been papers published showing that the proportion of DPD in bone changes very little between the ages of, say, 25 and 80. I accept the point, but I think that the changes in proportion of DPD in bone are actually very small. That's the least of our problems!

Epstein I have two questions. I know it may be discourteous to ask Graham (Russell), who gave the overview, as he should not be in the firing line, but he raised some incredibly provocative ideas. In osteoporosis it is known that obesity is protective, and do you think this is related to leptin and, particularly, not only to the osteoblasts, but to the oseotocytes, because I think Tim Skerry has been looking at neural transmission via glutamate receptors? The second question concerns looking at early signals for formation; I am sure that you are aware that Pamela Robey, Jane Aubin and Tom Einhorn, among others, have looked at transcription messages, and particularly Dr Einhorn, after fracture and callus formation, has been able to identify certain sequences of gene activity. Perhaps this may well be a way, in terms of gene transcription, to be able to identify what information we can derive for osteoblastic development and function.

Russell The leptin story is obviously pretty new, but in a sense it's going to turn the pharmaceutical approach on its head, because instead of people going after appetite and body size, there is now a need to look at antagonists, rather than agonists, for this system. I think it is very provocative. There are two angles. One is the lineage angle, which in a sense is somewhat separate and involves switching between adipocyte and osteoblast differentiation, probably using other molecules that will do that work through the YPAR receptor and so on. I think it is an area to watch. Whether it is useful in terms of the marker business is something else. The second is the apparent refilation of bone mass via leptin acting on the hypothalamus. Leptin deficiency appears to be associated with increased bone mass. Obviously, monitoring components of this pathway and looking for genetic variants is important. With regard to the efferent limb of this pathway from the hypothalamus to the skeleton, there is likely to be a resurrection of interest in neuropeptides, which have been known to have effects on bone for a long time, but have generally been overlooked. There is work on amylin, for example, that suggests it could be a bone former. So the efferent part of this pathway is going to be where the action is.

With regard to early signals in osteoblast generation, obviously with gene array techniques, you can now actually scan for 6000 or more genes that get turned on when you add stimulators such as PTH to osteoprogenitor cells, and then you have the problem of sorting out what you get. Tim Skerry's work, which you refer to, is one of the best illustrations I know of differential display actually leading to a discovery, in this case the glutamate signalling pathway, which is possibly involved in osteocyte interactions. So we are faced with some pretty powerful approaches for following these things. The challenge is trying to find markers that will monitor cell precursors—traditionally, cell surface markers have been best in relation to haematophoresis, and that's what we don't have for osteoblasts, or indeed for osteoclasts, as far as I know. I suppose that alkaline phosphatase stimulation is a measure of stimulating early osteoblast cells, in contrast with what you see with osteocalcin, which is a product of more mature osteoblasts.

Demers I guess what flies against the leptin hypothesis is the fact that sedentary young girls who are heavy have thicker bones than girls who are thin, and, although they may have menses, their bones are a lot thinner in terms of the developmental process. So I guess there is a lot that we need to understand before we accept that theory.

Russell Yes, that is exactly what the ob/ob mouse is. It is a fat mouse with dense bones, and it is leptin deficient.

Epstein I have got some other questions as well. I think most of us who look at bone markers in terms of clinical practice (and I don't think the panel addressed this specifically) do it in terms of either the disease or monitoring the disease, and what we do about, for example, diurnal variations, when we measure the various bone markers, particularly NTX and CTX; people have written about menstrual cycle and the changes in the bone markers. What I would really like, besides those two questions, to ask the panel is do they know of any good studies where histomorphometry has been sequentially followed in terms of the bone markers in order to see what the relationship is between the percentage decline and the decrease in activation frequency?

Demers I could just start by saying that I think most of the data we have in terms of the diurnal variation etc. come from urine assays, particularly with respect to the resorption markers. Now that picture might change a little bit with respect to serum assays, and may in fact change our approach to when exactly those assays should be collected in terms of the serum assay as a point in time as opposed to the urine assays, which really reflect an integrated production over a period of time. So I think we are going to have to re-look at those data with the serum assays if we indeed make the transition to the serum assays as the markers develop.

Gundberg When we looked at osteocalcin circadian variation, we found that the most stable levels occurred in the middle of the night, which is totally impractical in a true clinical situation.

Teisner With respect to PINP, there is no diurnal variation.

Price Can I move onto another question? I wanted to address a question primarily to the three speakers who were talking about the collagen cross-links. Increasingly, I think, we have been able to define either the key molecule, as in the case of DPD, or the peptide against which the antibody is raised, as in the case of NTX and CTX. These peptides and the DPD molecule could provide the primary standard; however, variation is seen either between methods or between batches of tests, due to the inability to transfer the calibration from a primary standard to the secondary reference material used in the test. How can we overcome this in the future?

Robins Well, I'm not sure that was directed at me. In terms of pyridinium cross-links, I made the point that you know precisely what you are measuring, and there is no problem as far as I can see in preparing calibrators and knowing what you are measuring in the assay. Now, for the peptide assays, I think it is less clear precisely what is being measured and whether all components have the same affinity for the antibodies; therefore, there are particular problems in standardization of the telopeptide assays as far as I can see. But Dr Leary may wish to comment.

Leary As far as the CTX is concerned, right now there is a monopoly of one company which is providing the antibody, so that can at least be somewhat controlled until somebody else gets into the picture. The same goes for NTX.

Ebeling For urinary NTX, it is derived from a monoclonal antibody against a urinary pool from a patient with Paget's disease, so it is not against an actual polypeptide. So there is probably less specificity for that particular monoclonal antibody for measuring a certain serum epitope. However, the serum NTX assay uses an antibody against the NTX peptide, rather than a pool of peptides.

Epstein I just want to ask Caren Gundberg something along the same lines. We all know that in Paget's there is almost a discrepancy between osteocalcin and alkaline phosphatase. Do you think this is related to the osteoblast being abnormal or do you think there may be different transcription or do you think it may be phases of maturation of the osteoblast?

Gundberg Several possibilities have been suggested, including differences in the activation and differentiation of the osteoblast.

Epstein So you would say that, in terms of monitoring therapy, probably alkaline phosphatase would still be the gold standard instead of osteocalcin?

Gundberg Absolutely. The question is: is total alkaline phosphatase just as good as bone-specific alkaline phosphatase?

Risteli In Paget's bone disease, PINP also shows very high circulating levels. With respect to Dr Teisner, I would like to comment on the instability of PINP. I work with the propeptide, and in my hands it is not unstable. When I found his paper, I tested its stability and kept one serum sample on the table for 1 week, and PINP did not denature. I think that the main issue is not in vivo denaturation, because it can be kept for a very long time at +37°C and the serum half-life is only about 1 h. So an 8-h half-life for the denaturation is too long for the in vivo situation.

Teisner Yes, but it is an in vivo phenomenon that is going on, because it takes place at 37°C. You never see a biological sample without these components in it, and it has nothing to do with degradation or denaturation.

Risteli Yes, but from the lessons with PINP, we know that there is also a lot of this monomeric form, and PINP cannot denature because it is stabilized by disulphide bridges. There we have thought that it is a degradation product of the whole procollagen, and I think that also in the case of PINP the whole procollagen can be degraded so that we get these monomeric chains. Actually, the cleavage of the propeptide is occurring, so that all three chains are cleaved at the same time, so this is a different origin.

Teisner We do not see any degradation of the α1 chain. We only see intact α1 chain, and that has been confirmed by multi-MS in the paper in *Clinical Chemistry*.

Risteli Yes, but you have studied amniotic fluid, and it seems that the amniotic fluid propeptide is much more unstable than the authentic propeptide from other sources.

Teisner We have performed studies on thermal stability on PINP, and we have assayed it in, so far, amniotic fluid, cord serum, normal human serum and purified material, and found the same transitions.

Raisz Dr Robins, I was interested in the possible role of muscle in cross-link execution. Could this confound our measurements?

Robins The turnover rate of myofibrillar proteins in muscle is indeed high, but animal studies have shown that intramuscular collagen turnover is much lower. The myofibrillar proteins appear to get replaced within the epi- and perimyceal envelopes, so the turnover of muscle collagen is actually not that high.

Raisz How high does it get?

Robins In inflammation it may well be much higher, as indeed it would be in any fibrotic situation; then, yes, the turnover rate will be much higher. There are no data from clinical studies as yet, but it is an important area to look at.

9

Biochemical markers of bone turnover: age, gender and race as sources of biological variability

Yvette M Henry and Richard Eastell

Summary • Effect of age • Effect of gender • Effect of race • Reference ranges • Conclusion • References

SUMMARY

The clinical use of biochemical markers of bone turnover is affected by biological variability. Biological variables such as age, gender and race influence all markers of bone formation and resorption. With increasing age, it is important to be aware of various factors that elevate bone markers during infancy, childhood and adulthood. Bone marker levels are increased in infants and children due to the effect of growth, and increased in women due to the menopause. The effect of gender on bone markers is age-dependent, whereby in infants the effect is small but in adults the effect is definite. It is beneficial to understand how markers of bone turnover are influenced by ethnicity. This chapter aims to discuss age, gender and race as sources of biological variability. The consequences of these biological variables in establishing an appropriate reference range are also considered.

EFFECT OF AGE

Infancy

Bone turnover markers are highest at birth and during infancy, compared to values during the rest of childhood. Urinary pyridinoline (U-PYD), deoxypyridinoline (U-DPD) and cross-linked N-telopeptides of type I collagen (U-NTX) as markers of bone resorption are several times higher at birth compared to values generally seen in older children and adults.[1–3] Similarly, bone alkaline phosphatase (ALP), C-terminal propeptide of type I procollagen (PICP) and serum osteocalcin as markers of bone formation parallel the trend seen in bone resorption markers.[4,5] Bone markers during this time seem to be more strongly associated with gestational age than chronological age. Mora et al[3] showed a significant association between NTX and gestational age in infants during the first month of life.

Infancy is the period of most rapid skeletal growth. Biochemical markers of bone turnover during growth reflect both modelling and remodelling of bone. Bone markers have been

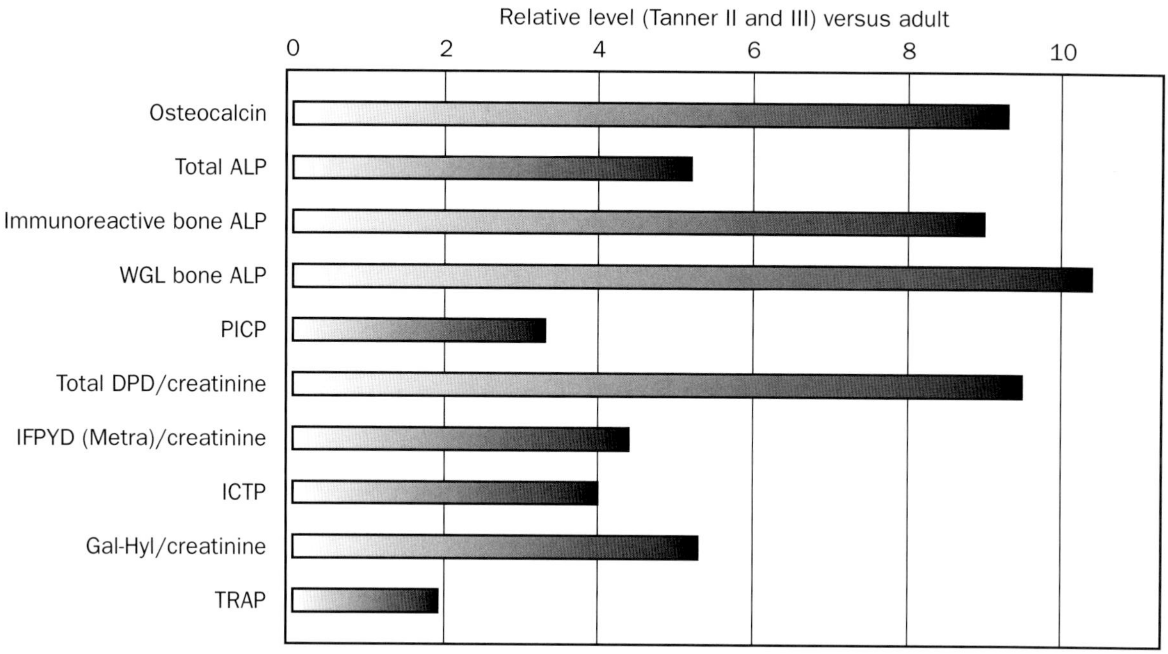

Fig. 9.1 Biochemical markers of bone formation and resorption in mid-puberty (Tanner stages II and III) relative to the mean level in adults. From Blumsohn et al.[10]

shown to increase at birth and peak at 1 month of life.[2] Growth velocity also increases during this time.[2,3,6] The accelerated bone turnover during infancy seems to reflect this increase in growth. Beyond 3 months of life, bone markers appear to decrease rapidly until about 2 years of age, after which there is a continued decrease, but at a slower rate, until levels stabilize during early childhood.[5,7] Therefore, it appears that the change in bone turnover markers with age during infancy reflect the patterns of postnatal growth.

Childhood

Adolescence is the period during which the growth spurt and the greatest accrual of bone mineral occurs. Biochemical markers of bone turnover appear to reflect these events. Several studies have examined the effect of age on bone turnover markers during childhood, and have demonstrated increases in markers of both bone formation and resorption. Generally, bone turnover markers in children are higher compared to adults, peak at mid-puberty and decrease thereafter. Work has shown bone ALP, PICP and osteocalcin to increase during puberty, reaching peak values between Tanner stages II and III and then decreasing in the later stages of pubertal development.[8–12] Markers of bone resorption (U-PYD, U-DPD, U-NTX, tartrate-resistant acid phosphatase (TRACP), galactosyl hydroxylysine (Gal-Hyl) and cross-linked C-terminal telopeptide of type I collagen (ICTP)) also show similar changes with age as seen in the bone formation markers.[10,13–15] Blumsohn et al[10] have reported that the magnitude of increase in bone marker levels differs between markers (Fig. 9.1). In this study, bone ALP increased 10-fold but PICP only increased 3-fold compared to the adult reference range. Taken together, these studies show that alterations in these bone markers with age parallel

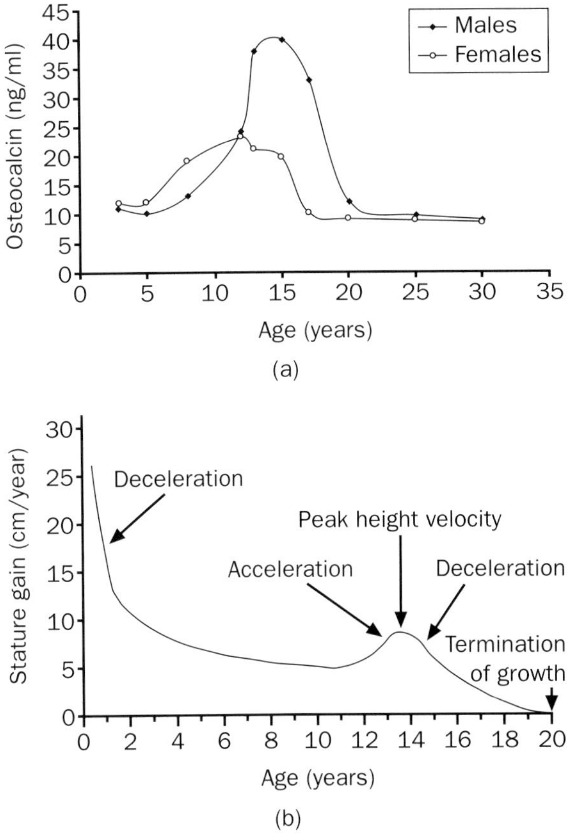

Fig. 9.2a, b Changes in serum osteocalcin levels during growth (a). The age-dependent phases of childhood skeletal growth velocity (b). These graphs illustrate similar patterns of changes in bone turnover markers and skeletal growth. 9.2a reproduced with permission from Hauschka et al.[16]
9.2b reproduced with permission from Malina.[17]

growth height velocity, with peaks in bone turnover around the time of the pubertal growth spurt, and decreasing with the deceleration of linear growth in late puberty. Fig. 9.2 shows the similarity between changes in osteocalcin during growth and the pattern of growth velocity.[16,17] After puberty, biochemical markers of bone turnover remain slightly elevated compared to levels in the mature adult. This elevation reflects the increase in bone mass that occurs after the closure of the growth plates ('consolidation').[18–20]

Adulthood and old age

In Caucasian women, frequent reports have suggested age-related increases in the rates of bone formation and resorption. Several large population-based studies have clearly shown age-related increases in serum osteocalcin, bone ALP, NTX, U-PYD and U-DPD.[21–25] Several other studies have illustrated age-related trends in bone markers analogous to those findings in the larger population-based studies.[18–20,26–33]

Some studies have also indicated a biphasic effect of age on bone turnover markers (Fig. 9.3). Markers of bone formation and resorption have been shown to decrease with age, reaching a nadir between 30 and 50 years, and increasing thereafter.[18,19,21,27,28] This biphasic pattern is likely to be the consequence of two factors. First, the initial decrease in markers probably reflects the completion of bone mass consolidation and fusion of the growth plates. Some of the growth plates (such as in the vertebral body) do not fuse until 25 years of age. Second, the increase in bone markers after age 50 is due to the effect of the menopause on bone turnover. It is well established that during the menopausal transition there is a dramatic increase in bone turnover, with an imbalance between bone formation and bone resorption which favours the latter. This altered bone turnover, which is associated with accelerated trabecular bone loss, is reflected in the level of bone turnover markers. It is evident that biochemical markers of bone turnover are markedly elevated in the first years following menopause. Compared to premenopausal women, levels of bone formation (bone ALP and osteocalcin)[22,34,35] and bone resorption (U-PYD and U-DPD)[23,34–37] are substantially higher in postmenopausal women. Also, Garnero et al[22] demonstrated a much greater increase (50–100%) in markers of bone resorption (U-NTX and C-terminal telopeptide of type I collagen (U-CTX)) compared to markers of bone formation (bone ALP and osteocalcin). Although this marked increase in the level of biochemical markers of bone turnover begins to decline in elderly women, a higher bone

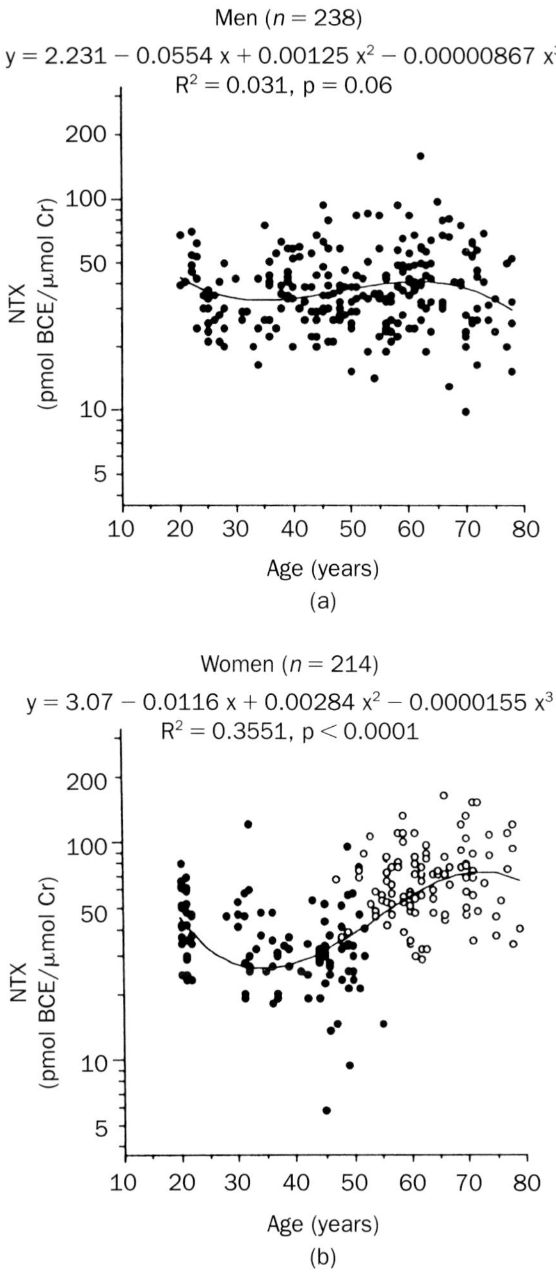

Fig. 9.3a, b Urinary NTX against age in men (a) and women (b). The curves for NTX were best described with a cubic model. In women, solid and open circles represent pre- and postmenopausal women respectively. Reproduced with permission from Sone et al.[30]

turnover state is maintained for many years after the menopause.[22] It appears that these biochemical markers have the ability to discriminate between different rates of bone turnover in pre- and postmenopausal women.

In contrast, some work has shown differing results for the effect of age on the propeptides of type I procollagen; a decrease with age in N-terminal propeptide of type I procollagen (PINP) in women[38] or no effect of age on PICP.[39] PICP and PINP are not specific to bone, in contrast to the established markers such as osteocalcin and bone ALP.

In men, the effect of age on biochemical markers of bone turnover remains controversial. Bone formation (osteocalcin, bone ALP, PICP and PINP) and bone resorption (U-PYD, U-DPD and U-NTX) have been reported to decrease with age,[19,32,38,40–42] and in some cases an additional small increase after the 5th decade has been seen (Fig. 9.3).[21,27,28,40] Although the levels of biochemical markers may increase somewhat in elderly men, the magnitude of this increase with age is of a much lesser degree compared to women.[30] On the contrary, other reports have shown an age-related increase in biochemical markers of bone turnover, without the initial decrease seen in the studies mentioned previously.[23,24,29,35,43] Furthermore, others have failed to show an effect of age on bone turnover markers; osteocalcin, bone ALP, PINP,[26,32,38] and U-NTX, U-CTX, U-PYD and U-DPD.[18,30,33] A plausible explanation for these discrepancies in the literature has yet to be given.

Biochemical markers of bone turnover, bone mineral density and fractures

Thus, men and women possess different age-related changes in rates of bone resorption and bone formation. These differences are concomitant with changes in bone mineral density (BMD) with age in both sexes, and the prevalence of osteoporotic fracture. In women, the elevated biochemical marker levels at the time of the menopause translate into an accelerated rate of bone loss at this time, which is primarily trabecular in nature and is much greater than in men. However, with ageing, trabecular bone loss slows in women due to loss of trabeculae but in men this continues at a steady rate. Therefore the absolute amount of trabecular bone loss is similar in men and women.[44–47] Age-related cortical bone loss is greater in women than in men because in men periosteal apposition compensates for endocortical resorption.[44,47] Therefore, the prevalence of osteoporotic fractures increases with age in both sexes but the increase is steeper in women; vertebral deformity,[49–51] hip fracture[52,53] and Colles' fracture.[54–56]

EFFECT OF GENDER

Infancy

Few studies have examined the effect of gender on bone turnover markers during infancy. However, there is some evidence to suggest that a gender effect on pyridinoline cross-links, U-CTX, U-NTX, osteocalcin and PICP[5,57,58] does not exist during the first few months of life. During the first 2 years of life, studies have reported similar levels of bone formation and resorption (PICP and U-DPD respectively) in boys and girls,[58] or greater levels in boys (PICP)[5] and greater levels in girls (U-CTX and U-NTX).[58] Thus the effect of gender on biochemical markers during infant life is small relative to the higher levels found at this time.

Childhood

Puberty is the time during childhood when the gender differences in biochemical markers of bone turnover are clear. In prepubertal children, the levels of biochemical markers seem to be similar in both girls and boys, but during puberty the increase in bone turnover markers occurs approximately 2 years earlier in girls.[11] Both markers of bone formation (osteocalcin) and resorption (U-NTX, U-PYD, U-DPD and ICTP) have been reported to increase earlier in girls than in boys.[11,13,14] Also, an earlier decrease in markers to adult levels (in late puberty) occurs in girls.[11–13,59] This reflects the earlier timing of peak growth velocity in girls. Consequently, in the earlier stages of pubertal development, girls tend to have higher levels of biochemical markers of bone turnover, but in late puberty the converse occurs.[11,12] Although boys exhibit a delayed increase in bone turnover compared to the girls, this increase is more prolonged and is associated with their longer duration of the pubertal growth spurt.[60,61]

Adulthood and old age

Studies comparing the effect of gender on biochemical markers of bone turnover have yielded discordant results. The effect of gender is age-dependent. During young adulthood, bone formation (bone ALP) and bone resorption (U-PYD and U-DPD) have been shown to be higher in men than in women.[18,19,62] This could be due to the later skeletal maturity, and hence a higher rate of skeletal metabolism, in men. In contrast to these reports, others have shown markers of bone turnover to be higher in men throughout adulthood and in the elderly.[27,28,35,63]

After the age of 50 years, markers of bone turnover are higher in women than in men, usually due to the menopause. It has been demonstrated that older postmenopausal women tend to have higher rates of bone formation (bone ALP and osteocalcin),[19,24,64,65] with stronger evidence for higher rates of resorption (U-NTX, U-PYD and U-DPD),[18,23,30,64–66] compared to men. Alternatively, others have shown women to have higher levels of markers throughout adulthood.[23,29,38,67] Additionally, there are studies that have not shown an effect of gender on the levels of bone markers during adulthood; in some cases, this could be a consequence of small sample size.[31,64–68]

EFFECT OF RACE

The prevalence of osteoporosis is lower in populations of African ancestry than in Caucasians.

Ethnic differences in biochemical markers of bone turnover have become apparent in studies investigating the aetiology of this difference. Reports investigating this issue have compared bone turnover markers in African-Americans, Mexican-Americans and West Africans with those in Caucasian groups, and reveal differing results.

Childhood

Studies have considered ethnic differences in bone turnover during childhood to try and understand when these differences are expressed. Work of this nature is scarce but has shown black children (ages 6–15 years) to have lower bone resorption (measured by TRACP and pyridinoline cross-links) and lower formation rates (measured by osteocalcin) compared to white children.[69,70] Some work has failed to demonstrate ethnic differences in bone turnover markers in children.[69,71] The reason for these discrepancies between studies is not known.

Adulthood

In adults, several studies have investigated ethnic differences in bone turnover. Bone formation has been examined in these studies by measuring primarily osteocalcin and bone ALP, and one study has measured PICP. Osteocalcin and bone ALP have been reported to be lower in African-Americans than in Caucasians.[72–78] Also, bone resorption measured by hydroxyproline, pyridinoline cross-links and U-NTX are in accordance with the ethnic differences seen in bone formation markers.[72,73,75,77,78] These differences have not been supported by all studies, with some workers reporting higher rates of bone turnover in blacks.[79] This study differed from those previously discussed, since the study population were West African blacks (from The Gambia or Nigeria) as opposed to African-Americans. Other reports have shown no effect of race on some bone turnover markers, and this may be due to some markers being measured in young adults, when bone turnover is relatively stable (Fig. 9.4).[73,78,80,81] Markers may be elevated in Mexican-Americans compared to Caucasians.

It is therefore evident from these results that race has a differential effect on bone markers between African populations, especially between West Africans and African-Americans. Consequently, it is important that, in epidemiological studies investigating skeletal metabolism, groups from different ethnic origins should be considered separately.

Biochemical markers of bone turnover, bone mineral density and fractures

The lower rate of bone turnover reported in some African-American children and adults has been associated with increased BMD. Slemenda et al[70] reported that lower rates of bone turnover in black children were associated with significantly higher BMD (approximately 10%) at the lumbar spine, proximal femur and radius. Others have shown similar skeletal advantages in black children at various skeletal sites.[71,82–85] Higher BMD in African-American adults compared to their white counterparts is also a consistent finding.[72,73,75,76,86–89] Greater bone size at some skeletal sites in African-Americans also explains the higher BMD in this group. This higher bone mass in blacks is thought to contribute to the lower incidence of osteoporotic fractures in this population.[52,53,90,91] However, in West African blacks the situation is different because BMD in this group tends to be either the same as or lower than in whites.[81,92–94] This parallels the higher rate of bone turnover seen in this group. Surprisingly, fracture rates are still significantly lower in West African blacks compared to their white counterparts,[94,95] and if BMD is similar in both groups, bone quality may play role in these different fracture rates.

REFERENCE RANGES

Age, gender and race greatly influence biochemical markers of bone turnover. This will

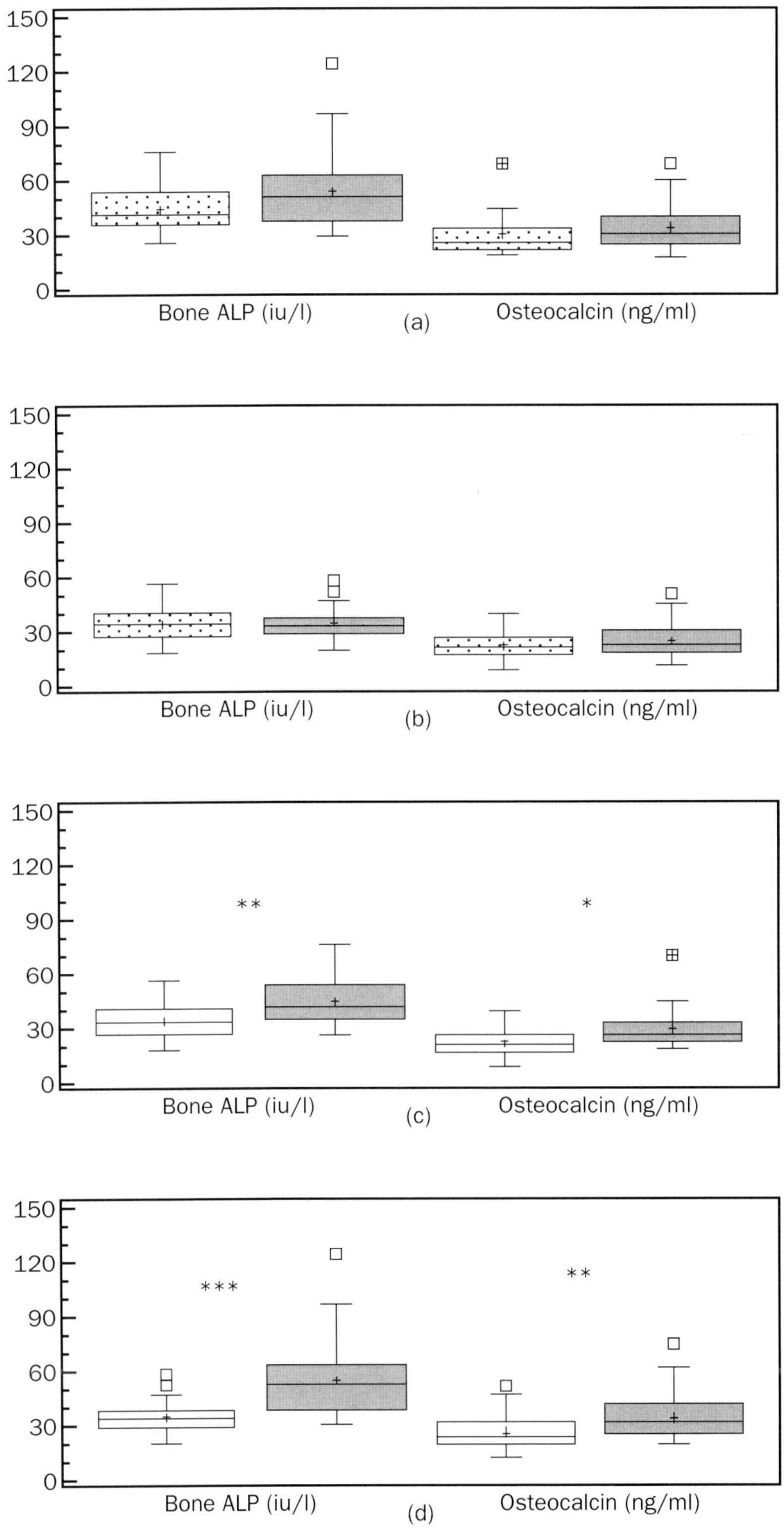

Fig. 9.4 Box and whisker plots showing race and gender effects on bone ALP and osteocalcin in African-Caribbean and Caucasian men and women. There were no race differences in markers in men (a) and women (b) (blacks shown in spotted boxes and whites in grey). Gender differences are shown in panels (c) (blacks) and (d) (whites) with women in open boxes and men in grey. Men had significantly higher bone ALP and osteocalcin in blacks (**$P < 0.01$, *$P < 0.05$ respectively) and whites (***$P < 0.001$, **$P < 0.01$ respectively). Based on Henry and Eastell.[96]

have important implications for the development of an appropriate reference range and in the interpretation of results. Higher markers should be expected in children and infants than in adults. In children, the timing of when markers vary with age is earlier in girls, and this should be taken into consideration. In adults, the differential effect of ageing and events such as the menopause in older women are important factors. As a result, it is essential to establish separate reference ranges for men and women over several age ranges. The best time during adulthood for the development of appropriate reference ranges would appear to be 30–45 years, when markers of bone turnover are relatively stable. It is recommended that the population should include at least 100 subjects and that these should be free of disease and drug therapy known to affect bone metabolism. Furthermore, in epidemiological studies comprising individual subjects from different ethnic backgrounds, the lower biochemical marker levels in some groups may warrant separate reference ranges.

CONCLUSION

Age, gender and race as sources of biological variability should be fully understood when measuring biochemical markers of bone resorption and bone formation. With increasing age, various factors affect markers of bone turnover. The rapid skeletal growth during infancy and puberty causes large increases in biochemical marker levels. Also, the menopause is a time when bone markers are elevated. The effect of age in men remains controversial and warrants further investigation. The effects of gender are not consistent but reflect differences in the timing of skeletal growth between boys and girls, and the differential effect of ageing in adults. The differing effect of race requires further study. In establishing appropriate reference ranges, it is advisable that men and women are treated separately and several age ranges are considered.

REFERENCES

1. Tsukahara H, Watanabe Y, Hirano S et al, Assessment of bone turnover in term and preterm newborns at birth: measurement of urinary collagen crosslink excretion. *Early Hum Dev* 1999; 53(3): 185–91.
2. Tsukahara H, Miura M, Hori C et al, Urinary excretion of pyridinium cross-links of collagen in infancy. *Metabolism* 1996; 45(4): 510–14.
3. Mora S, Prinster C, Bellini A et al, Bone turnover in neonates: changes of urinary excretion rate of collagen type I cross-linked peptides during the first days of life and the influence of gestational age. *Bone* 1997; 20(6): 563–6.
4. Crofton PM, Wheat-germ lectin affinity electrophoresis for alkaline phosphatase isoforms in children: age-dependent reference ranges and changes in liver and bone disease. *Clin Chem* 1992; 38(5): 663–70.
5. Lieuw-A-Fa M, Sierra RI, Specker BL, Carboxy-terminal propeptide of human type I collagen and pyridinium cross-links as markers of bone growth in infants 1 to 18 months of age. *J Bone Miner Res* 1995; 10(6): 849–53.
6. Lapillonne A, Travers R, Dimaio M et al, Bone remodeling assessed by urinary excretion of cross-linked N-telopeptides of type I collagen (NTx) in infants from birth to age one year. *J Bone Miner Res* 1996; Suppl 1: S149 (abstr).
7. Trivedi P, Risteli J, Risteli L et al, Serum concentration of the type I and III procollagen propeptides as biochemical markers of growth velocity in healthy infants and children and in children with growth disorders. *Pediatr Res* 1991; 30(3): 276–80.
8. Mora S, Pitukcheewanont P, Kaufman F et al, Biochemical markers of bone turnover and the volume and the bone density of bone in children at different stages of sexual development. *J Bone Miner Res* 1999; 14(10): 1664–71.
9. Cadogan J, Blumsoh A, Barker M, Eastell R, A longitudinal study of bone gain in pubertal girls: anthropometric and biochemical correlates. *J Bone Miner Res* 1998; 13(10): 1–11.
10. Blumsohn A, Hannon RA, Wrate R et al, Biochemical markers of bone turnover in girls during puberty. *Clin Endocrinol* 1994; 40: 663–70.
11. Johansen JS, Giwercman A, Hartwell D et al, Serum bone Gla-protein as markers of bone growth in children and adolescents: correlation with age, height, serum insulin-like growth

factor I, and serum testosterone. *J Clin Endocrinol Metab* 1988; 67(2): 273–8.
12. Crofton PM, Wade JC, Taylor MRH, Holland CV, Serum concentrations of carboxyl-terminal propeptide of type I procollagen, amino-terminal propeptide of type III procollagen, cross-linked carboxyl-terminal telopeptide of type I collagen, and their interrelationships in schoolchildren. *Clin Chem* 1997; 43(9): 1577–81.
13. Fujimoto S, Kubo T, Tanaka H et al, Urinary pyridinoline and deoxypyridinoline in healthy children with growth hormone deficiency. *J Clin Endcrinol Metab* 1995; 80(6): 1922–8.
14. Bollen AM, Eyre DR, Bone resorption rates in children monitored by the urinary assay collagen type I cross-linked peptides. *Bone* 1994; 15(1): 31–4.
15. Rauch F, Rauch R, Woitge HW et al, Urinary immunoreactive deoxypyridinoline in children and adolescents: variations with age, sex and growth velocity. *Scand J Clin Lab Invest* 1996; 56(8): 715–19.
16. Hauschka PV, Lian JB, Cole DE, Gundberg CM, Osteocalcin and matrix Gla protein: vitamin K-dependent proteins in bone. *Physiol Rev* 1989; 69: 990–1047.
17. Malina RM, Growth and Development: the first 20 years in man. Minneapolis: Burgess Publishing Company, 1975.
18. Ohishi T, Takahashi M, Kawana K et al, Age-related changes of urinary pyridinoline and deoxypyridinoline in Japanese subjects. *Clin Invest Med* 1993; 16(5): 319–25.
19. Kuwana T, Sugita O, Yakata M, Reference limits of bone and liver alkaline phosphatase isoenzymes in the serum of healthy subjects according to age and sex as determined by wheat germ lectin affinity electrophoresis. *Clin Chim Acta* 1988; 173: 273–80.
20. Kelly PJ, Pocock NA, Sambrook PN, Eisman JA, Age and menopause-related changes in indices of bone turnover. 1989; 69(5): 1160–5.
21. Khosla S, Melton III LJ, Atkinson EJ et al, Relationship of serum sex steroid levels and bone turnover markers with bone mineral density in men and women: a key role for bioavailable estrogen. *J Clin Endocrinol Metab* 1998; 83(7): 2266–74.
22. Garnero P, Sornay-Rendu E, Chapuy MC, Delmas PD, Increased bone turnover in late postmenopausal women is a major determinant of osteoporosis. *J Bone Miner Res* 1996; 11(3): 337–49.
23. Delmas PD, Gineyts E, Bertholin A et al, Immunoassay of pyridinoline crosslink excretion in normal adults and in Paget's disease. *J Bone Miner Res* 1993; 8(5): 643–8.
24. Gallagher JC, Kinyamu K, Fowler SE et al, Calciotropic hormones and bone markers in the elderly. *J Bone Miner Res* 1998; 13(3): 475–82.
25. Lewis LL, Shaver JF, Woods NF et al, Bone resorption levels by age and menopausal status in 5,157 women. *Menopause* 2000; 7(1): 42–52.
26. Resch H, Pietschmann P, Kudlacak S et al, Influence of sex and age on biochemical bone metabolism parameters. *Miner Electrolyte Metab* 1994; 20: 117–21.
27. Duda J, O'Brien JF, Katzmann JA et al, Concurrent assays of circulating bone Gla-protein and bone alkaline phosphatase: effects of sex, age, and metabolic bone disease. *J Clin Endocrinol Metab* 1988; 66(5): 951–7.
28. Vanderschueren D, Gevers G, Raymaekers G et al, Sex- and age-related changes in bone and serum osteocalcin. *Calcif Tissue Int* 1990; 46: 179–82.
29. Epstein S, McClintock R, Bryce G et al, Differences in serum bone Gla protein with age and sex. *Lancet* 1984; 1: 307–10.
30. Sone T, Miyake M, Takeda N, Fukunaga M, Urinary excretion of type I collagen crosslinked N-telepeptide in healthy Japanese adults: age- and sex-related changes and reference limits. *Bone* 1995; 17(4): 335–9.
31. Uebelhart D, Gineyts E, Chapuy MC, Delmas PD, Urinary excretion of pyridinium crosslinks: a new marker of bone resorption in metabolic bone disease. *Bone Mineral* 1990; 8: 87–96.
32. Tsai KS, Pan WH, Hsu SHJ et al, Sexual differences in bone markers and bone mineral density in normal Chinese. *Calcif Tissue Int* 1996; 59: 454–60.
33. Garnero P, Gineyts E, Riou J-P, Delmas PD, Assessment of bone resorption with a new marker of collagen degradation in patients with metabolic bone disease. *J Clin Endocrinol Metab* 1994; 79(3): 780–5.
34. Ebeling PR, Ately LM, Guthrie JR et al, Bone turnover markers and bone density across the menopausal transition. *J Clin Endocrinol Metab* 1996; 81(9): 3366–71.
35. del Pino J, Martin-Gómez E, Martin Rodriguez M et al, Influence of sex and age and menopause in serum osteocalcin (BGP) levels.
36. Uebelhart D, Schlemmer A, Johansen JS et al, Effect of menopause and hormone replacement

therapy on the urinary excretion of pyridinium cross-links. *J Clin Endocrinol* 1991; 72(2): 367–73.
37. Hassager C, Colwell A, Assiri MA et al, Christiansen C, Effect of menopause and hormone replacement therapy on urinary excretion of pyridinium cross-links: a longitudinal and cross-sectional study. *Clin Endocrinol* 1992; 37: 45–50.
38. Ebeling PR, Peterson JM, Riggs BL, Utility of type I procollagen propeptide assays for assessing abnormalities in metabolic bone disease. *J Bone Miner Res* 1992; 7(11): 1243–50.
39. Khosla S, Atkinson EJ, Melton III LJ, Riggs BL, Effects of age and estrogen status on serum parathyroid hormone levels and biochemical markers of bone turnover in women: a population-based study. *J Clin Endocrinol Metab* 1997; 82(5): 1522–7.
40. Wishart JM, Need AG, Horowitz M et al, Effect of age on bone mineral density and bone turnover in men. *Clin Endocrinol* 1995; 42: 141–6.
41. Fatayerji D, Eastell R, Age-related changes in bone turnover in men. *J Bone Miner Res* 1999; 14(7): 1203–10.
42. Orwoll ES, Bell NH, Nanes MS et al, Collagen N-telopeptide excretion in men: the effects of age and intrasubject variability. *J Clin Endocrinol Metab* 1998; 83(11): 3930–5.
43. Orwoll ES, Deftos LJ, Serum osteocalcin (BGP) levels in normal men: a longitudinal evaluation reveals an age-associated increase. *J Bone Miner Res* 1990; 5(3): 259–62.
44. Kalender WA, Felsenberg D, Louis D et al, Reference values for trabecular and cortical vertebral bone density in single and dual-energy quantitative computed tomography. *Europ J Radiol* 1989; 9(2): 75–80.
45. Maggio D, Pacifici R, Cherubini A et al, Age-related cortical bone loss at the metarpal. *Calcif Tissue Int* 1997; 60: 94–7.
46. Seeman E, The structural basis of bone fragility in men. *Bone* 1999; 25(1): 143–7.
47. Mazess RB, On aging bone loss. *Clin Orthop* 1982; 165: 239–52.
48. Ruff CB, Hayes WC, Sex differences in age-related remodelling of the femur and tibia. *J Orthop Res* 1988; 6(6): 886–96.
49. Burger H, van Daele PLA, Grashuis K et al, Vertebral deformities and functional impairment in men and women. *J Bone Miner Res* 1997; 12(1): 152–7.
50. O'Neill TW, Felsenberg D, Varlow J et al, The prevalence of vertebral deformity in European men and women: the European Vertebral Osteoporosis Study. *J Bone Miner Res* 1996; 11(7): 1010–18.
51. Cooper C, Atkinson EJ, O'Fallon WM et al, Incidence of clinically diagnosed vertebral fractures: a population-based study in Rochester, Minnesota, 1985–1989. *J Bone Miner Res* 1992; 7(2): 221–7.
52. Karagas MR, Grace L, Yao L et al, Heterogeneity of hip fracture: age, race, sex and geographic patterns of femoral neck and trochanteric fractures among the US elderly. *Am J Epidemiol* 1996; 143(7): 677–82.
53. Hinton RY, Smith GS, The association of age, race and sex with the location of proximal femoral fractures in the elderly. *J Bone Joint Surg* 1993; 75A(5): 752–9.
54. Owen RA, Melton III LJ, Johnson KA et al, Incidence of Colles' fracture in a North American community. *Am J Public Health* 1982; 72(6): 605–7.
55. Miller SVM, Evans JG, Fractures of the distal forearm in Newcastle: an epidemiological survey. *Age Ageing* 1985; 14: 155–8.
56. Mallmin H, Ljunghall S, Incidence of Colles' fracture in Uppsala. *Acta Orthop Scand* 1992; 63(2): 213–15.
57. Gfatter R, Braun F, Herkner K et al, Urinary excretion of pyridinium crosslinks and N-terminal crosslinked peptide in preterm and term infants. *Int J Clin Lab Res* 1997; 27(4): 238–43.
58. Zanze M, Souberbielle JC, Kindermans C et al, Procollagen propeptide and pyridinium cross-links as markers of type I collagen turnover: sex- and age-related changes in healthy children. *J Clin Endocrinol Metab* 1997; 82(9): 2971–7.
59. Magnusson P, Häger A, Larsson L, Serum osteocalcin and bone liver alkaline phosphatase isoforms in healthy children and adolescents. *Pediatr Res* 1995; 38(6): 955–61.
60. Theinz G, Buchs B, Rizzoli R et al, Longitudinal monitoring of bone mass accumulation in healthy adolescents: evidence of a marked reduction after 16 years age at the levels of lumbar spine and femoral neck in female subjects. *J Clin Endocrinol Metab* 1992; 75(4): 1060–5.
61. Kröger H, Kotaniemi A, Kröger L et al, Development of bone mass and bone density of the spine and femoral neck—a prospective study of 65 children and adolescents. *Bone Mineral* 1993; 23(3): 171–82.
62. Rosalki SB, Ying Foo A, Two new methods for

separating and quantifying bone and liver alkaline phosphatase isoenzymes in plasma. *Clin Chem* 1984; 30(7): 1182–6.

63. Sørsensen S, Wheat-germ agglutinin method for measuring bone and liver isoenzymes of alkaline phosphatase assessed in postmenopausal osteoporosis. *Clin Chem* 1988; 34(8): 1636–40.
64. Dresner-Pollak R, Parker RA, Poku M et al, Biochemical markers of bone turnover reflect femoral bone loss in elderly women. *Calcif Tissue Int* 1996; 59: 328–33.
65. Woitge HW, Scheidt-Nave C, Kissling C et al, Seasonal variation of biochemical indices of bone turnover: results of a population-based study. *J Clin Endocrinol Metab* 1988; 83(1): 68–75.
66. Greenspan SL, Dresner-Pollak R, Parker RA et al, Diurnal variation of bone mineral turnover in elderly men and women. *Calcif Tissue Int* 1997; 60: 419–23.
67. Beardsworth LJ, Eyre DR, Dickson IR, Changes with age in urinary excretion of lysyl- and hydroxylysyl pyridinoline, two new markers of bone collagen turnover. *J Bone Miner Res* 1990; 5(7): 671–6.
68. Behr W, Barnert J, Quantification of bone alkaline phosphatase in serum by precipitation with wheat-germ lectin: a simplified method and its clinical plausibility. *Clin Chem* 1986; 32(10): 1960–6.
69. Pratt JH, Manatunga AK, Peakcock M, A comparison of the urinary excretion of bone resorptive products in white and black children. *J Lab Clin Med* 1996; 127(1): 67–70.
70. Slemenda CW, Peacock M, Hui S et al, Reduced rates of skeletal remodeling are associated with increased bone mineral density during the development of peak skeletal mass. *J Bone Miner Res* 1997; 12(4): 676–82.
71. Gilsanz V, Skaggs DL, Kovanlikaya A et al, Differential effect of race on the axial and appendicular skeletons of children. *J Clin Endocrinol Metab* 1998; 83: 1420–7.
72. Ettinger B, Sidney S, Cummings SR et al, Racial differences in bone density between young adult black and white subjects persist after adjustment for anthropometric, lifestyle, and biochemical differences. *J Clin Endocrinol Metab* 1997; 82(2): 429–34.
73. Meier DE, Luckey MM, Wallenstein S et al, Racial differences in pre- and postmenopausal bone homeostasis: association with bone density. *J Bone Miner Res* 1992; 7(10): 1181–9.
74. Bell NH, Greene A, Epstein S et al, Evidence for alteration of the vitamin D endocrine system in blacks. *J Clin Inv* 1985; 76: 470–3.
75. Kleerekoper M, Nelson DA, Peterson EL et al, Reference data for bone mass, calciotropic hormones and biochemical markers of bone remodeling in older (55–75) postmenopausal white and black women. *J Bone Miner Res* 1994; 9(8): 1267–76.
76. Perry III HM, Horowitz M, Morley JE et al, Aging and bone metabolism in African American and Caucasian women. *J Clin Endocrinol Metab* 1996; 81(3): 1108–17.
77. Aloia JF, Mikail M, Pagan CD et al, Biochemical and hormonal variables in black and white women matched for age and weight. *J Lab Clin Med* 1998; 132(5): 383–9.
78. Bikle DD, Ettinger B, Sidney S et al, Differences in calcium metabolism between black and white men and women. *Miner Electrolyte Metab* 1999; 25(3): 178–84.
79. Baca EA, Ulibarri VA, Scariano JK et al, Increased serum levels of N-telopeptides (NTx) of bone collagen in postmenopausal Nigerian women. *Calcif Tissue Int* 1999; 65: 125–8.
80. Cosman F, Morgan DC, Nieves JW et al, Resistance to bone resorbing effects of PTH in black women. *J Bone Miner Res* 1997; 12(6): 958–66.
81. Dibba B, Prentice A, Laskey MA et al, An investigation of ethnic differences in bone mineral, hip axis length, calcium metabolism and bone turnover between West African and Caucasian adults living in the United Kingdom. *Ann Hum Biol* 1999; 26(3): 229–42.
82. Nelson DA, Simpson PM, Johnson CC et al, The accumulation of whole body skeletal mass in third- and fourth-grade children: effects of age, gender, ethnicity, and body composition. *Bone* 1997; 20(1): 73–8.
83. Gilsanz V, Roe TF, Mora S et al, Changes in vertebral bone density in black and white girls during childhood and puberty. *N Engl J Med* 1991; 325: 1597–600.
84. Bell NH, Shary J, Stevens J et al, Demonstration that bone mass is greater in black that in white children. *J Bone Miner Res* 1991; 6(7): 719–23.
85. Bachrach LK, Hastie T, Wang MC et al, Bone mineral acquisition in healthy Asian, Hispanic, Black and Caucasian youth: a longitudinal study. *J Clin Endocrinol Metab* 1999; 84(12): 4702–12.
86. Liel Y, Edwards J, Shary J et al, The effects of race and body habitus on bone mineral density of the

hip, spine and radius in premenopausal women. *J Clin Endocrinol Metab* 1988; 66(6): 1247–50.
87. Nelson DA, Jacobsen G, Barondess DA, Parfitt AM, Ethnic differences in regional bone density, hip axis length, and lifestyle variables among healthy black and white men. *J Bone Miner Res* 1995; 10(5): 782–7.
88. Bell NH, Stevens GJ, Shary SJ, Demonstration that bone mineral density of the lumbar spine, trochanter and femoral neck is higher in black than in white young men. *Calcif Tissue Int* 1995; 56: 11–13.
89. Aloia JF, Vaswani A, Yeh JK, Flaster E, Risk for osteoporosis in black women. *Calcif Tissue Int* 1996; 59: 415–23.
90. Farmer ME, White LR, Brody JA et al, Race and sex differences in hip fracture incidence. *Am J Public Health* 1984; 74: 1374–80.
91. Baron JA, Barrett J, Malenka D et al, Racial differences in fracture risk. *Epidemiology* 1994; 5(1): 42–7.
92. Daniels ED, Pettifor JM, Schnitzler CM et al, Differences in mineral homeostasis, volumetric bone mass and femoral neck axis in black and white South African women. *Ost Int* 1997; 7: 105–12.
93. Prentice A, Shaw J, Laskey A et al, Bone mineral content of British and rural Gambian women ages 18–80+ years. *Bone Mineral* 1991; 12: 210–14.
94. Solomon L, Bone density in aging Caucasian and African populations. *Lancet* 1979; ii: 1326–30.
95. Adebajo AO, Cooper C, Grimley Evans J, Fractures of the hip and distal forearm in West Africa and the United Kingdom. *Age Ageing* 1991; 20: 435–8.
96. Henry YM, Eastell R, Ethnic and gender differences in bone mineral density and bone turnover in young adults: effect of bone size. *Osteoporos Int* 2000; 11: 512–17.

10

Circadian rhythm studies of serum bone resorption markers: implications for optimal sample timing and clinical utility

William D Fraser, Mark Anderson, Christine Chesters, Brian Durham, Aftab M Ahmad, Paula Chattington, Jiten Vora, Christine R Squire and Michael J Diver

Summary • Introduction • Subjects, patients and methods • Results • Discussion • Acknowledgements • References

SUMMARY

A pronounced circadian rhythm was detected for the serum bone resorption markers C-terminal telopeptide (CTX) and free deoxypyridinoline (fDPD), but not for serum N-terminal telopeptide (NTX), in healthy young normal men (mean age 26 years, range 23–31 years, $N = 10$). The circadian rhythm profile for CTX demonstrates an acrophase at 05:18, 24-h mean concentration 0.30 ng/ml, with the greatest change in concentration observed between 07:00 and 11:00. Elderly men (mean age 68 years, range 55–75, $N = 6$) had a significantly lower 24-h mean CTX (0.23 ng/ml, $P < 0.001$), and persistence of the circadian rhythm, with a reduced amplitude (0.10 versus 0.20 ng/ml), and the acrophase was earlier, at 04:52. In young men, the fDPD 24-h mean was 2.06 nmol/l, the amplitude was 0.28 nmol/l and the acrophase was 04:20. In thyrotoxic patients there was a significant increase in 24-h mean CTX (0.89 ng/ml) and fDPD (4.19 nmol/L) (both $P < 0.001$ compared to young men). The circadian rhythm for CTX was retained in five thyrotoxic patients (acrophase 02:27, amplitude 0.31 ng/ml) but absent for fDPD in four (amplitude 0.08 nmol/l). Following treatment with a block-and-replace regimen, euthyroid patients had a significant decrease in mean CTX (0.58 ng/ml) and fDPD (1.83 nmol/l, both $P < 0.001$ compared to toxic). The circadian rhythm was present for CTX in all euthyroid patients (acrophase 04:13, amplitude 0.27 ng/ml) and for fDPD in four patients (acrophase 01:33, amplitude 0.23 nmol/l). Postmenopausal women with osteoporosis ($N = 8$) had a significantly higher mean CTX (0.72 ng/ml) compared to young men ($P < 0.001$), and blunting of the circadian rhythm was noted in five subjects (acrophase 04:21, amplitude 0.23 ng/ml). Samples obtained for serum CTX and fDPD between 14:00 and 17:00 provided the best discrimination between normal young men and the disease states studied.

INTRODUCTION

Osteoclast resorption of bone releases collagen degradation products, including the C- and N-terminal telopeptides (CTX and NTX) that are subsequently metabolized to release the free cross-links pyridinoline (fPD) and deoxypyridinoline (fDPD). The measurement of CTX, NTX and free cross-links in urine[1,2] has provided valuable information on bone metabolism during normal bone turnover and in many metabolic bone diseases. Measurement of analytes in urine can be problematic, and urinary markers of bone resorption have a high combined biological and analytical variability.[3] There is also variability in urine creatinine secretion, so expressing results as a ratio to creatinine is not the complete solution to the problem. The problems of variability in urine measurements may be overcome by measuring serum bone resorption markers. Immunoassays for serum bone resorption markers have recently been developed, and assays are available on automated analysers, which significantly reduces their analytical variability.

It is well known that the urine bone resorption markers demonstrate a significant circadian rhythm, with a peak of excretion between 03:00 and 08:00 and a nadir at 14:00–23:00.[4,5] There appear to be minimal effects of age, menopause, posture, or osteopenia[6] on the circadian nature of urine cross-link excretion, but in men the nocturnal excretion in cross-links is 10% lower than that of postmenopausal women.[7] If serum resorption markers are to be useful clinically, then the biological variability, including the circadian variation, must be assessed and the optimal sampling procedures defined.

We have performed detailed studies on the variation of serum CTX and fDPD throughout a 24-h period using a standardized sampling technique in normal volunteers and in patients with osteoporosis and thyrotoxicosis.

SUBJECTS, PATIENTS AND METHODS

All subjects and patients were sampled in an identical manner. They were all admitted to a metabolic ward at 13:00 and were hospitalized for 25 h. A sampling cannula was inserted in a vein in the antecubital fossa and flushed with 1 ml sterile heparinized saline. Half-hourly or hourly sampling commenced at 14:00, and at each time point the initial 2.5 ml was discarded and a 5-ml blood sample obtained. Samples were separated immediately to obtain serum and plasma. Routine biochemical analysis was performed on fresh samples, and other aliquots were stored at −50°C prior to analysis for biochemical markers of bone metabolism. A standard hospital diet was taken at 08:00, 12:00, 18:00 and 21:00, and all subjects were freely ambulant but avoided strenuous exercise throughout the 24-h sampling period. They lay down to sleep at 23:00 and awoke at 07:30. All subjects demonstrated a significant increase in serum prolactin, consistent with deep sleep.

After obtaining full informed written consent, the following subjects and patients were studied:

1. Ten healthy young men (age range 23–31 years, mean 26 years) not suffering from any illness, with normal routine biochemistry and not on any medication.
2. Six healthy elderly men (age range 55–75 years, mean 68 years) not suffering from any illness, with normal routine biochemistry and not on medication.
3. Six patients (three women, three men, age range 20–62 years, mean 43 years) diagnosed as thyrotoxic on the basis of clinical symptoms, examination and abnormal thyroid function tests (thyrotrophin (TSH) <0.01 mU/l and total thyroxine (TT4) >180 nmol/l in all cases).
4. The same six patients 3 months after treatment with 20 mg twice daily carbimazole and 100 μg daily thyroxine as a block-and-replace treatment regimen. All patients were considered to be clinically euthyroid.
5. Eight postmenopausal women (age range 60–79 years, mean 65) with a mean lumbar spine bone mineral density 71% (range 62–82%) of their age-matched population,

not taking medication known to interfere with calcium metabolism, and who had not received any therapy for osteoporosis.

Routine biochemical analysis was performed using standard chemistries on a Hitachi 747 analyser (Roche, Lewes, UK).

Serum CTX was measured using an immunoelectrochemical assay (IECA) on an Elecsys 2010 immunoanalyser (Roche, Lewes, UK). The assay is a double antibody sandwich technique that employs one biotinylated monoclonal antibody specific for an epitope of an eight amino acid sequence of the C-terminal portion of type 1 collagen telopeptide and another monoclonal antibody directed against a separate epitope of the telopeptide labelled with ruthenium. The total assay incubation time was 18 min, and measurement was made via the electrochemical signal generated by the ruthenium-labelled sandwich complex. Inter-assay coefficient of variation (CV) was <9% between 0.2 and 0.5 ng/ml. The assay sensitivity (replicates of the zero standard) was <0.01 ng/ml.

Serum fDPD was measured using an in-house radioimmunoassay (RIA). A double-antibody RIA has been developed. DPD–TME conjugate was labelled with radioiodine using the chloramine-T method and purified using C18 solid-phase extraction columns. Samples (25 μl) and ^{125}I-labelled DPD (100 μl) were incubated overnight (18 h) at 2–8°C with sheep anti-DPD (50 μl) first antibody. Separation of the antibody-bound DPD was by PEG and microgranular cellulose-assisted precipitation with donkey anti-sheep second antibody. Inter-assay CV was 7.9–10.0% across the range 1.6–6.2 nmol/l. Sensitivity of the assay (22% CV) was 0.6 nmol/L. Serum NTX was measured using a commercial enzyme-linked immunosorbent assay (ELISA) (Ostex International Inc., Seattle, USA). Samples and horseradish peroxidase-labelled monoclonal antibody were incubated in microtitre plates coated with NTX for 90 min. After washing, bound antibody was estimated by generation of peroxide substrate. Inter-assay CV was 7% across the working range of the assay. Bone densitometry was performed by dual-energy X-ray absorptiometry (DXA) using a Lunar DPX densitometer (Lunar Corp., USA). Measurement was made at both the lumbar spine (L2–L4) and neck of femur and compared to the UK Lunar database. Typical precision with this method is 0.5% CV at L2–L4 and 1.05% at the dominant femoral neck.

COSIFIT software, a program that provides iterative non-linear least-squares analysis of biological rhythm data using Marquardt's modification of the Gauss–Newton algorithm, was used to analyse the circadian rhythms. The program provides both parametric and non-parametric estimates of goodness-of-fit, and the statistical differences between parameter values of curves were ascertained by ANOVA.[8] Student's *t*-test was used to analyse the differences between mean concentrations of analytes.

The appropriate local ethics committee approved all studies.

RESULTS

A circadian rhythm in serum CTX and fDPD, but not NTX, was demonstrated in normal young men (Fig. 10.1; Table 10.1). The greatest rates of change are observed between 07:00 and 11:00 in CTX and between 05:00 and 10:00 for fDPD, with a secondary increase in fDPD noted in 6 out of 10 individuals mid-morning (10:00–13:00). A tight distribution of results is seen between 14:00 and 17:00 for CTX. Serum NTX showed little variation throughout the 24-h (Table 10.1), but certain values were significantly lower than the mean in each profile (Fig. 10.1c). Investigation of these results indicated that a significant edge effect was observed using the serum NTX measurement system that resulted in lower values at particular positions on the microtitre plates. In view of these findings, further investigation of 24-h profiles using this assay was not performed.

Elderly males had a significant reduction in 24-h mean CTX and rhythm amplitude, and an earlier acrophase compared to young men (Table 10.1; Fig. 10.2).

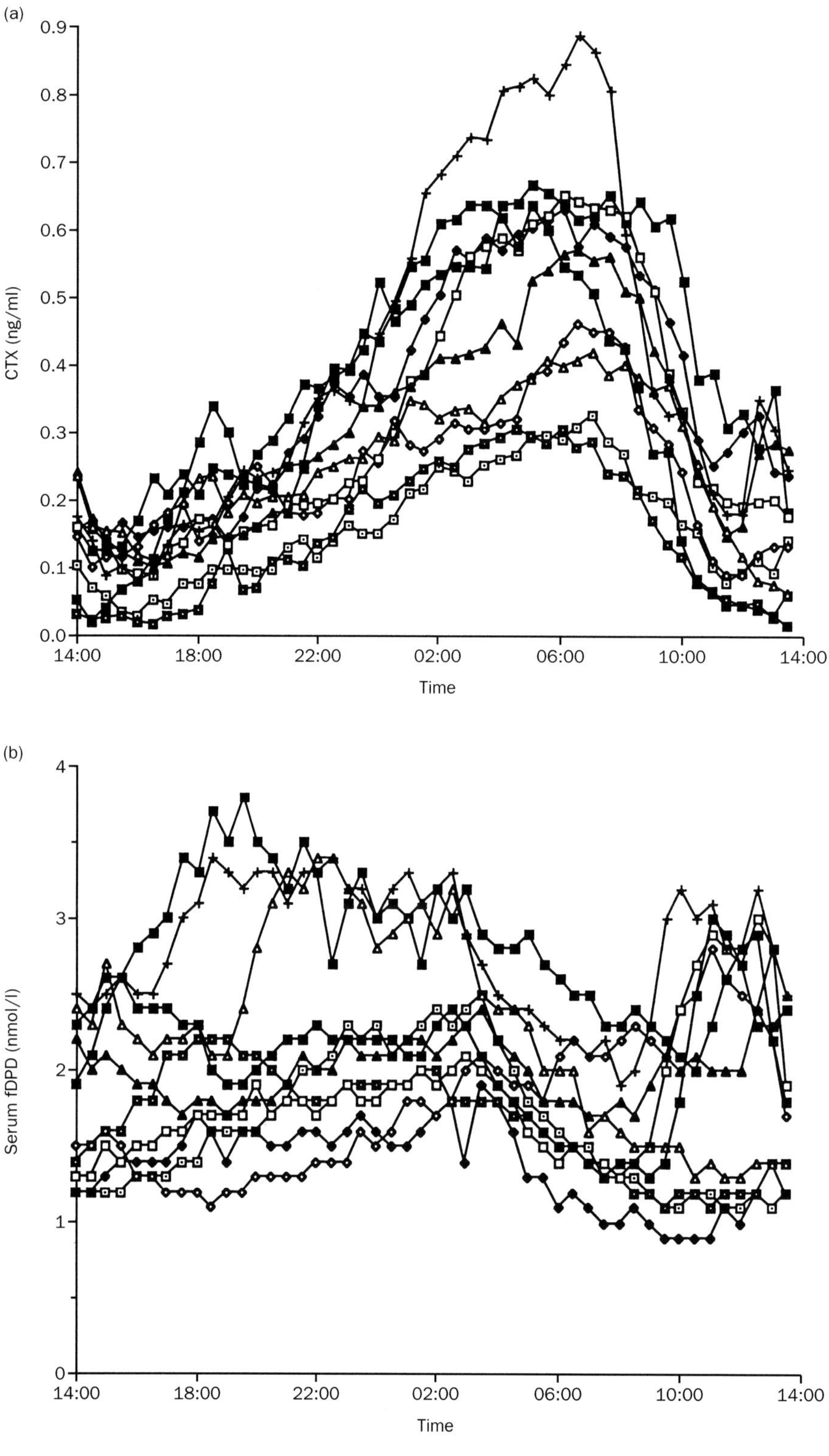

Fig. 10.1 The variation in serum (a) CTX, (b) fDPD (10 cases each), and (c) NTX (6 cases), in young, normal men during a 24-h sampling period.

(c)

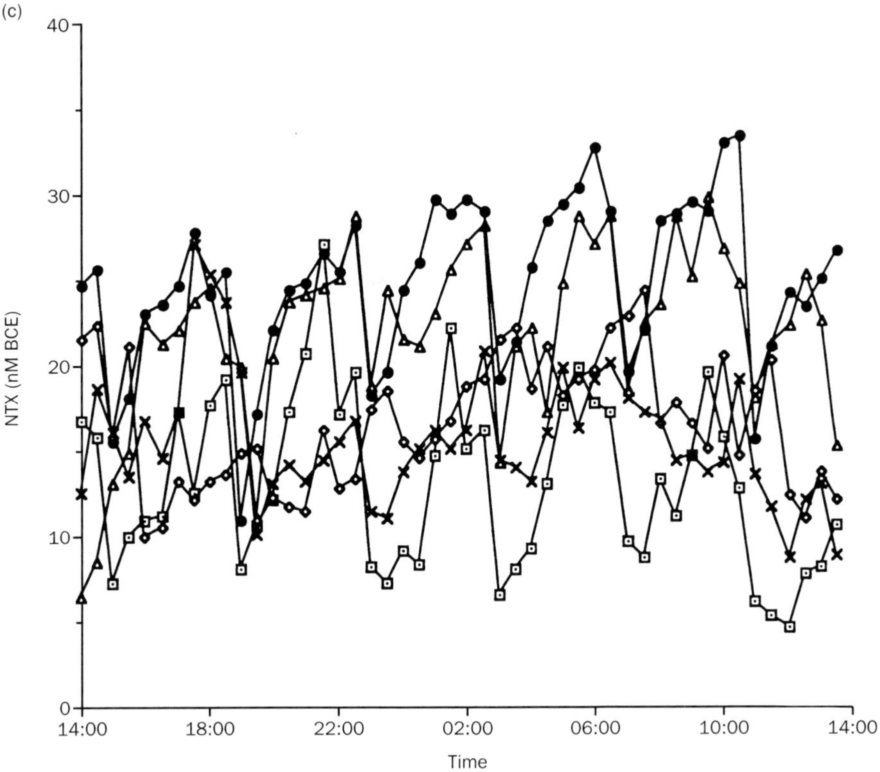

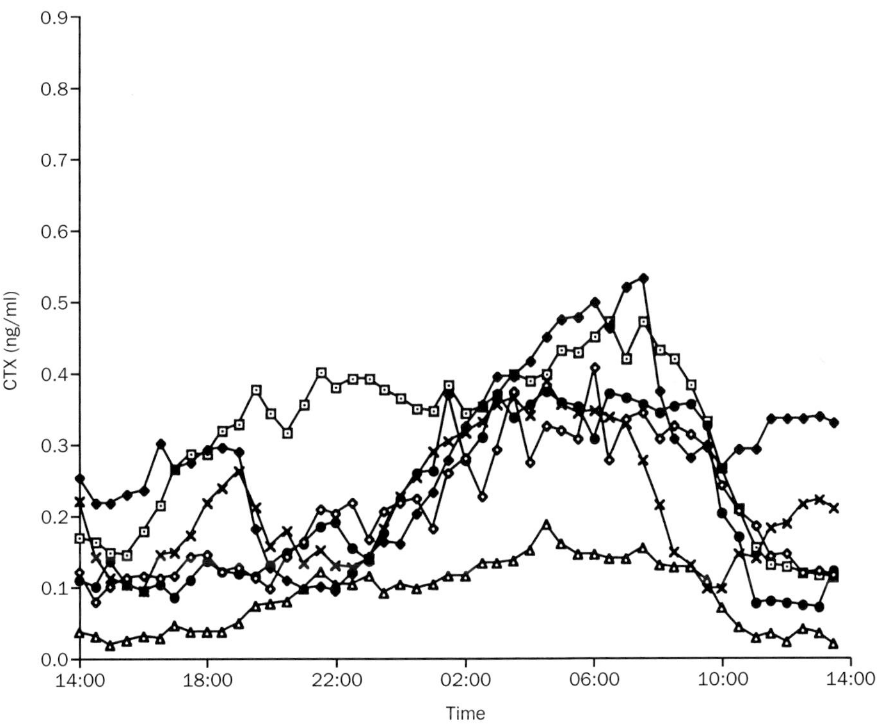

Fig. 10.2 The variation in serum CTX in six normal elderly men during a 24-h sampling period.

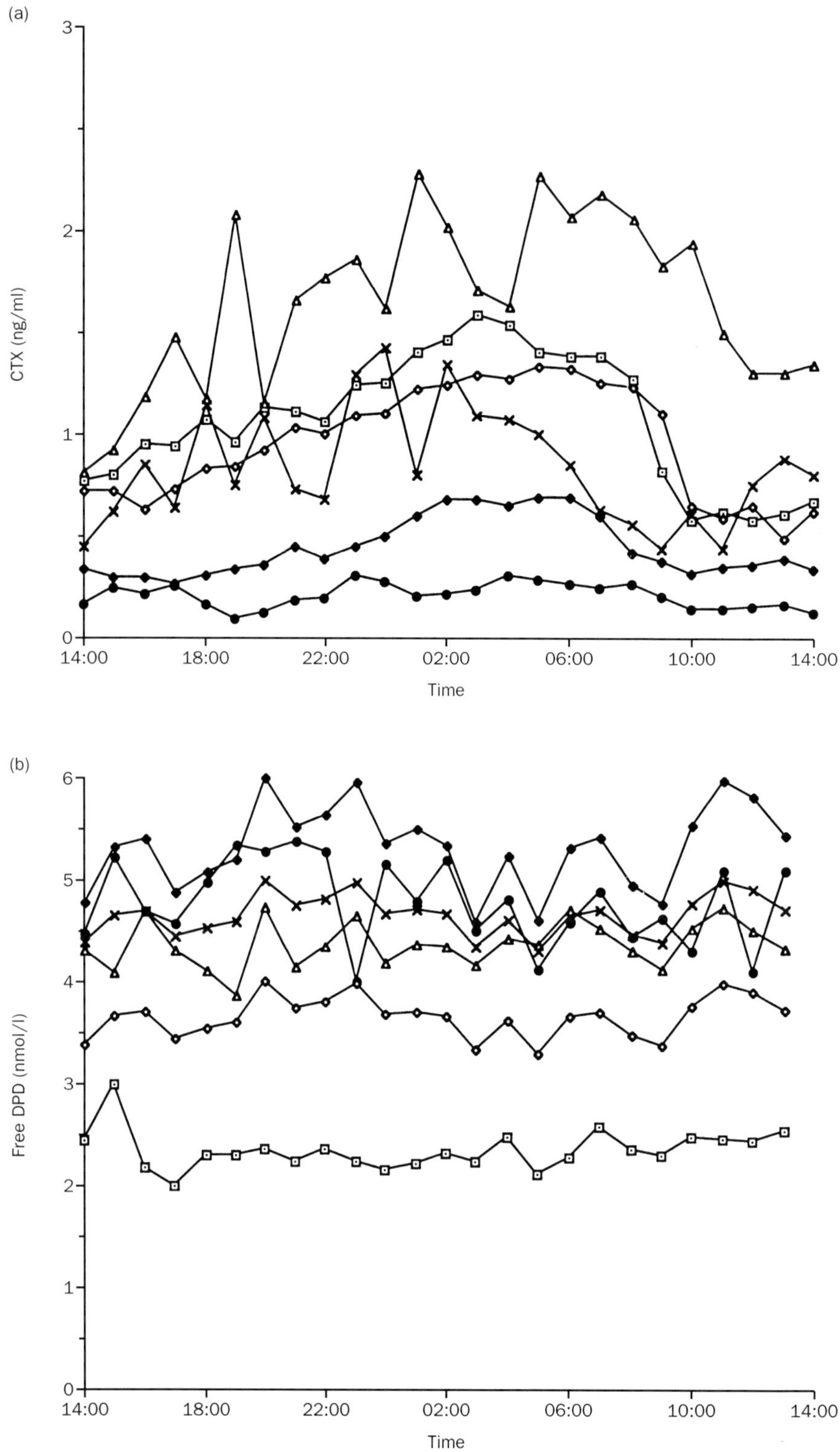

Fig. 10.3 The variation in serum CTX and fDPD during a 24-h sampling period in six thyrotoxic patients before (a,b) and after (c,d) treatment with carbimazole and thyroxine.

(c)

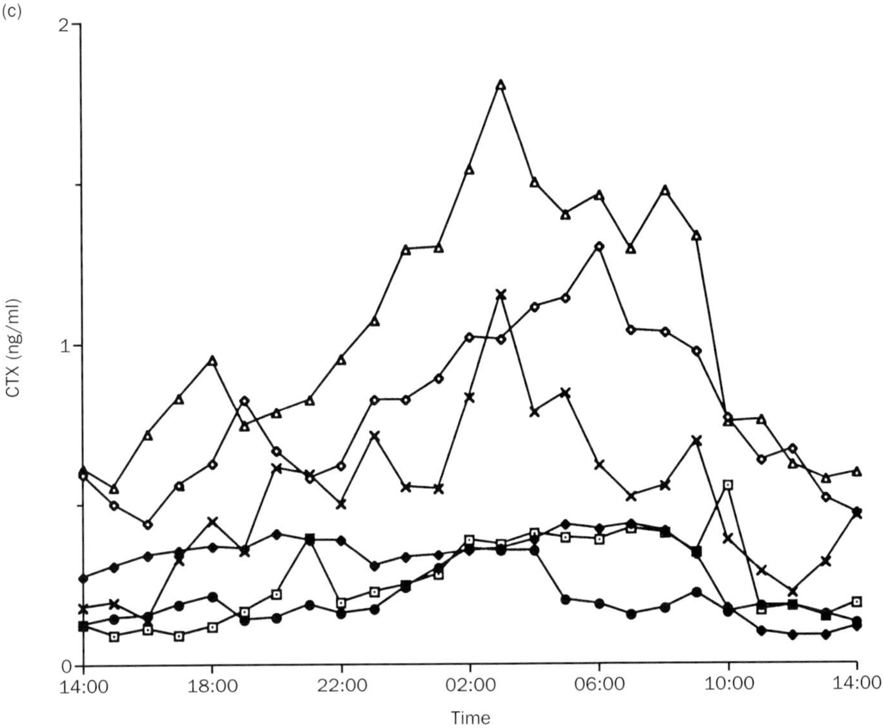
CTX (ng/ml)
2
1
0
14:00
18:00
22:00
02:00
06:00
10:00
14:00
Time

(d)

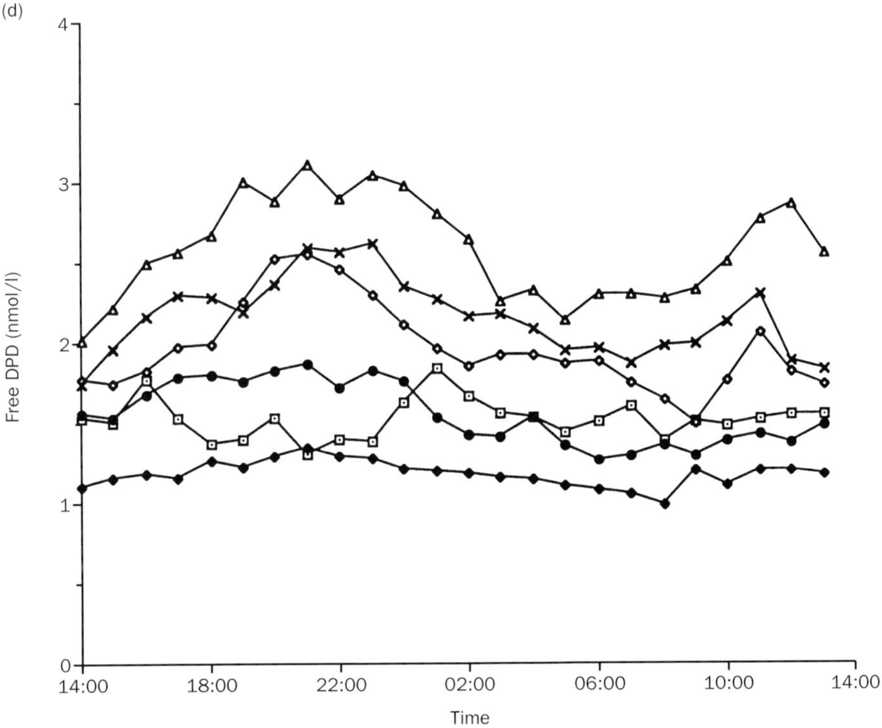
Free DPD (nmol/l)
4
3
2
1
0
14:00
18:00
22:00
02:00
06:00
10:00
14:00
Time

Table 10.1 Circadian rhythm analysis.

	Mesor	Amplitude	Acrophase	Correlation coefficient
CTX Young men	0.30 ± 0.008	0.20 ± 0.01	05:18	0.94
CTX Elderly men	0.23 ± 0.006	0.10 ± 0.007	04:52	0.90
CTX Osteoporosis	0.72 ± 0.009	0.23 ± 0.01	04:21	0.96
CTX Thyrotoxic	0.89 ± 0.025	0.31 ± 0.021	02:27	0.96
CTX Euthyroid	0.58 ± 0.021	0.27 ± 0.023	04:13	0.93
fDPD Young men	2.06 ± 0.01	0.28 ± 0.02	04:20	0.87
fDPD Thyrotoxic	4.19 ± 0.04	0.08 ± 0.059	19:59	0.24 (Poor fit. No significant rhythm)
fDPD Euthyroid	1.83 ± 0.02	0.23 ± 0.018	01:33	0.94
NTX Young men	18.50 ± 0.47	1.88 ± 0.534	04:32	0.47 (Poor fit. No significant rhythm)

All comparisons between groups for mesor, amplitude and acrophase are significantly different ($P < 0.001$), except fDPD thyrotoxic versus euthyroid amplitude ($P < 0.05$), CTX young men versus osteoporosis, CTX thyrotoxic versus euthyroid, fDPD young men versus euthyroid amplitude (all not significant), CTX young men versus elderly men acrophase ($P < 0.05$) and CTX young men versus osteoporosis acrophase (not significant).

In patients with thyrotoxicosis, 24-h mean serum CTX and fDPD were significantly elevated compared to the young men (Table 10.1). In the majority of thyrotoxic patients, the circadian rhythm was retained for CTX, but the acrophase was shifted. No significant rhythm was detected for fDPD. The greatest change in serum CTX was between 06:00 and 10:00 (Fig. 10.3a) and for fDPD between 22:00 and 03:00 (Fig 10.3b). Following treatment, there was a significant decrease in mean CTX and fDPD, and the circadian rhythm was present for CTX in all patients and for fDPD in four patients (Table 10.1). The greatest rate of change in CTX results was between 07:00 and 11:00, and in fDPD between 20:00 and 02:00 (Fig. 10.3c,d).

Patients with osteoporosis had a significant increase in mean serum CTX, and 5/8 had a blunted circadian rhythm (Table 10.1). In those patients who had retained a circadian rhythm, the percentage increase in CTX was lower than that seen in thyrotoxic patients or normal males (Fig. 10.4).

Comparisons were made between the discrimination in CTX values, for each subject studied, taken between 08:00 and 11:00 and between 14:00 and 17:00 for each group studied (Figs 10.5 and 10.6). It is clear that the greatest variability and hence poorest discrimination exists between 08:00 and 11:00 while greater discrimination is seen between the groups at 14:00–17:00. A significant difference is seen between thyrotoxic patients before and after treatment on samples taken between 14:00 and 17:00 ($P < 0.01$) that is not detected between 08:00 and 11:00.

DISCUSSION

Our studies have detected a significant circadian rhythm in both serum CTX and fDPD in the majority of subjects studied. An increase in bone resorption is observed in young men overnight, with the acrophase for CTX at 05:18 and that for fDPD at 04:20. The broader shape

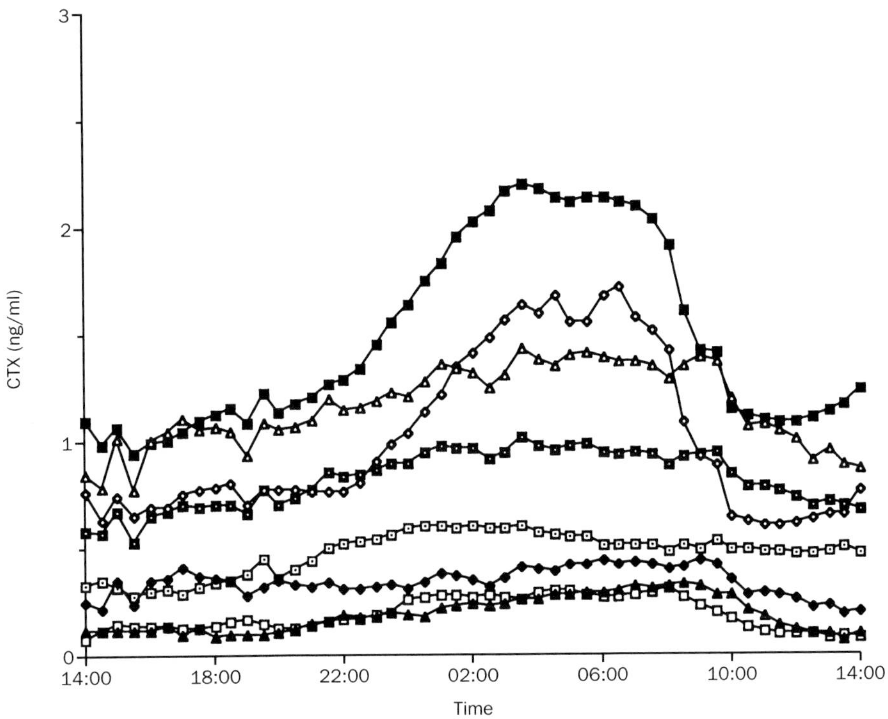

Fig. 10.4 The variation in serum CTX in eight women with postmenopausal osteoporosis during a 24-h sampling period.

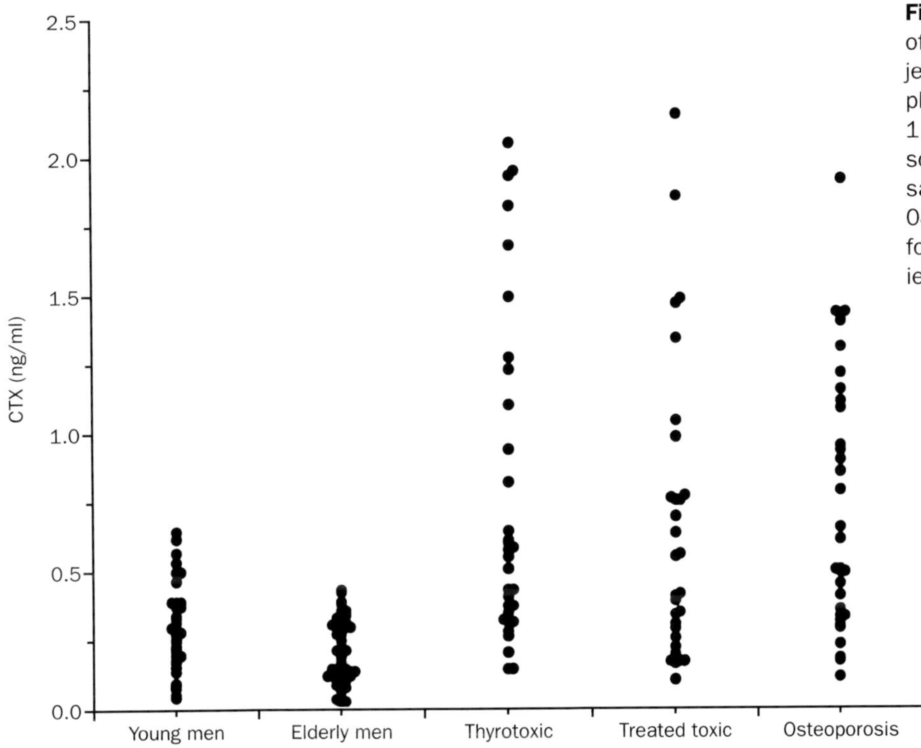

Fig. 10.5 Distribution of CTX results for all subjects studied when sampled between 08:00 and 11:00. The values in the scattergram include all sample points during 08:00–11:00 obtained for each individual studied.

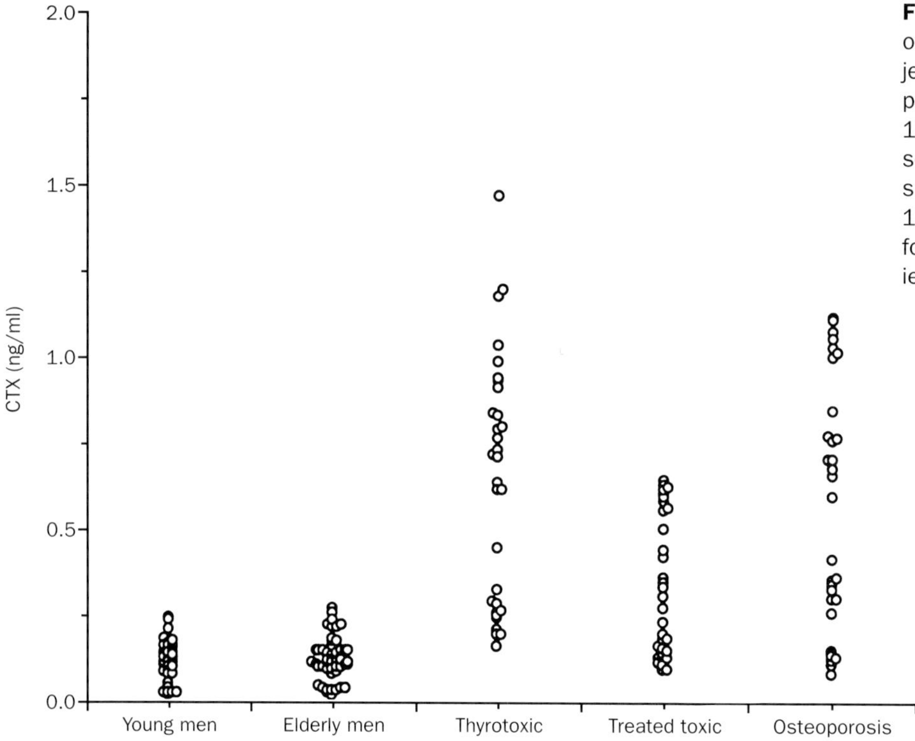

Fig. 10.6 Distribution of CTX results for all subjects studied when sampled between 14:00 and 17:00. The values in the scattergram include all sample points during 14:00–17:00 obtained for each individual studied.

to the fDPD profile may reflect a combination of release from bone and subsequent metabolism of telopeptides at the kidney, which is postulated to result in production of free cross-links. There is a consistency to the CTX rhythm that persists in all normal subjects studied and in many patients with thyrotoxicosis. Blunting of the rhythm was observed for serum CTX in a high percentage of postmenopausal patients with osteoporosis and some patients with thyrotoxicosis. There was absence of a significant rhythm in fDPD in thyrotoxic patients. The factors regulating the circadian rhythm are not immediately obvious, but strong similarities exist with the loss of the circadian rhythm for parathyroid hormone (PTH) and phosphate that has been observed in women with established osteoporosis.[9]

Our data suggest that the circadian rhythm observed in urinary bone resorption markers[4–6] is a direct reflection of the filtered load of telopeptide and free cross-links, with the highest excretion in urine following immediately behind the highest serum concentration. The role of renal metabolism in the generation of total urinary CTX would appear to be small.

There was no significant circadian rhythm for serum NTX detected in normal men. This was surprising, since previous reports have demonstrated that a circadian rhythm exists for urine NTX.[5,10] The variability in urine NTX is greater than either urine CTX or fDPD throughout a 24-h period[5] and this may contribute to an apparent dampening of the urine NTX circadian rhythm compared to urine fDPD and CTX.[5,10] It is possible that the serum NTX assay we used has insufficient sensitivity to detect the subtle changes in NTX, at low concentrations, that may take place over a 24-h time period in normal men. A significant diurnal variation in serum NTX has been reported in patients with osteoporosis before and after treatment with alendronate when sampled at 4-h intervals.[11] In these patients with osteoporosis, the NTX concentration was higher before and after treatment than in the normal men, and the differences in sampling frequency may also have contributed to detection of a diurnal rhythm. If the obviously discrepant values are excluded from analysis of our NTX results, then a difference ($P < 0.05$) in mean value between daytime and nighttime in the young normal men is observed which is consistent with the previous report.[11] The analytical (edge) effect that was observed may have prevented us from obtaining consistent results, which may have masked the presence of a circadian rhythm.

A significant increase in 24-h mean serum CTX and fDPD was observed in patients with osteoporosis and thyrotoxicosis compared to normal men. This reflected an overall increase in bone resorption in these patients. It is well known that thyrotoxic patients are predisposed to a significant reduction in bone mass and subsequently osteoporosis with fractures.[12] Urinary indices of bone resorption are significantly elevated in thyrotoxicosis and normalize within 2 months of treatment, whereas markers of bone formation remain elevated in these patients for up to 8 months after commencement of treatment.[13,14] Normalizing DPD was shown to be a good predictor of increase in bone mineral density.[14] It is interesting to speculate whether the additional loss of the circadian rhythm is a poor prognostic factor or whether this merely reflects the changes in bone metabolism initiated by the disease process. The fact that the circadian rhythm in resorption markers can be restored rapidly by treatment of thyrotoxicosis suggests that overproduction of thyroid hormones drives the resorption process in these patients, eventually overcoming the natural control of the circadian rhythm. In osteoporosis, a very large number of circadian profiles would need to be measured in order to establish the prognostic implications of loss of the circadian rhythm. Treatments directed at restoring this circadian rhythm, however, may prove to be beneficial in the long term, and it is interesting that intermittent PTH therapy is known to be anabolic to bone, while continuous infusion of PTH is catabolic.[15,16] The circadian rhythm of serum crosslinks may therefore reflect a beneficial effect of a combination of factors required to maintain skeletal integrity by initiating the bone remodelling cycle in an intermittent fashion.

The circadian rhythm of serum bone resorption markers has significant implications for their clinical utility. In all of the profiles, the greatest variability in serum CTX and a high variability in fDPD was observed between 07:00 and 11:00. This is the time when the majority of hospital samples are taken and when phlebotomy usage is at its highest. We have shown that the best discrimination between the groups studied, for CTX, was seen in samples taken between 14:00 and 17:00, when the distribution of results in normal males was tightest and the effect of the circadian nature of bone resorption was lowest. Preliminary studies have indicated that the ability of serum CTX measurements to predict fracture is significant and strongest when samples are obtained in the afternoon or when fasting late morning rather than early morning.[17,18] These findings would be explained by the circadian rhythm and early morning variability in serum CTX and fDPD that we have demonstrated, the greater morning variability in CTX leading to poorer discrimination and poorer predictive value.

In conclusion, we have shown that a significant circadian rhythm exists for serum CTX and fDPD which is lost in some patients with thyrotoxicosis and osteoporosis. The greatest variability in concentration of serum resorption markers is seen in the early morning (07:00–11:00), and this means that venesection should be performed, for the best clinical discrimination, between 14:00 and 17:00.

ACKNOWLEDGEMENTS

All reagents for CTX and NTX were supplied by Roche Diagnostics Ltd. A dedicated team of doctors and nursing staff assisted in the sampling and care for patients on the metabolic ward.

REFERENCES

1. Calvo MS, Eyre DR, Gundberg CM, Molecular basis and clinical application of biological markers of bone turnover. *Endocr Rev* 1996; 17: 333–68.
2. Fraser WD, The collagen crosslinks pyridinoline and deoxypyridinoline: a review of their biochemistry, physiology, measurement, and clinical applications. *J Clin Ligand Assay* 1998; 21: 102–10.
3. Panteghini M, Pagani F, Biological variation in urinary excretion of pyridinium cross-links: recommendations for the optimum specimen. *Ann Clin Biochem* 1996; 33: 36–42.
4. Schlemmer A, Hassager C, Jensen SB et al, Marked diurnal variation in urinary excretion of pyridinium crosslinks in premenopausal women. *J Clin Endocrinol Metab* 1992; 74: 476–80.
5. Ju HJ, Leung S, Brown B et al, Comparison of analytical performance and biological variability of three bone resorption assays. *Clin Chem* 1997; 43: 1570–6.
6. Schlemmer A, Hassager C, Pedersen BJ et al, Posture, age, menopause and osteopenia do not influence the circadian variation in the urinary excretion of pyridinium crosslinks. *J Bone Miner Res* 1994; 9: 1883–8.
7. McLaren AM, Isdale AH, Whiting PH et al, Physiological variations in the urinary excretion of pyridinium crosslinks of collagen. *Br J Rheumatol* 1993; 32: 307–12.
8. Teicher MH, Barber NI, COSIFIT: an interactive program for simultaneous multioscillator cosinor analysis of time-series data. *Comput Biomed Res* 1990; 23: 283–95.
9. Fraser WD, Logue FC, Christie JP et al, Alteration of the circadian rhythm of intact parathyroid hormone and serum phosphate in women with established postmenopausal osteoporosis. *Osteoporos Int* 1998; 8: 121–6.
10. Bullen AM, Martin MD, Leroux BG, Eyre DR, Circadian variation in urinary excretion of bone collagen crosslinks. *J Bone Miner Res* 1995; 10: 1885–90.
11. Gertz BJ, Clemens JD, Holland SD et al, Application of a new serum assay for type 1 collagen cross linked N-telopeptides: assessment of diurnal changes in bone turnover with and without alendronate treatment. *Calcif Tissue Int* 1998; 63: 102–6.
12. Adams PH, Jowsey J, Kelly PJ et al, Effects of hyperthyroidism on bone and mineral metabolism in man. *Q J Med* 1967; 36: 1–15.
13. Nagasaka S, Sugomoto H, Nakamura H et al, Antithyroid therapy improves bony manifestations and bone metabolic markers in patients with Graves' thyrotoxicosis. *Clin Endocrinol* 1997; 47: 215–21.
14. Siddiqi A, Burrin JM, Noonan K et al, A longitudinal study of markers of bone turnover in Graves' disease and their value in predicting bone mineral density. *J Clin Endocrinol Metab* 1997; 82: 753–9.
15. Podbesek R, Edonard L, Meunier P et al, Effects of two treatment regimens with synthetic human parathyroid hormone fragment on bone formation and the tissue balance of trabecular bone in greyhounds. *Endocrinology* 1983; 112: 1000–6.
16. Reeve J, Davies UM, Hesp R et al, Treatment of osteoporosis with human parathyroid peptide and observations on effect of sodium fluoride. *Br Med J* 1990; 301: 314–18.
17. Chapurlat RD, Garnero P, Bréart G et al, Afternoon sampled serum cross-laps predicts hip fracture in elderly women: The EPIDOS study. *J Bone Miner Res* 1999; 14(suppl): Abstract 1118.
18. Bauer DC, Black DM, Ensrud K et al, Serum markers of bone turnover and fracture of the hip and spine: a prospective study. *J Bone Miner Res* 1999; 14(suppl): Abstract 1058.

11

Sources of biological bone marker variability

Charles H Chesnut III

Summary • Introduction • Previous observations • Current observations • Conclusions • References

SUMMARY

The biological variability of bone turnover markers includes the diurnal (circadian) variability (typically over 24 h), as well as the longer-term intra-subject variability (over days to months). Factors which might affect bone markers include such nutrients as dietary calcium, lifestyle factors such as smoking, exercise, and bed rest, and medications such as hormone replacement treatment or bisphosphonates. The effect of such factors on bone marker variability should be differentiated from their direct effect on bone markers, exclusive of their effect on variability. Previous observations on diurnal variability include: a magnitude for urinary N-terminal cross-linking telopeptide of type I collagen (NTX) of about 40%, a magnitude for serum C-terminal cross-linking telopeptide of type I collagen (CTX) of about 65%, and an approximate 50% reduction in the magnitude of diurnal variability with bisphosphonates such as alendronate. More recent data demonstrate a short-term (3-day) and long-term (2-month) variability of urinary NTX of 13.1% and 15.6% respectively, with short-term variability of 6.3% and long-term variability of 7.5% for the serum NTX. Smoking, bisphosphonates such as alendronate, hormone replacement therapy and dietary calcium intake modestly influence marker variability, but such influences are of questionable clinical significance.

INTRODUCTION

Assessment of the biological variability of bone turnover markers includes diurnal (circadian) variability, usually over a 24-h period, and intra-subject variability (over longer periods of time, days to months). Factors which may affect bone turnover markers include such nutrients as calcium,[1] lifestyle factors such as smoking, exercise,[2,3] and bed rest,[4] and use of medications such as hormone replacement treatment or bisphosphonates. These factors would obviously have a direct effect on markers, but the question of whether such factors also affect the diurnal or longer-term intra-subject variability of markers could be examined.

PREVIOUS OBSERVATIONS

Previous data on diurnal (circadian) variability include those of Greenspan et al,[5] demonstrating in 28 male and female subjects (75–80 years of age) a 'magnitude' of diurnal variability of serum osteocalcin and serum bone-specific alkaline phosphatase of approximately 10–20%, and

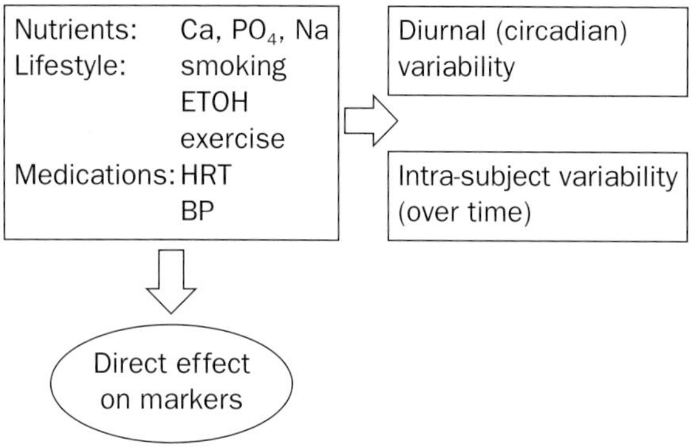

Fig. 11.1 Factors affecting bone markers. Ca: calcium; PO_4: phosphate; Na: Sodium; ETOH: ethanol (alcohol); HRT: hormone replacement therapy; BP: bisphosphonates.

of urinary N-terminal cross-linking telopeptide of type I collagen (NTX) of approximately 40%. Also, variability was significantly higher in females than in males. A second study[6] in six healthy males of approximately 25 years of age demonstrated the 'magnitude' of diurnal variability of serum C-terminal cross-linking telopeptide of type I collagen (CTX) to be ~60% to 66%. Finally, Gertz et al[7] demonstrated in 38 osteopenic females, aged approximately 69 years, an effect of alendronate in significantly reducing the 'magnitude' of diurnal variability by about 50% as compared to placebo, but a physiological, circadian pattern of bone resorption was still maintained. In this paper it was also noted that there was an approximately 25% suppression of serum NTX due to alendronate as compared to placebo, and an approximately 30% suppression of urinary NTX due to alendronate as compared to placebo. There is a paucity of previous information on the effect of lifestyle factors such as smoking, exercise, and bed rest on the biological variability of markers. One study[1] has, however, noted a significant effect (suppression) of evening calcium supplementation on the circadian rhythm of bone resorption markers.

CURRENT OBSERVATIONS

A recent paper by Eastell et al[8] was designed to evaluate the short-term (over 3 days) and long-term (over 2 months) intra-subject variability of urine and serum NTX in 277 healthy postmenopausal women, at four geographical sites. Also, this study aimed to relate NTX variability to age, hormone replacement therapy, bone mineral density, diet, exercise, and smoking.

In this study, it was found that the short-term (3-day) and long-term (2-month) variability were 13.1% and 15.6% respectively for urinary NTX; for serum NTX, short-term variability was 6.3% and long-term variability 7.5%. Short-term variability for urine NTX was affected by smoking and geographical site; long-term variability of urinary NTX was affected by hormone replacement therapy use ($P = 0.05$) and by dietary calcium ($R = -0.22$, $P < 0.01$). Short-term variability of serum NTX was affected by smoking (coefficient of variation 6.5% for never/former smokers, and 5.5% for current smokers, $P < 0.03$) and by geographical site.

A calculation of the signal-to-noise ratio to compare two NTX measures to determine the effect that variability has on the ability to measure a change in bone resorption was done; for the urine NTX, this was 2.9, and for the serum NTX it was 2.8. These ratios are therefore similar for urinary and serum NTX, indicating similar diagnostic value. Also, the least significant change can be calculated, which incorporates variability into laboratory test interpretation. Calculating the least significant change indicates that a decrease of 31% in urinary NTX, and of 14% in the serum NTX, is required to achieve 90% confidence that a decrease between two sequential NTX measurements after initiation of therapy is clinically relevant, and not due to variability alone.

CONCLUSIONS

1. Significant diurnal (circadian), or longer, variability exists for markers of bone turnover; to control for such variability,

consistent timing of sample collections is mandatory.
2. Factors such as bisphosphonates or hormone replacement therapy, dietary calcium intake and smoking may influence marker variability of the NTX.
3. The above influences are, however, modest, and are of questionable clinical significance.

REFERENCES

1. Blumsohn A, Herrington K, Hannon RA et al, The effect of calcium supplementation on the circadian rhythm of bone resorption. *J Clin Endocrinol Metab* 1994; 79: 730-5.
2. Woitge HW, Friedmann B, Suttner S et al, Changes in bone turnover induced by aerobic and anaerobic exercise in young males. *J Bone Miner Res* 1998; 13: 1797–804.
3. Fujimura R, Ashizawa N, Watanabe M et al, Effect of resistance exercise training on bone formation and resorption in young male subjects assessed by biomarkers of bone metabolism. *J Bone Miner Res* 1997; 12: 656–62.
4. Zerwekh JE, Ruml LA, Gottschalk F, Pak CY. The effects of twelve weeks of bed rest on bone histology, biochemical markers of bone turnover, and calcium homeostasis in eleven normal subjects. *J Bone Miner Res* 1998; 13: 1594–601.
5. Greenspan S, Dresner-Pollak R, Parker R et al, Diurnal variation of bone mineral turnover in elderly men and women. *Calcif Tissue Int* 1997; 60: 419–23.
6. Wichers M, Schmidt E, Bidlingmaier F, Klingmüller D, Diurnal rhythm of Cross-Laps in human serum. *Clin Chem* 1999; 45: 1858–60.
7. Gertz B, Clemens J, Holland S et al, Application of a new serum assay for type I collagen cross linked n-telopeptides: assessment of diurnal changes in bone turnover with and without alendronate treatment. *Calcif Tissue Int* 1998; 63: 102–6.
8. Eastell R, Mallinak N, Weiss S et al, Biological variability of serum and urinary n-telopeptide of type I collagen in postmenopausal women. *J Bone Miner Res* 2000; 15: 594–8.

12

The effects of fracture or disease on biochemical markers of bone remodelling

Michael Kleerekoper

Summary • Introduction • Case 1 • Case 2 • Fractures and bone remodelling • Immobilization and bone remodelling markers • Non-osteoporosis skeletal disease and bone remodelling markers • Conclusions • References

SUMMARY

Biochemical markers of bone remodelling have comprised an important adjunct to clinical examination and radiographic studies in the diagnosis of metabolic bone diseases for at least 50 years. They have also been the mainstay of monitoring progression or regression of metabolic bone diseases. The newer markers of remodelling currently employed in patients with osteoporosis, where bone remodelling abnormalities are less pronounced than in other metabolic bone diseases, have improved sensitivity over the older markers but offer no apparent improvement in diagnostic specificity. Skeletal fractures heal by local remodelling, and the extent to which this remodelling is reflected in biochemical markers is most probably related to the extent of the fracture. Specific questions that cannot yet be fully answered from the available data include the following: (1) Are there levels of markers that indicate the presence of a secondary cause of bone loss or metabolic disease other than osteoporosis? (2) Are there patterns of abnormality in markers that offer differential diagnostic information? (3) To what extent and for how long do common osteoporotic fractures influence levels of markers? (4) Is there a minimum amount of immobilization that is likely to affect levels of markers?

INTRODUCTION

For decades, clinicians have relied on biochemical markers of bone remodelling in the management of patients with metabolic bone diseases such as Paget's disease of bone, osteomalacia, and hyperparathyroidism. In most of these patients, the remodelling abnormality was so marked that sensitivity and specificity of the marker assays was not a practical problem. As attention has turned to the more subtle remodelling changes seen in the primary osteoporoses, newer markers of bone remodelling with improved sensitivity and specificity have been developed. Increased utilization of these newer markers has brought to light more subtle changes in remodelling in metabolic bone diseases other than osteoporosis that might have gone unnoticed with older assays. Even more subtle are the changes in remodelling that accompany normal fracture healing. This is a localized skeletal phenomenon which might be very robust at the site of fracture but has limited impact on levels of markers that reflect more global skeletal activity. This chapter will focus on changes in these newer markers in skeletal disease other than osteoporosis and on the changes that accompany fracture healing. Two recently evaluated patients are presented to underscore how knowledge of biochemical markers of bone remodelling has impacted on diagnosis and management.

CASE 1

A 64-year-old woman was referred by an endocrinologist because 2 years of alendronate therapy had had no impact on her bone mineral density (BMD). Initial evaluation revealed a urine N-telopeptide (NTX) level of 146 bone collagen equivalents (BCE)/mM creatinine (normal 5–65). The patient claimed that she had been compliant with medication except for three brief episodes when she experienced diarrhoea. A diagnostic work-up revealed no cause for the diarrhoea, which was then attributed to the alendronate therapy. Anti-diarrhoeal medication controlled her symptoms, and she restarted alendronate therapy. She had been fully compliant with this therapy for the 3 months prior to the NTX measurement. She reported weight loss and difficulty in sleeping, and, when questioned, agreed that she had been less troubled by cold weather. Physical examination revealed no goitre or eye changes of hyperthyroidism, but she did have a tachycardia and brisk deep tendon reflexes. Laboratory studies confirmed the diagnosis of hyperthyroidism with suppressed thyroid-stimulating hormone (TSH) and elevated free thyroxine (FT4).

Thyroxine has effects on bone remodelling similar to those of parathyroid hormone, such that it can be quite difficult to distinguish these effects histologically. Much has been written about the potential role of thyroid hormone therapy as a predisposing factor for postmenopausal osteoporosis, but no consensus opinion prevails on this point. Frank hyperthyroidism is known to affect the biochemical markers of bone remodelling, with an increase in both resorption and formation markers. Usually, this poses no clinical problem, since the diagnosis of hyperthyroidism is easily made on clinical and biochemical evaluation. It is uncertain why the diagnosis had not been made earlier in this patient, but the major clue when she was eventually referred for a specific skeletal evaluation was the very high level of NTX despite antiresorptive therapy.

Siddiqui et al[1] reported their findings in 17 thyrotoxic patients studied over 1 year of therapy. Before treatment, in 10 of 17 patients bone-specific alkaline phsophatase (bone ALP) was elevated above the mean 2.3-fold. In these same patients, osteocalcin (OC) was elevated 2.2-fold. The traditional total alkaline phosphatase (total ALP) was only elevated 1.2-fold, even though the correlation between bone ALP and total ALP was excellent ($R = 0.95$, $P < 0.001$), indicating the improved sensitivity of bone ALP. The changes in markers of bone resorption were even more pronounced. Urine deoxypyridinoline (DPD) was elevated in all patients 3.5-fold. Serum DPD was less sensitive, with elevated values in only 8 of the 13 patients in whom the measurement was made, and the mean increase being only 1.1-fold. Serum pyridinoline (PYD) was more affected, being elevated in 13 of 14 patients to a mean of 2.3-fold. Pretreatment FT4 correlated only with urine DPD ($R = 0.61$, $P < 0.02$). On therapy, bone ALP rose by 2 weeks in 15 of 16 patients (one patient was excluded from on-treatment analyses because of failure of the hyperthyroidism to respond to therapy) and peaked at 4 weeks, coincident with attainment of the euthyroid state. Bone ALP remained elevated in eight patients during the 1 year of follow-up. A rise in OC was apparent after 2 weeks of therapy, but levels were restored to normal by 12 weeks of therapy. By

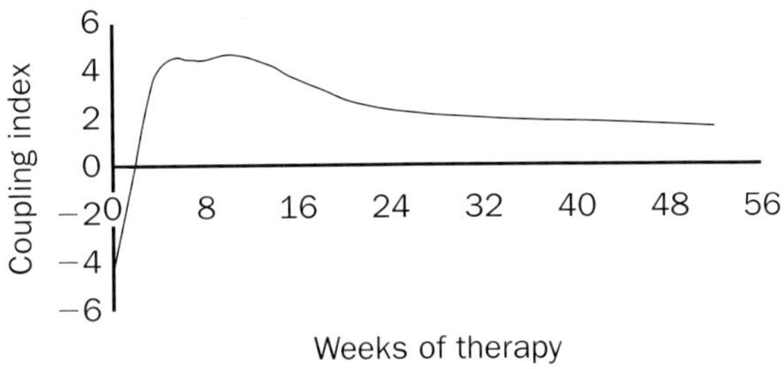

Fig. 12.1 The coupling index[2] provides an assessment of the balance between bone resorption and bone formation. A negative value indicates that the remodelling balance is in favour of resorption. This figure demonstrates the change in coupling index during 52 weeks of therapy for hyperthyroidism. Adapted from Siddiqui et al[1] with permission.

week 8, all of the markers of bone resorption had been restored to within the normal range, with the exception of four patients in whom normal values were not attained until 24 weeks of therapy. Using the Eastell coupling index,[2] resorption predominated at baseline, but a change in favour of formation was apparent by 4 weeks, and this remained the case through 1 year of therapy (Fig. 12.1).

CASE 2

A 66-year-old woman was referred by a haematologist 7 months after she had sustained three thoracic vertebral fractures without antecedent trauma. The initial evaluation had revealed a kappa light chain monoclonal gammopathy, but multiple myeloma had been excluded. Prior to her fractures, the patient had been working at a manual labour job. She denied medical problems apart from excess alcohol intake, and she had been abstinent for the preceding 2 months. She had undergone a temporary colostomy for unremembered reasons 15 years earlier. Since her fractures, she had been unable to work because of pain, and had experienced anorexia and an 18-kg weight loss without change in bowel habits or other gastrointestinal symptoms. There was no relevant family history or priory history of fractures, and a detailed history seeking possible secondary causes of bone loss was non-contributory. On examination, she was an African-American woman with typical spine changes of the vertebral fractures. There was decreased range of motion of the spine but no skeletal tenderness. She had some features of hypercortisolism (truncal obesity, peripheral wasting). Serum calcium, phosphorus and ALP were normal, as were hepatic and renal function. The haematology profile was normal. The results of 24-h urine studies at presentation and repeated after 2 months of therapy was calcediol 20 μg three times a week together with 1500 mg calcium are given in Table 12.1.

This patient, in whom a definitive diagnosis other than osteoporosis has not been established despite extensive evaluation, poses several questions:

- What is the magnitude of the effect of her vertebral fractures on the bone resorption indices, and how long does such an effect last?
- Did her relative immobilization since her fractures contribute to the increased levels of bone resorption markers? How much immobilization is needed before any effect on these markers is seen?
- Is there a level of bone resorption marker that indicates that a secondary cause for accelerated bone loss is operative? If so, what is that level?
- Is there a pattern of bone resorption markers that indicates or suggests a specific secondary cause for bone loss?

FRACTURES AND BONE REMODELLING

There have been no published reports of the acute effect of vertebral fractures on markers of bone remodelling. This is not particularly surprising, since many acute vertebral fractures are not identified as being related to osteoporosis by most clinicians, and few patients with vertebral fractures require admission to hospital as a

Table 12.1 24-hour urine collection results in patient Case 2.

	Baseline	After 2 months of calcifediol and calcium
Creatinine (mg)	600	800
Sodium (mmol)	120	113
Calcium (mg)	44	55
NTX (nmol BCE/mmol creatinine)	148	147
PYD (nmol/mmol creatinine)	233	167
DPD (nmol/nmol creatinine)	61	41

Reference interval: NTX 5–65; PYD 22–89; DPD 4–21.

direct result of the fracture. Recently, Ingle et al reported on the acute changes in markers following forearm[3] and ankle[4] fractures. There were statistically significant increases in formation markers of 20–50% seen as early as 2 weeks following fracture and persisting at 52 weeks following the fracture. In contrast, while the increases in resorption markers were a little more pronounced, the differences were not statistically significant, and baseline values were evident 52 weeks after the fracture. Boonen[5] studied 40 elderly men within 18 h of a hip fracture and, compared to matched controls, found significant increases in the bone resorption markers PYD and DPD. The mean increase in PYD was 71% ($P < 0.004$), and in DPD it was 92% ($P < 0.008$). There was no significant difference in serum OC between patients and controls. The fracture cases also had significantly lower values for calcediol and calcitriol, as well as total serum testosterone. Multivariate analysis demonstrated that the greatest increases in the resorption markers were found in those men with either hypogonadism or hypovitaminosis D or both. In sum, there are limited data on the magnitude of the increase in biochemical markers of bone remodelling following fracture and on the time course of this change following the fracture.

IMMOBILIZATION AND BONE REMODELLING MARKERS

The adverse skeletal effects of immobilization have been studied extensively. Initially, these studies related to the immobilization following acute spinal cord injury, and denervation–immobilization experiments in animals have been conducted in an attempt to find a suitable animal model for osteoporosis. Most recently, attention has focused more directly on the skeletal losses associated with the anti-gravity effects of space flight, where rates of bone loss exceed most of those seen under full-gravity conditions. Inoue et al[6] performed a rigorous study in healthy young male volunteers strictly immobilized for protracted periods in a 6° downwards head tilt posture. Levels of bone resorption markers rose dramatically within a few weeks, reaching peak values some two-fold greater than baseline and remaining elevated while the experiment continued. Almost identical changes, albeit some time later, were observed with the formation markers. Recovery after mobility was restored was rapid but not complete. Zerwekh et al[7] also studied formation and resorption markers during 12 weeks of bed rest and post-recovery. They demonstrated a rapid rise in resorption markers (hydroxypro-

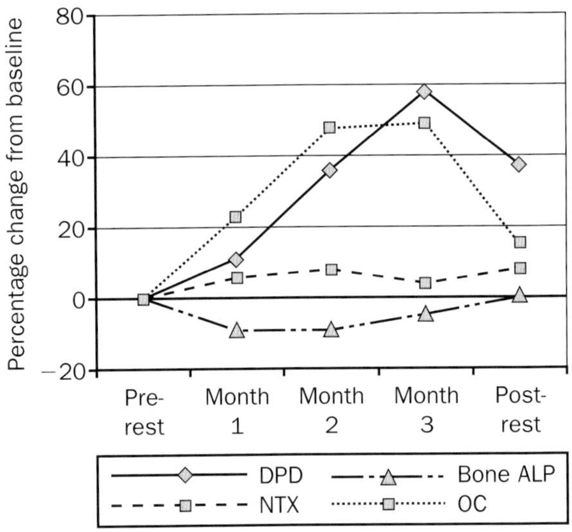

Fig. 12.2 Changes in markers of bone formation (bone ALP and OC) and bone resorption (DPD and NTX) before bed rest, during three 4-week periods of rest, and following recovery from bed rest. Adapted from Zerwekh et al[7] with permission.

line, DPD and NTX) but not in the formation markers osteocalcin or bone ALP (Fig. 12.2). It is unlikely that the relatively short periods of incomplete immobilization during a period of recovery from acute vertebral or hip fracture would raise levels of markers to that extent. This is particularly so in an older population most likely to sustain such fractures.

NON-OSTEOPOROSIS SKELETAL DISEASE AND BONE REMODELLING MARKERS

Substantial literature exists on the abnormalities in both resorption and formation markers in several metabolic and metastatic bone diseases. In general, there are significant increases in all markers in most of these reports, but the magnitude and pattern of change vary substantially from one report to the next. The highest levels of markers are generally seen in Paget's disease of bone and in patients with skeletal metastases. It has been suggested that bone remodelling, and markers of remodelling, are more related to the number of skeletal sites involved with Paget's disease than to the severity at a particular site. Thus a patient with polyostotic Paget's disease is likely to have a higher level of any marker than a patient with profound involvement at a single site. Similar data are not available for metastatic bone disease, but it is likely that the level of marker reflects the degree of bone tumour burden. The levels of markers in primary hyperparathyroidism also reflect the severity of disease, as evidenced by the degree of hypercalcaemia and of bone involvement. Case reports of patients with severe primary hyperparathyroidism and overt osteitis fibrosa document very high levels of resorption markers.[8] In mild primary hyperparathyroidism, the situation is quite different. Silverberg et al[9] recently reported a 10-year follow-up of 121 patients with mild disease, approximately half of whom met NIH criteria for surgical cure.[10] The mean levels of total ALP, OC, PYD and DPD were normal in both groups. Following parathyroidectomy, there was a significant decrease in total ALP but results for other markers were not reported. In those with mild disease not subject to surgery, all parameters of disease, including serum calcium, serum parathyroid hormone and BMD, were stable during follow-up. While not specifically reported, it is reasonable to assume that levels of markers of remodelling were stable during that same 10-year period. A slightly different conclusion was drawn from an earlier study by Guo et al.[11] This group did note minimal increases in formation markers (bone ALP and the C-terminal propeptide of type I procollagen) and more marked increases in the resorption markers (NTX and galactosyl hydroxylysine). More importantly, they demonstrated increases in bone ALP, NTX and galactosyl hydroxylysine during 2 years of follow-up. A recent series of metabolic and malignant bone diseases was published by Woitge et al,[12] and the results are summarized in Fig. 12.3. As anticipated, there was substantial variability in the degree of elevation in all studied markers in the various

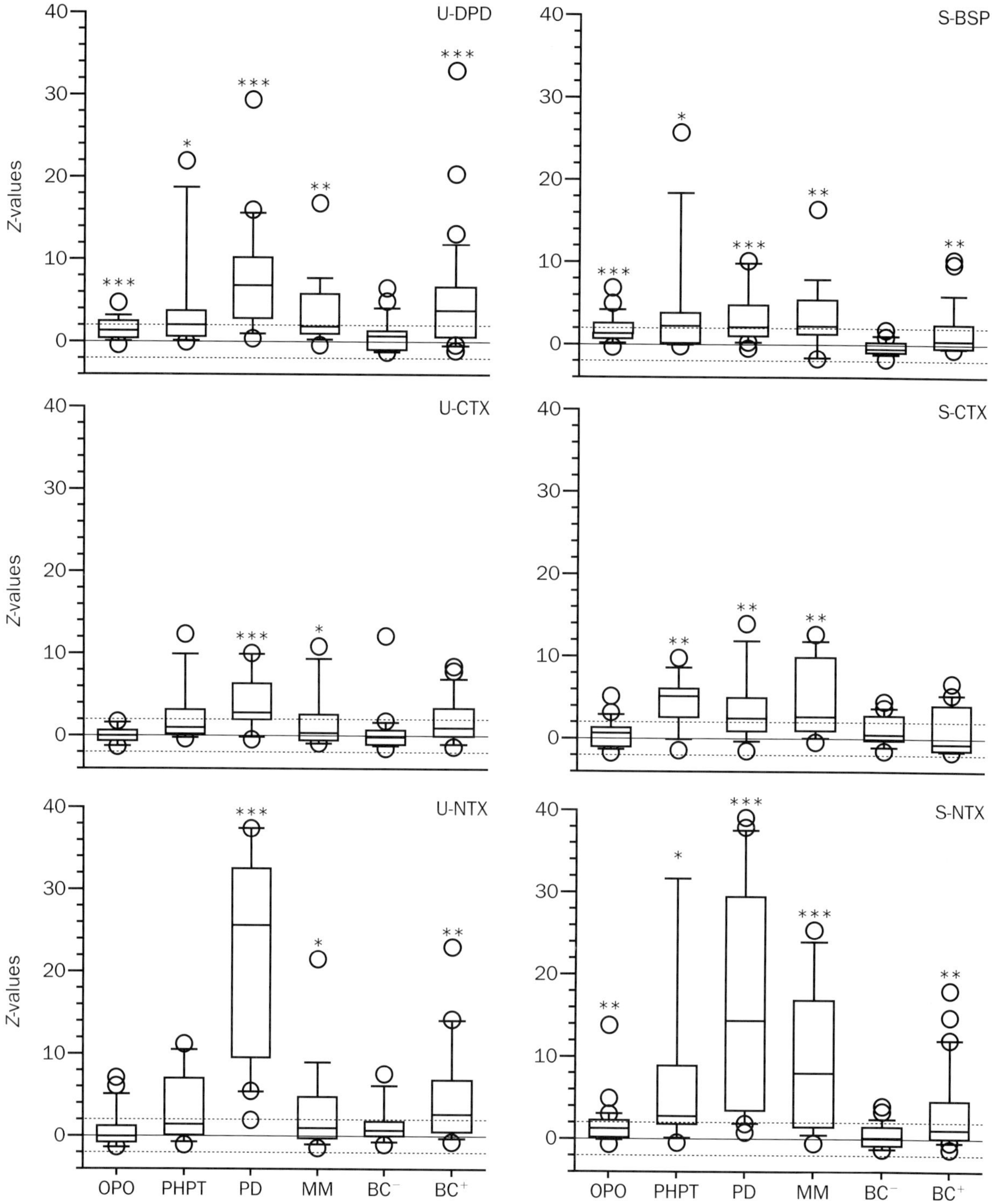
U-DPD
S-BSP
U-CTX
S-CTX
U-NTX
S-NTX
Z-values
0
10
20
30
40
OPO
PHPT
PD
MM
BC⁻
BC⁺

diseases. They confirmed the very high levels in Paget's disease, and in breast cancer patients with skeletal metastases. Not demonstrated in Fig. 12.3 but reported in the text, high levels of markers were also seen in patients with humoral hypercalcaemia of malignancy. Patients with multiple myeloma had increases in the serum markers of resorption but not in the urine markers.

Each of the diseases already mentioned represents non-subtle changes in bone metabolism, usually easy to distinguish from a primary osteoporosis by history, physical examination, and appropriate laboratory studies. More subtle is mild secondary hyperparathyroidism due to malabsorption or malnutrition. In these patients, serum calcium may be normal, there may or may not be mild hypophosphataemia, and total ALP may be only mildly elevated on a routine biochemical profile. Keaveny et al[13] studied 19 newly diagnosed patients with celiac sprue and found that 25% had hypovitaminosis D and 19% had secondary hyperparathyroidism with an elevated serum parathyroid hormone. Both bone resorption and formation markers were elevated at presentation. The poor vitamin D status of many elderly citizens without overt gastrointestinal disease is receiving appropriate increased attention throughout the world. The adverse impact of this on biochemical markers of bone resorption and formation was well documented in the Chapuy study.[14] The direct impact of this mild secondary hyperparathyroidism on the pathogenesis of hip fractures and the rapid reduction in hip fracture incidence following correction of this with supplemental calcium and vitamin D[15] underscores the importance of assessing bone remodelling parameters in the elderly.

Fig. 12.3 Relative values for biochemical markers of bone remodelling in several diseases known to affect the skeleton. Values are reported in standard deviation units (*Z*-score) relative to the mean value in healthy premenopausal women. OPO, osteoporosis; PHPT, primary hyperparathyroidism; PD, Paget's disease of bone; MM, multiple myeloma; BC^-, breast cancer without metastases; BC^+, breast cancer with metastases; U-DPD, urine deoxypyridinoline; U-CTX, urine C-terminal telopeptide; U-NTX, urine N-terminal telopeptide; S-BSP, serum bone sialoprotein; S-CTX, serum CTX; S-NTX, serum NTX, Reproduced from Woitge et al[12] with permission.

CONCLUSIONS

Any disease that affects bone remodelling either locally or generally is likely to have a demonstrable effect on the levels of biochemical markers of bone remodelling. With few exceptions, these changes will be apparent more quickly, but not necessarily more profoundly, in the resorption markers than in the formation markers. The level of increase appears to reflect the degree of skeletal involvement, as evidenced by observations in Paget's disease of bone, primary hyperparathyroidism, and, to a lesser extent, hyperthyroidism. However, as has been known for generations, there is no evidence that a particular pattern of change in markers has any differential diagnostic value. Even the relatively non-specific marker, total ALP, will reflect metabolic bone disease if the disease is severe enough. But an elevated total ALP level alone cannot distinguish Paget's disease of bone from severe primary or secondary hyperparathyroidism, rickets or osteomalacia, or even metastatic bone disease. It seems reasonable to conclude that there is a level of elevation in markers that increases the likelihood of metabolic bone disease other than primary osteoporosis. What that level might be remains unknown.

REFERENCES

1. Siddiqui A, Burrin JM, Noonan K et al, A longitudinal study of markers of bone turnover in Graves' disease and their value in predicting bone mineral density. *J Clin Endocrinol Metab* 1997; 82: 753–9.
2. Eastell R, Robins SP, Colwell T et al, Evaluation of bone turnover in type I osteoporosis using

biochemical markers for both bone formation and bone resorption. *Osteoporosis Int* 1993; 3: 255–60.

3. Ingle BM, Hay SM, Bottjer HM, Eastell R, Changes in bone mass and bone turnover following distal forearm fracture. *Osteoporosis Int* 1999; 10: 399–407.
4. Ingle BM, Hay SM, Bottjer HM, Eastell R, Changes in bone mass and bone turnover following distal ankle fracture. *Osteoporosis Int* 1999; 10: 408–15.
5. Boonen S, Vanderschueren D, Cheng XG et al, Age-related (type II) femoral neck osteoporosis in men: biochemical evidence for both hypovitaminosis D and androgen deficiency induced bone resorption. *J Bone Miner Res* 1997; 12: 2119–26.
6. Inoue M, Tanaka H, Moriwake T et al, Altered biochemical markers of bone turnover in humans during 120 days of bed rest. *Bone* 2000; 26: 281–6.
7. Zerwekh JE, Ruml LA, Gottschalk F, Pak CY, The effects of twelve weeks of bed rest on bone histology, biochemical markers of bone turnover, and calcium homeostasis in eleven normal subjects. *J Bone Miner Res* 1998; 13: 1594–601.
8. Kulak C, Bandeira C, Voss D et al, Marked improvement in bone mass after parathyroidectomy in osteitis fibrosa cystica. *J Clin Endocrinol Metab* 1998; 83: 732–5.
9. Silverberg SJ, Shane E, Jacobs TP et al, A 10-year prospective study of primary hyperparathyroidism with or without parathyroid surgery. *N Engl J Med* 1999; 341: 1249–55.
10. Consensus Development Conference Panel, Diagnosis and management of asymptomatic primary hyperparathyroidism. *Ann Intern Med* 1991; 114: 593–7.
11. Guo CY, Thomas WE, al-Dehaimi AW et al, Longitudinal changes in bone mineral density and bone turnover in postmenopausal women with primary hyperparathyroidism. *J Clin Endocrinol Metab* 1996; 81: 3487–91.
12. Woitge HE, Pecherstorfer M, Li Y et al, Novel serum markers of bone resorption: clinical assessment and comparison with established urinary markers. *J Bone Miner Res* 1999; 14: 792–801.
13. Keaveny AP, Freaney R, McKenna MJ et al, Bone remodeling indices and secondary hyperparathyroidism in celiac disease. *Am J Gastroenterol* 1996; 91: 1226–31.
14. Chapuy MC, Schott AM, Garnero P et al, Healthy elderly French women living at home have secondary hyperparathyroidism and high bone turnover in winter. EPIDOS Study Group. *J Clin Endocrinol Metab* 1996; 81: 1129–33.
15. Chapuy MC, Arlot ME, Delmas PD, Meunier PJ, Effect of calcium and cholecalciferol treatment for three years on hip fractures in elderly women. *BMJ* 1994; 309: 193.

13

Variation in biochemical markers of bone turnover: sources, quantification, minimization

G Russell Warnick

Summary • Introduction • Components and sources of variation • Intra-individual within-day variation • Intra-individual among-day variation • Variation related to specimen, collection and processing • Analytical variation • Total variation • Least significant change • Index of individuality • Considerations in minimizing confounding biological factors in urine • Other considerations in interpretation of results • Considerations in decreasing variability • References

SUMMARY

Variation in the biochemical markers of bone metabolism can compromise their ability to characterize disorders of bone metabolism. Variation can be categorized into pre-analytical and analytical sources. Pre-analytical variation includes biological components of intra- and inter-individual differences as well as variation from specimen collection, processing and storage. The intra-individual variation is composed of within-day and among-day components, the latter including both short-term and longer-term factors. The within-day intra-individual variation is primarily related to diurnal changes associated with bone remodelling. The diurnal cycle has usually been described in terms of amplitude or magnitude, the peak value in relation to nadir. Although uncommon in bone literature, description of the diurnal cycle in terms of statistical distribution, corresponding to approximately 10% CV, facilitates comparisons with other sources. Studies have suggested that the full range of among-day intra-individual variation, averaging approximately 20% CV, can be observed within a week. Inter-individual or group variation results from characteristic patterns of bone turnover at different ages, between genders and depending on the physiological status. Analytical sources of variation arise from the measurement process and can range to over 15% or more. An index of individuality can be used to determine whether a reference range based on a particular group is applicable to the individual. When the inter-individual variation in the group is relatively large, a reference range derived from the group has little utility in characterizing the individual. Age- and gender-specific ranges can be used to narrow the group, decreasing effective variation and improving characterization of the individual. Calculation of least significant change

(LSC = 1.96 × $2^{1/2}$ × [CV total variability]) indicates whether an observed change, e.g. response to therapy, can be considered meaningful. In general, the greater the variability, the less predictive is the marker. Thus, predictive value can be improved by decreasing variation. Sampling at the same time each day, preferably in the morning when the value is highest, can minimize the diurnal component. The effect of the largest contributor, among-day variation, can be decreased by averaging results over 2 or 3 days, cost-effectively by pooling collections for analysis as an alternative to analysing separate samples. Analytical variation can be reduced by choosing appropriate markers, using reproducible assay methods and performing, if necessary, replicate analyses. With relatively modest changes in practice, variation can be substantially decreased.

INTRODUCTION

Biochemical markers of bone turnover are considered to have utility in diagnosis and management of patients with bone disorders (for recent review, see Watts[1]). In practice, however, the benefit of biochemical marker testing has been limited by variability from both biological and analytical sources. Biochemical markers have been clearly useful in identifying and monitoring treatment of conditions that result in large changes in bone turnover. Their use in conditions with more subtle changes, such as postmenopausal osteoporosis, has been somewhat controversial.

Laboratory measurements have inherent advantages in clinical decision-making. Measurements are objective and generally quantitative, facilitating comparisons with accepted reference ranges or cut-points. Determination of a biochemical marker either in the laboratory or at the point of care can be relatively inexpensive and can often be performed simultaneously with other necessary laboratory testing on an existing specimen. The incremental cost for each successive test can be quite low. Ever-improving technologies available in the modern clinical laboratory facilitate accurate measurements by highly automated and computer-linked instruments. Bone marker values can be interpreted in conjunction with other relevant laboratory tests in the same report. Technology for point-of-care measurement of biochemical parameters is advancing rapidly, and bone marker tests are beginning to appear on such devices, a capability that could facilitate mass screening and initial diagnosis.

Biochemical markers of bone turnover, especially those of bone resorption, respond rapidly to treatment, providing a relatively early indication of efficacy. In contrast, measurements of bone density are quite accurate in establishing the status of the bone at a point in time, but the slow rate of change in bone mass requires a much longer interval before a treatment effect can be reliably detected by densitometry. Provided the variability issues can be appropriately addressed, biochemical markers used in conjunction with bone densitometry could offer considerable synergy; densitometry to determine the current state of the bone, and biochemical markers to signal the rate of change.

COMPONENTS AND SOURCES OF VARIATION

From the perspective of the laboratorian, variation can be categorized into pre-analytical and analytical sources (Table 13.1). Analytical components of variation are related to the measurement process. Pre-analytical components include those occurring before the specimen is analysed; biological variation within the individual, within-day and among-day variation, and group variation among individuals, as well as the variation associated with specimen collection, processing, transport and storage. Each of these components will be considered subsequently.

INTRA-INDIVIDUAL WITHIN-DAY VARIATION

Bone turnover follows a well-characterized diurnal or circadian pattern that can be moni-

Table 13.1 Variation: components/sources.

Pre-analytical
Intra-individual
Within-day (diurnal)
Among-day
Inter-individual
Group variation
Age/gender etc. differences
Specimen/collection/processing
Analytical
Measurement

tored based on the release of molecules associated with bone formation and resorption.[2] Measured in urine, resorption markers peak in the early morning and reach a nadir in the late afternoon to early evening, with a peak to nadir ratio averaging approximately 1.5. The diurnal change has generally been reported in terms of amplitude, e.g. ratio or percentage increase of peak over nadir or percentage of the deviation of each peak and nadir from the 24-h mean. Because sources of variability other than diurnal are reported in terms of statistical distribution, i.e. coefficient of variation (% CV = standard deviation/mean), conversion of amplitude to distribution is useful in making comparisons. One study used a statistical package to determine that a peak to nadir ratio of 1.5 corresponds to a CV of approximately 10%.[3]

The observed daily cycle could be expected to be most extreme with frequent sampling of a marker with a short residence time. The formation markers, bone alkaline phosphatase and osteocalcin, with longer half-lives, exhibit lesser amplitudes. Because the release of the small-molecule markers into urine essentially pools or time-averages the signal between voids, smoothing of the actual cycle is expected. Less frequent sampling or longer pooling should also be expected to decrease the observed cycle.

Considering physiological factors, even though the output of cross-links in urine varies with conditions, the amplitude of the cycles seems to be reasonably consistent on a relative basis.[4,5] Early and elderly postmenopausal and elderly osteopenic women all demonstrated similar relative amplitudes in their diurnal patterns.[5] The diurnal cycle is reportedly unaffected by oestrogen and other hormonal changes during the menstrual cycle.[5] Small differences were observed in amplitude between men and women.[6] Children demonstrated a diurnal pattern similar to those of adults.[7] Even after 5 days of bed rest with total cross-link output decreasing significantly, by 28%, the relative amplitude of the cycles remained the same.[5]

INTRA-INDIVIDUAL AMONG-DAY VARIATION

The among-day variation observed in a variety of studies seems to be about the same for periods of days as well as weeks and months (Table 13.2). For pyridinoline (PYD), among-day variation with daily collections for 5–21 days averaged 16% (ranging from 12% to 21% in five collection conditions/studies) and for deoxypyridinoline (DPD) it averaged 17% (range 5–24% in six combinations).[8–10] Variation was similar in weekly collections made for up to 5 weeks, averaging 16% for PYD and 17% for DPD.[11,12] Variation in DPD with monthly collections over 5 months in both males and females averaged 17%.[3] For N-terminal cross-linking telopeptide of type I collagen (NTX) similar observations have been reported; among-day variation was about the same over 2 weeks as over 8 years, approximately 19%.[13] Thus, factors contributing to variation over months seem not to outweigh shorter-term factors. In fact, one study reported that three collections over 3–5 days captured the full range of biological variation.[10]

VARIATION RELATED TO SPECIMEN, COLLECTION AND PROCESSING

Biochemical markers of bone formation have been measured in blood plasma or serum. The

Table 13.2 Variation: intra-individual among-day (CV).

	Day to day	Week to week	Month to month
PYD[a]	16%	16%	
DPD[a]	17%	17%	17%
NTX[13]	19%		18%

[a]Six studies/183 subjects.
PYD, pyridinoline; DPD, deoxypyridinoline; NTX, N-terminal cross-linking telopeptide of type I collagen.

collagen degradation markers of resorption can be measured in either serum or urine, each fluid having merits. Urine has the advantage that the analytes are concentrated approximately 10 times compared to serum in a less complex matrix, because many serum constituents are filtered by the kidney. Specimen collection procedures for urine are minimally invasive. But modulation by the kidney from the blood compartment to urine may change relative concentrations of free compared with conjugated cross-links,[14] affecting some assay procedures.

Quantitative measurements made in spot urines are corrected for volume fluctuations based on creatinine output, even though creatinine contributes to analytical variation and is also subject to biological variation, with a diurnal cycle nearly counter-cyclical to that of the resorption markers, high in the afternoon and low in the morning.[15] Lifestyle and other factors such as diet, stress, exercise and renal failure can also affect creatinine output.[16] In older and immobilized patients, creatinine output may be low, resulting in overestimation of resorption markers. Nevertheless, correction by creatinine is still considered to improve prediction.[11]

In studies of different modes of urine collection,[9,11,12] the second morning void (SMV) appeared to be slightly more predictive than a 24-h collection, which was a little better than a first morning void (FMV). The differences were slight, but considering the considerable inconvenience of the 24-h collection, either a SMV or perhaps FMV seems preferable. The SMV might have some advantage in clinical practice, in that patients who urinate frequently during the night might better accommodate the SMV, and some patients scheduled in the morning might be able to provide the SMV during the office visit.

There are also aesthetic and logistical disadvantages with urine; for example, the preferred morning void collection may require that patients receive collection supplies for use at home. Because serum is upstream from urine, there have been suggestions that biological variation might be less, although the evidence to date is not compelling. Serum values do not require correction for dilution, as does urine, eliminating the variability from creatinine. Serum or plasma, the most common specimens for laboratory analysis, facilitate measurement of the markers in a specimen often already collected for other analytes. Nevertheless, collection of serum is invasive and still subject to diurnal and other sources of variation.

Considering storage effects, markers can deteriorate during processing and storage. The proteins tend to be unstable, and PYD and DPD are especially sensitive to UV light, requiring that specimens be stored in opaque containers. Alternatively, urine collection containers can be acidified.

ANALYTICAL VARIATION

The total analytical or measurement variation includes within- and among-run components. Variation is described as precision or imprecision, generally in relative terms as CV. Within-run variation is primarily a function of the reproducibility of the analytical system, e.g. pipetting and optical systems. Among-run variation adds components related to the long-term stability of the instrument and reagent, lot-to-lot changes and calibration-related fluctuations. Chemical assays, which usually employ highly specific enzymatic reactions, tend to be more precise than immunoassays, which, especially in manual assays, are often run in duplicate or triplicate to compensate. Effective variability falls by the square root of the number of replicate analyses averaged; for example, with two replicates the variability of the reported value is decreased by the factor 1.4 (square root of 2), with three replicates by 1.7, and with four replicates to half that obtained on a single measurement.

Analytical variation or imprecision can be determined from replicate analyses of patient specimens. More often, quality control materials are used to determine whether measurement runs are acceptable and provide the basis for estimates of imprecision. In most cases, the procedures required to create a control material at reasonable cost and of acceptable stability alter the material so that it is no longer commutable, i.e. emulates an actual patient specimen. Such analyte or specimen matrix alterations are termed matrix effects and can compromise the ability of a control material to predict imprecision or bias on actual patient specimens. Imprecision usually varies with analyte concentration; measurements are often less precise at lower levels.

Discerning the actual analytical imprecision of a method from published reports or making comparisons between methods is often difficult because of differences in terminology, control materials, analyte levels, matrix effects, etc. Many reports do not specify the number of replicate analyses included in each reported value. Analyte levels may not be specified, and descriptions often do not identify whether the stated imprecision represents within-run, among-run or total analytical variation. A review of published reports describing analytical variation for various biochemical bone markers did not reveal any obvious patterns; however, the newer automated instruments are often more precise than manual measurements. Total analytical variation (CVa) in most published studies seemed to range between approximately a high of 15% to a low of 7%, with some outliers. Analytical variation should be considered together with pre-analytical components in establishing requisite performance criteria.

TOTAL VARIATION

In estimating total variation within the individual; the variances, CVs of each component squared, are additive, e.g. total intra-individual variation CVt is estimated by combining biological within-day (CVbwd), among-day (CVbad) and total analytical (CVa) by the relationship:

$$\mathrm{CVt} = (\mathrm{CVbwd}^2 + \mathrm{CVbad}^2 + \mathrm{CVa}^2)^{1/2}$$

Total intra-individual variation represents the confounding 'noise' factor in making reliable interpretations from results for biochemical markers.

LEAST SIGNIFICANT CHANGE

The least significant change (LSC), sometimes termed 'minimum significant change' or 'critical value', can be used to determine if an observed change, e.g. in monitoring therapy between pre- and post-treatment values, is meaningful based on the total intra-individual variation in each of the two measurements.[17] The LSC is calculated by the following relationship:

$$\mathrm{LSC} = 1.96 \times 2^{1/2} \times \mathrm{CVt}$$

The factor 1.96 converts the CV, derived from one standard deviation increment, to a 95%

Table 13.3 Least significant change.

Marker	LSC (%)	Percentage change (HRT 24 weeks)
Osteocalcin[a]	21	−26
PINP[a]	21	−40
Bone ALP	26	−22
f DPD/Cr[a]	26	−27
t DPD/Cr[a]	48	−48
NTX/Cr	70	−38

[a]Change is significant.
PINP, procollagen type I N propeptide; ALP, alkaline phosphatase; DPD, deoxypyridinoline; NTX, N-terminal cross-linking telopeptide of type I collagen; Cr, creatinine.

certainty interval. The factor 1.4 recognizes that variation is experienced in both pre- and post-treatment measurements. LSC was estimated for various formation and resorption markers to determine if the observed changes were meaningfully predictive of response to hormone replacement therapy (Table 13.3).[17] Other studies have reached conflicting conclusions about the predictive value of the various markers.

INDEX OF INDIVIDUALITY

An index of individuality, which compares the variation in the individual to the variation in a reference group, is useful in determining whether a reference range based on the group is appropriate for a particular analyte. When the individual variation is small compared to the group variation (intra/inter < 0.6), a reference range based on the group has little utility. On the other hand, when the ratio is >1.4, the reference range is considered useful. An evaluation of bone PYD measured in urine from men and women concluded that gender-specific reference ranges, which decrease the applicable group variation, are necessary for reliable prediction.[17]

CONSIDERATIONS IN MINIMIZING CONFOUNDING BIOLOGICAL FACTORS IN URINE

Because of the diurnal cycle, collecting serial specimens from each patient as close as possible to the same time each day should minimize the within-day component of variation. Because levels are highest in early morning, a morning (SMV or FMV in the case of urine) collection can be expected to obtain the best signal-to-noise ratio, thereby decreasing effective analytical variation as well as being more convenient than a night-time or 24-h urine collection. In routine clinical practice, some collections, for convenience, will probably be made during office visits scheduled throughout the day. In this instance, scheduling each patient requiring serial collections at the same time each day could minimize the confounding effects of diurnal variation.

Estimates of total variation under various conditions are presented in Table 13.4. Assuming a typical diurnal variation of 10%, among-day variation of 20% and analytical variation of 15%, specimen collections made at random times during the day would be expected to result in total intra-individual variation, includ-

Table 13.4 Minimizing intra-individual variation.

	CV (%)			
	Within-day (bwd)	Among-day (bad)	Analytical (a)	Total (t)[a]
Worst case[b]	10	20	15	27
Morning sample	0	20	15	25
+precise assay	0	20	7	21
+3 days averaged	0	12	7	14

[a] $CVt = (CVbwd^2 + CVbad^2 + CVa^2)^{1/2}$.
[b] Random specimen collection on one day with poor analytical precision.

ing the analytical component, of approximately 27%. Minimizing the within-day component with successive collections at the same time of day could be expected to decrease the effective variability to 25%.

Improving the total analytical imprecision from 15% to 7% could decrease the effective variation further to about 21% (Table 13.4). Increasing the number of analysis replicates or using a more precise analytical system can be expected to improve analytical imprecision. However, clearly, the among-day component is by far the largest contributor to overall variation; thus, replicate collections should have considerably more effect than replicate analyses. Physiological changes and lifestyle factors that might affect the bone markers or the normalizing creatinine in urine measurements such as calcium supplementation, diet factors and exercise level should be maintained as constant as is practical from one collection to the next, or at least considered in the interpretation. However, by far the greatest improvement could be derived from averaging multiple collections made over several days. There is precedent for such a practice in bone markers. A study in children recommended multiple collections to improve reliability of classification.[8] A study of bone markers in young adults suggested that urine collections made over 3 days is sufficient to capture the full range of among-day variability.[10]

The usual approach might be to collect and analyse individually two or three successive specimens collected over 3–5 days. However, the assay cost is an issue that might limit replicate analysis on a single specimen or analysis of each of multiple specimens. An alternative approach, possibly more cost-effective, might be to pool multiple specimen collections made over several days for analysis. For example, the patient could pool three morning void urine specimens collected over a week. Combining and averaging three successive days' collection would be expected to result in the largest decrease in total variation to 14%, half that of the simulated worst-case scenario (Table 13.4).

OTHER CONSIDERATIONS IN INTERPRETATION OF RESULTS

Reference ranges stratified appropriately by gender and age will certainly improve interpretation of bone marker values. The use of stepped cutpoints could also facilitate interpretations. There are precedents for this approach. Bone mineral density (BMD) measurements are stratified stepwise as osteopenic and osteoporotic. Similarly, interpretation of total and LDL cholesterol values are made based on stepped cut-points defining borderline and high risk with graded response.[18] Use of stepped cut-points with a graded response for bone markers could reduce the effects of unavoidable variability in making interpretations, facilitating appropriate follow-up. Considering the biochemical markers as risk factors rather than diagnostic would also seem to be more appropriate in the context of unavoidable variability. Clinicians using the biochemical markers in patient management would be more likely to interpret results appropriately, not as definitive alone, but within the context of other information, such as BMD measurements and clinical observations.

CONSIDERATIONS IN DECREASING VARIABILITY

Recent improvements in the clinical utility of the lipid/lipoprotein analytes offer a useful precedent for the biochemical markers of bone turnover.[19] An essential component is standardization of the analytes, in this context not of the methods, but of analytical results. Standardization requires two primary elements, an accepted target, usually a reference method for each analyte, and a mechanism for feedback, commonly high-quality reference materials in conjunction with proficiency testing programmes. Initially, standardization of results among research laboratories facilitates cross-comparisons and pooling of data. When the analytes move into general clinical practice, a mechanism must be established for transferring the accuracy base to the clinical laboratories.

With standardization programmes in place among the research laboratories and consequent improvements in laboratory performance, clinical studies with the various formation and resorption markers become more meaningful. Standardization facilitates comparison studies of marker clinical utility in various bone disorders. Estimates of individual and population variation are less likely to be compromised by laboratory variation. Broadly applicable reference ranges can be derived from population studies, and eventually health-based rather than population norm-based cut-points can be established for interpretation of the biochemical markers. Standardization also facilitates establishing realistic goals for analytical performance based on attainable performance and the requisite needs for clinical decision-making. Guidelines for patient preparation, specimen collection and result interpretation to minimize the confounding effects of variation can be developed. Standardization of results is thus an essential step in improving the clinical utility of the markers.

REFERENCES

1. Watts NB, Clinical utility of biochemical markers of bone remodeling. *Clin Chem* 1999; 45: 1359–68.
2. Schlemmer A, Hassager C, Alexandersen P et al, Circadian variation in bone resorption is not related to serum cortisol. *Bone* 1997; 21: 83–8.
3. Ju HJ, Leung S, Brown B et al, Comparison of analytical performance and biological variability of three bone resorption assays. *Clin Chem* 1997; 43: 1570–6.
4. Eastell R, Calvo MS, Burritt MF et al, Abnormalities in circadian patterns of bone resorption and renal calcium conservation in type I osteoporosis. *J Clin Endocrinol Metab* 1992; 74: 487–94.
5. Schlemmer A, Hassager C, Pedersen BJ et al, Posture, age, menopause, and osteopenia do not influence the circadian variation in the urinary excretion of pyridinium crosslinks. *J Bone Miner Res* 1994; 9: 1883–8.
6. Stone PJ, Beiser A, Gottlieb DJ, Circadian variation of urinary excretion of elastin and collagen crosslinks. *Proc Soc Exp Biol Med* 1998; 218: 229–33.
7. Fujimoto S, Kubo T, Tanake H et al, Urinary pyridinoline and deoxypyridinoline in healthy children and in children with growth hormone deficiency. *J Clin Endocrinol Metab* 1995; 80: 1922–8.
8. Marowska J, Kobylinska M, Lukaszkiewicz J et

al, Pyridinium crosslinks of collagen as a marker of bone resorption rates in children and adolescents. *Bone* 1996; 19: 699–77.

9. Sarno M, Powell H, Tjersland G et al, A collection method and high-sensitivity enzyme immunoassay for sweat pyridinoline and deoxypyridinoline cross-links. *Clin Chem* 1999; 45: 1501–9.
10. Ginty F, Flynn A, Cashman K, Inter and intra-individual variations in urinary excretion of pyridinium crosslinks of collagen in healthy young adults. *Eur J Clin Nutr* 1998; 52: 71–3.
11. Panteghini M, Pagani F, Biological variation in urinary excretion of pyridinium crosslinks: recommendations for the optimum specimen. *Ann Clin Biochem* 1996; 33: 36–42.
12. Leino A, Impivaara O, Kaitsaari M, Measurements of deoxypyridinoline and hydroxyproline in 24-h, first morning, and second morning urine samples. *Clin Chem* 1996; 42: 2037–9.
13. Gerrits MI, Vecht-Hart IM, Oldenhave A et al, Comparison of urinary bone resorption markers in women of 40–70 years; day-to-day and long-term variation in individual subjects. *Maturitas* 1998; 16: 247–55.
14. Colwell A, Eastell R, The renal clearance of free and conjugated pyridinium cross-links of collagen. *J Bone Miner Res* 1996; 11: 1976–80.
15. Bollen AM, Martin MD, Leroux BG et al, Circadian variation in urinary excretion of bone collagen cross-links. *J Bone Miner Res* 1995; 10: 1885–90.
16. James IT, Walne AJ, Perrett D, The measurement of pyridinium crosslinks: a methodological overview. *Ann Clin Biochem* 1996; 33: 397–420.
17. Hannon R, Blumsohn A, Naylor K et al, Response of biochemical markers of bone turnover to hormone replacement therapy: impact of biological variability. *J Bone Miner Res* 1998; 13: 1124–33.
18. The Expert Panel, Report of the National Cholesterol Education Program expert panel on detection, evaluation, and treatment of high blood cholesterol in adults. *Arch Intern Med* 1988; 148: 336–69.
19. Myers GL, Cooper GR, Henderson LO et al, Standardization of lipid and lipoprotein measurements. In: Rifai N, Warnick GR, Dominiczak MH, eds. *Handbook of Lipoprotein Testing*, lst edn. Washington, DC: AACC Press, 1997: 223–50.

Discussion paper B

Panel: *Charles H Chesnut III, Richard Eastell, William D Fraser, Yvette Henry, Sundeep Khosla, Michael Kleerekoper, Lawrence G Raisz, René Rizzoli and G Russell Warnick*

Eastell First, we'll run through a number of questions which were predefined and ask the opinion of both the panelists and the speakers on these issues. Then we will open the discussion.

The first question which we posed was about the whole issue of reference ranges. Is there really a need when a laboratory is setting up assays to establish its own reference range or can it use a manufacturer's reference range? And if it does set up its own reference range, should it just be for premenopausal women or does it really need to do reference ranges for postmenopausal women and for men? I would ask my co-panelists to address this first, and then we can throw questions over to the speakers if we need to.

Raisz Well, clearly we need reference ranges. Let's start with a broader question that we are using markers in two different ways: we are using markers for studies and we are using markers for patients. With respect to the use for patients, the reference range is clearly necessary. With respect to the studies, each study is its own control and you are comparing two different groups—and there the reference ranges are important but not critical. You heard a good deal of information on reference ranges—I think that there is a bit of good news in that the racial and ethnic difference so far don't look very large. We have had similar results looking at smaller populations in the USA. I believe that a complex series of multiple reference ranges is impractical for physicians. Therefore, if you are going to have reference ranges, you've got to pick what you think is the best. And since we believe that some of the changes in markers are going to be related to an overall change in bone metabolism related to some ideal person in their 30's or 40's, that would be the reference range that would be used. One issue that has not been discussed very much and is important to develop—particularly in northern climates, where the range of vitamin D changes with season, and in parts of the world where the intake of calcium and physical activity change with the season—is that you need to be sure that the reference ranges aren't substantially affected by vitamin D and calcium intake and physical activity. I would love to hear some comments on that from any of the panelists.

Eastell Can I take two of the points there and throw them over to Bill (Fraser) to comment on seasonal changes in a moment? But the other one I would like to put to Mike (Kleerekoper). In your case, you chose the reference range for osteoporotic women as being the appropriate reference to compare your subjects with, and so do we really need to have a reference range for osteoporotic women? Would it be sufficient to have a reference range for postmenopausal women, or could we just use the premenopausal reference range and have, say, a cut off—such as 2 SD above the mean? So what do you think is a practical approach to using a reference range in the setting of your case?

Kleerekoper Well, I'm glad you raised the question in that way. I am not sure that I fully agree that we need reference ranges. I think that the first thing you have to ask is: what is it you want from the marker? If you want the marker to make a diagnosis, then clearly you need a reference range. If you want the marker to be something with which to monitor progression or regression of disease, then I am less satisfied that you need a reference range. My understanding, and one of the messages that I tried to get across in my presentation, maybe not well, is that there is no diagnostic value in a marker. We've known this for decades. Alkaline phosphatase is high in Paget's disease, in parathyroid disease, and in metastatic disease, for example. So I am not as satisfied as you are, Larry, that we actually need a reference range.

Raisz I guess I'll rebut it, as I think that we are talking about different things. We need a reference range if we want to decide whether the value is high or not. The fact that it's high does not give us any diagnostic information, and in fact normal values for markers don't give diagnostic information either. So the issue of reference ranges is simply to decide whether the value is high or not. If it's high, we then have to interpret it according to whatever we have learned about high values for markers. What you've said is absolutely correct, that these high markers can be from many different causes. I agree with you wholeheartedly that a low, normal number can be used as the starting point for a therapeutic response. The studies with alendronate are clear in that regard—it doesn't make any difference what the starting value is for NTX; you can get a therapeutic response with alendronate. I think those two issues are clearly differentiable and we can understand them—I don't find that a big problem.

Eastell We'll leave the concept of the reference range and the high value perhaps to this afternoon, when we are going to be discussing whether high values mean an increased risk of fracture and so forth. Bill (Fraser), can I ask you to comment on the issue of season?

Fraser In terms of season, we have to look at children and adults differently and, as you know, I have done some very large studies on both aspects. In adults, what we tend to see is a change in the resorption markers in the winter, where there is a significant rise in the marker. In our own studies, we see a 25% increase in the resorption marker coming into the winter period, and then a return back down during the summer. This is seen in elderly adults. In our young adults, we didn't see the same change in the resorption marker. Interestingly, formation markers didn't change very much in either of those two studies. But in children, and this comes back to one of Larry's points, the differences are quite dramatic. In fact, what we saw in children was a marked rise in resorption and formation markers into the spring and summer. We've got a paper that's going to be published on this, where we linked the markers, just like the study did which linked markers to growth, and the growth changes that happened in late spring and summer. These growth changes correlated very strongly with the changes in the markers that we saw in children. However, there is a problem with children, in that the markers can change by, on average, 100% from day to day. I used my own daughters for some of these studies. My youngest daughter is part of the National Ballet and does a lot of exercise, and her resorption markers change dramatically after she has been at ballet classes, and they increase by 200% after she has had a period of this marked activity. The markers of my elder daughter, who is a bit of a couch potato like me, change by no more than 20–25% on day-to-day variability. So I think that these two things do have quite marked variability, and the seasonal variability is different in children than in adults.

Eastell Thank you very much. Rene (Rizzoli), you had a point you wanted to make.

Rizzoli Yes, I would like to ask the panelists at what age do you think that the steady state is reached in a growing individual—at which age the premenopausal range could be applicable, 20, 25, 18?

Eastell Yvette (Henry), can I get you to comment on that?

Henry Yes, in my opinion, markers may still be elevated into the third decade of life. Some work has suggested that some of the growth plates in the vertebra don't fuse until around age 25 years, so there will still be some growth and markers may still be elevated. As a result of that, the most ideal time to establish an appropriate reference range would be between 30 and 45 years.

Eastell Right, can we move onto the next question that we have here? It is actually related to the last question, so we will get a hard time here from Mike (Kleerekoper). There has been a proposal with bone markers that we should think of them in the same way that we think of bone density, and talk about T scores and Z scores. Thus, in the case described by Mike, should we compare a women's bone turnover to a postmenopausal or premenopausal range? I would like to hear if any of the panelists have got any further comments on T scores and Z scores. I know, Sundeep (Khosla), that you have got some comments that you would like to make about this question.

Khosla Actually, it's for practising physicians who aren't necessarily in the bone area that I think this is potentially very useful, because they are just learning how to understand bone density in terms of T and Z scores—so a sort of analogous interpretation of bone markers could be very useful. To the extent that we'll talk about later, increases over premenopausal levels might be independent predictors of fracture risk; I think this would be very useful. The other advantage would be that, because the units for different markers are all different, having some sort of consistent T and Z score scale would make it a little easier to deal with all the markers that are available. An important issue to deal with, however, is that, unlike bone density, many of the markers are not normally distributed, and is a T and Z score really valid in that setting—so that may require some mathematical transformations and so forth in order to be able to do that.

Chesnut III I think we need to be a bit pragmatic here as well. We certainly know that one of the concerns of practitioners regarding bone mineral density is confusion with T's, Z's and every other alphabet soup letter that we can think of. We are aware of their great concern regarding comparing normal reference databases across machines, across skeletal sites, etc. While we have obviously an excellent technology with bone densitometry, for practitioners it is frequently a black hole and we are not able, at this time I think, to really give them a standardized bone quantity unit across machines—although hopefully that is coming very soon—or, indeed, a standardized referent database across the various technologies for determining bone quantity. Now we have the opportunity with bone markers, still very much in their infancy, to do it right the first time around. Bone density has been available for 10–15 years, and there are still many problems of understanding for the practitioner. All I'm saying is, let's be aware of the difficulties encountered with densitometry and not go down the same route of confusion for markers for the day-to-day management of osteoporosis. Let's define with meetings of this sort exactly how we want to do this and do it right this time.

Kleerekoper I would not like to see the T and Z scores for a number of reasons, and clinicians are not terribly troubled by different reference values for different circumstances—when you order an LH or an FSH, for example, you don't get a number back, you get a number that relates to whether it is a prepubertal or postpubertal, follicular phase, luteal phase, postmenopausal, male—that's what clinicians are used to, so I see no reason to confuse them by adding Z scores and T scores. I don't think they will add anything.

Eastell I think that the argument I've heard is that in a postmenopausal woman, the advantage in comparing it to the premenopausal reference range to give a T score is the concept that T score >2 relates to increased fracture risk—that's been the argument, and that is why I think it is relevant to the discussion this afternoon that for fracture risk it seems to relate to the data that have been produced by one or two groups that a bone turnover above the reference range for premenopausal women appears to be associated with increased fracture risk.

Fraser Richard (Eastell), I'd like to come in on this. You've raised this several times and you know my feeling on it. What we need is an intelligent interpretation of the number—not displaying it as nmol, mg, T's, Z's. What the practising physician does when he gets the bone density back is to look at what the interpretation says on that result. That is what we need, and it comes back to Mike Kleerekoper's point of knowing what the numbers actually mean. I've sent in a paper with over 4500 results where we tried to do this, saying, when does a value above a certain point mean that you have a patient with cancer or thyrotoxicosis, or that you may have myeloma? I think we have got to know what the numbers mean, and that is far more important than changing how we present them. This is important for clinical chemists in the audience and those who are providing these results, making intelligent interpretation.

Warnick I have a comment that relates to both these questions. The first comment is about reference ranges; if the measurement values in a clinical or research laboratory are not accurate, then a reference range derived in another laboratory or by the manufacturer is not applicable. The second point is that if the result is not accurate, it doesn't matter whether you report it in actual units or in some statistical measure; the result is still not reliable. I think that the key issue is to make the measurements accurate.

Kleerekoper Two comments: first of all, there are two members of the editorial board of *Clinical Chemistry* in the audience, and I know that neither one of us reviewed that manuscript. But, more importantly, give us the answer—you said that you had the data. Can you give us the answer now—can you use the markers to make a decision—is this primary osteoporosis, is this secondary, what disease is it?

Fraser If I look at the HPLC values for DPD, looking at over 4500 measurements, and I see a value on HPLC >10 nmol, then I know that it is not likely to be involutional osteoporosis, and I look for a secondary cause. I can tell you where the cut-off is for myeloma. I can tell you what the cut-off is for the hypercalcaemia. You commented that you don't know why the hypercalcaemia of malignancy had such high values; well, it's a result of PTH-RP, interleukin-6 and the stimulation of osteoclast resorption. These patients get very high resorption as a result of production of these factors. So, yes, you can look at the resorption marker value and say, not diagnostically, this is X, but it is likely to be the following secondary causes depending on the level of deoxypyridinoline.

Kleerekoper There's nothing like that in the literature, is there?

Eastell It would be good to see such a paper. Can we move on to the next question, which was the recommendation about the timing of sampling? Russ, you raised the issue of morning fasting, second morning void, and 24-h urine collections, and said that it wasn't really so clear; probably, on balance, the second morning void came out as being the least variable of the three. Does anybody have any comments that they would like to add to Russ Warnick's point?

Raisz The striking finding that serum levels would be best done in the afternoon gives us a quandary, and I think we are going to come back again to the difference between research studies and clinical studies. I can conceive of a research study in which we get afternoon blood samples, although, knowing how clinical research operates, since there are other times when you want to have a fasting morning sample, a full day stay in the clinical research centre is just not going to occur for most of our studies. So I think that we are stuck with picking a time which is reasonable. This puts us in an almost impossible situation with regard to using serum CTX in morning values. I would like some further discussion of that. Should we therefore give up on serum CTX or could there be something going on with that measurement that changes more than you expect, that we could fix by changing the measurement?

Eastell Can I comment on that? I agree with Bill Fraser's point that when you look at the circadian rhythm, serum CTX is most stable in the afternoon, and that is probably true for the other resorption markers. But the critical issue is the signal-to-noise ratio, so if we want to use the biochemical markers to monitor therapy, we need to know the size of the decrement you have in the morning. Until we have that, I don't think we can make a firm recommendation that the afternoon is the best time to sample.

Raisz I agree with you. The question is, if you do a study in which the patient gets her sample (it's usually her) around 7.30 and the next sample is taken at 9.30, is that going to destroy the study with respect to the serum? I guess we have to get empirical answers to that.

Eastell We've covered most of our questions here, so I'll take a point from Mike (Kleerekoper) and we'll then move out to the floor.

Kleerekoper I have a very specific question for Russ (Warnick). Most of my patients that come with osteoporosis get a 24-h urine collection because I am interested in the calcium excretion. Am I wrong in getting the markers in that same collection? That is question number 1. Question number 2—once I have committed myself to the baseline being a 24-h collection, am I committed to all follow-up samples being 24-h samples or can I chop and change?

Warnick With regard to the first question, I think there are always trade-offs that have to be considered, and in this instance a 24-h collection might overall be more convenient and efficient. Values tend to be lower with the 24-h collection, perhaps by 20%, so you would have to be consistent in the way you sample, and if you are comparing against a reference range, then you have to consider how the reference range was obtained.

Seibel As we are up for a consensus: do we have that on how many samples we are going to take before treatment? For example, as baseline samples, is one enough?

Eastell Well, the quick answer to that is that it depends on how much you want to reduce the variability. I thought that Russ Warnick's example was a really nice illustration of the magnitude of the effect of multiple sampling in contrast to improving the assay performance and in contrast to taking the sample at a fixed time of day. I actually find that really useful to take two or more samples but that is just my opinion. Can I ask the panel members first, and then the speakers, what they think about one versus two versus three measurements, say at baseline at the initial evaluation of a patient with osteoporosis?

Rizzoli Ideally it would be perfect, but with feasibility it is obviously not possible, but everything depends on the variability of the measurement. Everything that has been presented today is from clinical research groups with very good technical aspects, with determination run in batches and so on. It's markedly different from daily practice, and for this reason—one point that has not been touched on so far—is the quality control adequate in private laboratories and in the daily practice? Because for the practitioner this is a major point—how much can he be confident in the data received?

Fraser I am fascinated by Larry's comment that you cannot get a woman to give a sample from 2 p.m. to 5 p.m., but we are going to try to get patients to come back and give two, three or four urine samples. It must be reflecting a difference between American and European culture, because we have phlebotomists who work all day, and we have clinics in the mornings and afternoons. Most of our general practitioners, our primary care physicians, have practice nurses who sample from 11.00 to 16.00, and the bulk of our samples that come in for analysis come in the afternoon. So, in fact, it's like changing your practice about cholesterol—nobody would think about doing a non-fasting cholesterol/triglyceride sample these days to look for hyperlipidaemia. Now you know that the signal-to-noise ratio is much better, because you saw the graph. I don't have to calculate the figures between morning and afternoon sampling to show you that the signal-to-noise ratio is better—it can be seen to be better.

Eastell But you didn't show us the signal, you only showed us the noise.

Fraser Well, you know that the signal-to-noise ratio is going to be better. Clearly, the discrimination is better. We need more data, I fully agree with that, but I cannot believe that we cannot say to patients, 'Come and have your blood sample at 14.00.' The alternative would be late morning or fasting.

Epstein I've got a question which pertains to Dr Henry's presentation about race. As you know, we had done this about 10 years ago, and showed that the African-Americans had lower osteocalcins, this was confirmed by Luckey, Cosman etc., and the only problem is that you cannot take this in isolation, as if you are going to use African-Americans in trials, their levels of 2,5-OH-D are lower and their PTHs are higher, and you know that in many osteoporosis trials these patients may get calcium and vitamin D. Norman Bell has shown that this can influence the levels in African-Americans, and I just wanted to know if your African-Caribbeans are subject to the same perturbations.

Henry In the study that I did looking at race and gender differences, we also measured 25-hydroxyvitamin D and it was lower in that particular population. So yes, in the UK the trends are the same. In terms of the differences in markers in terms of race, there are obviously no set rules; that was just a guide to say that those were the trends that have been illustrated so far in the markers in terms of race.

Raisz A very quick question; the issue of myeloma being diagnosed by ratio of resorption to formation hasn't been brought up. Would you like to comment on it, Bill (Fraser) or anyone?

Kleerekoper I alluded to it by pointing out that this woman had very high resorption markers but a normal ALP.

Raisz Did you do that analysis?

Fraser We presented some preliminary data in a paper about 2 years ago, where we looked at bone ALP against the resorption marker. Of course, what happens in myeloma is that the bone-specific alkaline phosphatase is suppressed more than you would have expected for the level of resorption. We have also looked at a very much bigger group of patients with myeloma and followed them up after transplant, looking at formation and resorption markers. It is a good way of looking at myeloma—it isn't a good way of diagnosing the disease, but a good way of indicating that you've got a myeloma.

Leary At the risk of being redundant, I wish to

re-emphasize the importance of having precise assays. Without these, it is not possible to study reference ranges, biological variability, or signal-to-noise ratio, so the first thing we need to have is really better procedures, and then we can answer all these questions we are addressing here.

Eastell Thank you.

Bauer Two quick questions. One, there were some data floating around at the last ASBMR about the attenuation of the circadian rhythm of serum CTX when fasting, and I wonder if anyone has any follow-up on that data. The second thing is, I wonder if Russ (Warnick) will comment on the total variability of the markers that we've talked about, both between persons and within person, compared to cholesterol, and I wonder if you would also like to comment about the NCEP recommendation that we not make treatment recommendations on less than two and preferably three cholesterol measurements.

Eastell Yvette is going to show some data we have on the effect of fasting. While she is getting them ready, Bill (Fraser) you were going to make some comments on the Christiansen study.

Fraser We've looked at the question of fasting and non-fasting, because it was suggested to me that you could get round it this way. We've matched up pairs of individuals and have followed individuals through prolonged fasts for 72 h, and what is quite clear is that if you are going to sample at 9, 10 and 11 o'clock in the morning, the difference between 9 o'clock to 10 o'clock to 11 o'clock is far greater than and more significant than the difference you get between fasting and non-fasting in those individuals. If you fast, you can increase the CTX value, and I think that is the slide that you are going to see here—you can actually increase the value that you get for the serum CTX but the spread stays exactly the same. So I don't think you improve discrimination by sampling non-fasting in the morning compared to taking the sample at 14.00 to 17.00.

Eastell The published data that you were referring to were from the Christiansen study, which shows an attenuation of the circadian rhythm with fasting. Here are the data on fasting. We had observed, like Bill (Fraser) and other people, that there is a reduction of about 50% before and after a meal, and we wondered how much of this was a meal effect, and how much of it was a time of day effect. On the slide, the NF refers to non-fasting and the F refers to fasting in the same 20 women, with everyone sampled at 9 o'clock in the morning. The only difference is therefore the fasting and non-fasting state. You can see that the individuals who are fasting are about 20% higher than non-fasting individuals for serum CTX. So about 20% of the effect is diet and the remaining 30% must be time of day and the endogenous rhythm. I don't think we disagree, Bill, I think we are just saying that there is a contribution from both. [The reader is referred to the opening paragraph of Discussion paper C. Data from Christgau et al examines the issue of fasting and its influence on diurnal variation of CTX.]

Any comment on the second question, the one about cholesterol? Should we follow similar guidelines and have two or three measurements of bone turnover to characterize someone's status? The answer, Doug, from me is, of course, yes, but is anyone going to disagree and say that we should have just one marker?

Raisz Were you talking about one marker repeated three times or different markers? Those are two different questions.

Eastell One marker three times.

Raisz In the case of cholesterol, the test is not only done several times, but there are three different things you're measuring. We are nowhere near that in the bone marker field.

Bauer Correct; I'll just pick on Mike Kleerekoper's example, that he was going to make diagnostic decisions or perhaps a therapeutic efficacy decision on whether vitamin D and calcium had been useful on the basis of a single follow-up measurement, whereas I think the current understanding or the current recommendations for cholesterol are that you actually not make treatment decisions based on a single measurement.

14

The use of biochemical markers of bone turnover for monitoring treatment of osteoporosis

Pierre D Delmas

Summary • Introduction • Effects of treatments of osteoporosis on bone markers • The association between changes in bone markers and changes in BMD under treatment • The association between changes in bone markers and fracture risk under treatment • References

SUMMARY

Most antiresorptive therapies for osteoporosis produce a significant increase in bone mineral density (BMD). The relatively low signal-to-noise ratio of this technique does not allow us to detect rapidly (within months) responders and non-responders to therapy. Conversely, sensitive markers of bone resorption and formation rapidly decrease under hormone replacement and bisphosphonate therapies, with a more favourable signal-to-noise ratio. Several studies have shown a significant inverse correlation between the short-term decrease in bone turnover markers and the 2-year increase in BMD at various skeletal sites. Cut-offs for these decreases can be defined for a given marker under treatment that will identify responders and non-responders (based on the BMD change) with adequate sensitivity and specificity. Because BMD change may not be an ideal surrogate marker of treatment efficacy, further studies should analyse the potential relationship between changes in bone turnover markers and the probability of fractures under treatment.

INTRODUCTION

The goal of treatment of osteoporosis is to reduce the occurrence of fragility fractures, but their incidence is low and the absence of events during the first year(s) of therapy does not necessarily imply that treatment is effective. Pain is clearly not an adequate index of treatment efficacy. Thus, it is of critical importance for the physician to have a biological tool to monitor the effects of the treatment of such a chronic disease. Measurement of bone mineral density (BMD) by dual energy X-ray absorptiometry (DXA) is a surrogate marker of treatment efficacy that has been widely used in clinical trials. Antiresorptive drugs used in postmenopausal osteoporosis induce a significant increase in BMD from baseline, the magnitude of which varies substantially according to the agent and to the skeletal site, contrasting with a small decrease in untreated women that is variable according to age and to the skeletal site. The use of DXA in the monitoring of treatment efficacy in the individual patient, however, has not been prospectively validated and raises methodological issues. Such monitoring requires the definition of the least significant change of BMD, i.e. the smallest detectable difference that is not related to organic changes of BMD (such as treatment effect) but to the precision error of the machine, depending on device errors, technician variability, differences in patient positioning, and other technical artefacts. Although there are few reliable data on the long-term precision error of DXA, several studies have determined the short-term precision error expressed as the coefficient of variation (CV). The CV ranges from 0.9% at the lumbar spine and total area of the hip in early postmenopausal women to 1.9% and 2.5% at the spine and total hip, respectively, in elderly women. Various cut-offs for individual significant changes in BMD have been suggested, based on different statistical models. At the lumbar spine, the skeletal site that shows the largest increase in BMD under antiresorptive therapy, these cut-offs vary from 2% to 5% in women with postmenopausal osteoporosis.[1,2] How do these cut-offs apply to the monitoring of treatment efficacy? A potent bisphosphonate such as alendronate induces a large increase in BMD at the spine, reaching 6–7% at 2 years and 8–9% at 3 years, contrasting with a spontaneous small decrease or small increase in early and elderly postmenopausal women respectively.[3,4] Thus, repeating signal BMD after 2 years is likely to identify responders to alendronate therapy. In contrast, raloxifene induces a small 2–3% increase in BMD at the spine and hip,[5] and nasal calcitonin an even smaller increase that is significant at the spine but not at other skeletal sites.[6] Clearly, DXA is not appropriate for the individual management of patients treated with these drugs. In addition, DXA may not be appropriate to detect responders to therapy after 1 year, as the change in BMD is likely to be within the variability of the technique and may be influenced, in cases of large changes, by a statistical artefact called the regression to the mean.

Table 14.1 Considerations on the use of bone turnover markers to monitor treatment of osteoporosis.

Number and schedule of measurements
Which marker(s) should be used?
Which criteria should be used to assess the bone marker response to therapy?
Least significant change of a marker (based on variability)
Prediction of BMD change under therapy
Prediction of probability of fracture under therapy
Factors that will influence the cut-off value of a marker in detecting responders/non-responders:
Variability of the marker
Type of treatment (hormone replacement therapy, SERM, bisphosphonate)
Use of percentage change from baseline or of absolute value of a bone marker under treatment
Sensitivity versus specificity

Failure to respond to treatment may be due to non-compliance (probably the most important single factor), to poor intestinal absorption (i.e. bisphosphonates), to other factors contributing to bone loss, or to other unidentified factors. Monitoring may improve compliance, although this needs to be proven for osteoporosis treatment. As discussed below, several studies suggest that markers of bone turnover may be used for monitoring treatment of osteoporosis. We will discuss their clinical utility in the management of the individual patient. Some of the critical issues are listed in Table 14.1.

EFFECTS OF TREATMENTS OF OSTEOPOROSIS ON BONE MARKERS

Effects of hormone replacement therapy on bone markers

Oestrogen deficiency induces a rapid and sustained increase in skeletal remodelling, which is reflected by a 50–100% mean increase in formation/resorption markers that is sustained throughout life.[7] This increase is more pronounced for more specific and therefore sensitive markers of bone turnover than for conventional ones. Hormone replacement therapy (HRT) induces a decrease in bone markers that is dose-dependent and that reaches a plateau maintained for the duration of treatment. In most studies using adequate doses of oestrogens, the mean level of bone markers at plateau is within 1 SD of the mean of healthy premenopausal women that represents the normal range of bone turnover. Cessation of HRT induces a rapid increase in resorption markers towards baseline values, followed by an increase in formation markers. The pattern of these changes according to the route of administration and dose of HRT and according to the marker used has been widely documented in many studies.[8–21] Under adequate doses of either oral, percutaneous, transdermal or nasal oestrogens, the mean decrease is 40–60% for urinary C-terminal cross-linking telopeptide of type I (CTX) and N-terminal cross-linking telopeptide of type I (NTX), slightly less for serum CTX, and 20–30% for urinary free deoxypyridinoline (DPD).

The decrease in bone formation markers is delayed, probably reflecting the coupling of bone formation to pre-existing bone resorption, and the plateau is only reached after 6–12 months of HRT. In comparative studies, the decrease occurs earlier with serum osteocalcin and procollagen type I N propeptide (PINP) than with serum bone alkaline phosphatase (ALP). The pattern of decrease during the first 3 months may differ according to the route of administration, and appears to be delayed following parenteral HRT as compared to oral HRT. These differences between routes of administration in their effects on the kinetics of decrease of bone formation markers could be related to the first-pass effect of oral oestrogen on liver metabolism, especially on the production of IGF-1.

Once the plateau is reached, the mean decrease in serum osteocalcin ranges from 10% with 25 µg to 45% with 100 µg for transdermal 17β-oestradiol, and from 25% to 50% with oral oestrogens. The decrease is of similar magnitude for serum PINP and bone ALP. The decrease in formation markers is less pronounced when oestrogens are combined with norethisterone acetate (NETA). The cyclical use of NETA results in a transient increase in bone formation markers, probably reflecting the unique stimulating activity of this progestin on osteoblastic activity.[22]

Effects of bisphosphonates on bone markers

Oral daily treatment of postmenopausal women with or without osteoporosis with various bisphosphonates, including alendronate and ibandronate, induces a dose-dependent decrease of bone turnover that follows a pattern comparable to that with HRT.[23–30] These changes have been extensively studied under alendronate treatment. A significant decrease in resorption markers is already seen after 1 month of treatment and reaches a plateau from 3 months onwards. The magnitude of the decrease is the largest (60–70%) for the cross-link-related peptides (i.e. urinary CTX and NTX and serum CTX), substantial for the total excretion of DPD (50%) and marginal (or even non-significant) for the urinary free cross-links and for serum ICTP. The decrease in bone formation markers is delayed, reaching a plateau after 6–12 months of treatment, depending on the marker and on the dose. With 10 mg of alendronate daily, the maximum decrease is about 50% for osteocalcin, bone ALP and PINP, and 30–40% for procollagen type I C propeptide (PICP). Once treatment is stopped, bone turnover increases rapidly, with a significant increase in resorption markers

3 months after withdrawal, followed by an increase in formation markers.[30,31] Whether bone turnover returns to the level of untreated patients or to a lower level is not yet clear. Intermittent bisphosphonates such as cyclical oral etidronate and risedronate, and intravenous pamidronate, produce a different pattern of bone marker changes. There is a rapid decrease in resorption markers, followed within a few weeks by a slow increase that may or may not reach baseline value at the time of the second course of bisphosphonate, depending on the potency and dose of the bisphosphonate and on the interval between the two courses. The decrease in formation markers is slower and less pronounced and shows little fluctuation after a few courses of bisphosphonates.

Effects of other antiresorptive agents on bone markers

Oral raloxifene, 60 mg/day, produces a sustained decrease in bone turnover of smaller magnitude than that with most HRT regimens, with a 30–40% reduction in urinary CTX and a 20–30% reduction in bone formation markers.[5] The reduction in bone turnover observed with daily nasal calcitonin is of small magnitude and appears to be consistently significant only for doses of 200 IU or higher. In summary, most effective antiresorptive treatments, where given continuously, induce a decrease in bone turnover that reaches a plateau within a few weeks or months, earlier for resorption than for formation markers, at a level that depends on the potency of the drug and on the marker used.

THE ASSOCIATION BETWEEN CHANGES IN BONE MARKERS AND CHANGES IN BMD UNDER TREATMENT

The role of baseline bone turnover in response to treatment

Patients with high-turnover osteoporosis have a greater remodelling space than patients with low turnover. Inhibiting bone resorption will result in a larger transient increase in bone mass than in low-turnover patients, and therefore high-turnover patients are likely to respond better to antiresorptive therapy than those with low turnover. When bone turnover was assessed by the 24-h whole-body retention of a technetium-99m-labelled bisphosphonate, injectable calcitonin and transdermal HRT induced a greater increase in spinal BMD in osteoporotic patients with high turnover as compared to those with low turnover.[32,33] When bone turnover was assessed by biochemical markers, a similar finding was reported in patients treated with oral or transdermal oestrogens. A significant relationship between baseline bone turnover—assessed by the 24-h whole-body retention of labelled bisphosphonate or by bone markers—and the BMD response to alendronate has also been reported.[24–34] In fact, in most of these studies, the overlap between the BMD responses of both groups is important, and most patients with low bone turnover treated with alendronate or with high doses of HRT will show an increase in BMD at the spine. It is therefore not clear what practical implications the assessment of baseline bone turnover could have for treatment decisions. In addition, there is no prospective study indicating that low-turnover patients should be treated with higher doses of antiresorptive drugs than low-turnover ones. The potential role of bone turnover in the subsequent response to treatment of osteoporosis—with antiresorption and bone-forming agents—remains an important question for future research. An association between baseline bone marker levels and fracture probability should be looked for, after adjustment for confounders, in large controlled trials.

The association between bone turnover changes and BMD changes (Table 14.1)

The decrease in bone markers under antiresorptive therapy, usually expressed as a percentage of the initial values, is strongly related to the increase in BMD, especially at the lumbar spine, which is the most responsive skeletal site to treatment. All studies of HRT except one have shown in the past 10 years that the short-term (3–6 months) decrease in bone turnover markers is significantly correlated with the long-term increase in BMD. Most studies were performed in early postmenopausal women, with BMD measurement at the spine or radius. In a large double-blind placebo-controlled study of different oestrogen and oestrogen/progestin regimens, including the oral and transdermal routes of administration, Johansen et al[8] found a significant negative correlation between the changes in osteocalcin and changes in BMD at the forearm and whole body during a 24-month period ($R = -0.40$ and -0.39 respectively, $P < 0.01$), i.e. the larger the decrease in osteocalcin, the larger the increase in bone mass. This association has been confirmed in several other studies using other markers, such as urinary free DPD, NTX and CTX, for bone resorption. The association between bone turnover and BMD changes is usually stronger with the most precise BMD measurements (DXA versus DPA, spine versus hip) and stronger with the most sensitive and specific bone markers. Correlations are stronger with increasing interval between the two measurements of bone turnover, as shown in the study of Riis et al.[11] The use of a combination of several markers slightly increases the prediction of BMD changes.

In postmenopausal women with or without osteoporosis treated with alendronate, a similar association has been reported between the short-term changes in bone turnover markers and the 2-year increase in BMD.[23–25] In 75 women with osteoporosis treated with alendronate 10 mg/day or 5 mg/day or with placebo, the 2-year increase in spine BMD was correlated with the 3-month decrease in NTX and total urinary DPD, with R values of 0.53 and 0.48 respectively ($P < 0.0001$ for both). The correlation was weaker and not significant for free pyridinoline (PYD) and serum ICTP. Although formation markers had not yet reached their lowest value at 3 months, their decrease was highly correlated with the BMD increase, with R values of 0.67 for bone ALP and PICP, and of 0.63 for osteocalcin.[23] Finally, significant correlations, with R values ranging from -0.29 to -0.47, were found between the change in either total or bone ALP, osteocalcin or urinary CTX and the change in BMD measured at the spine, hip and total body in 141 women treated for 1 year with various doses of oral ibandronate daily.[29] When two or more markers were combined, the R values increased significantly, up to -0.55.

Thus, all these studies indicate that a marked decrease in markers is associated with a positive subsequent BMD response to either HRT or bisphosphonates, while non-responders show little or no change in bone markers. These findings suggest that bone markers, especially sensitive and precise ones, can be used to monitor treatment efficacy, especially within the first 6 months of initiating therapy, at a time when BMD changes are still too small to be used clinically. In most studies, however, the coefficients of correlation are relatively low, between 0.3 and 0.7, indicating that early changes in bone markers can explain only 15–30% of the variance of BMD changes induced by treatment. Correlation coefficients increase markedly if the precision error of assessing the rate of bone loss by DXA is taken into account. If bone markers are to be used clinically, adequate cut-offs need to be defined and validated. A few studies have addressed this issue, and these are reviewed below.

Prediction of BMD changes with bone markers in the individual patient

For the clinician, the goal of treatment monitoring is to identify early responders and non-responders. If BMD changes measured by DXA are considered to constitute the 'gold standard' of response to treatment, it is critical to have an accurate definition of a positive/negative BMD response. Some studies have considered positive and negative BMD changes over 2 years as the criteria for defining responders and non-responders respectively. These criteria may lead to a substantial number of patients being misclassified, as any change within the analytical coefficient of variation of the technique could correspond to a true responder or non-responder. An alternative approach is to exclude from the analysis patients with an ambiguous BMD response, i.e. within the CV of the DXA technique. Different statistical models have been used to identify responders and non-responders according to the bone marker response to therapy. One approach is to consider the least significant change of a bone marker, based on the short-term or long-term within-subject variability, regardless of the BMD response.[15] Most studies have attempted to define the minimum marker change associated with a positive BMD response as previously defined. The optimal threshold of bone marker changes can be defined using ROC curves, or by using logistic regression models. The percentage change from baseline and/or the absolute value of the marker under treatment can be used, and cut-off values can be obtained with a prespecified sensitivity or specificity or by searching for the best trade-off between sensitivity and specificity. The sensitivity is the percentage of BMD responders that will be identified as responders according to the bone marker response, and the specificity is the percentage of BMD non-responders correctly identified by bone markers. The positive predictive value is the percentage of patients identified as responders by the test (bone marker change) who are responders according to the reference criteria (BMD change in this case). It is inversely related to the proportion of false positives. The negative predictive value is the percentage of patients identified as non-responders by the test who are truly non-responders according to BMD changes. It is inversely related to the proportion of false negatives. In 387 women treated with alendronate 5 mg, a cut-off of −40% for the percentage change of NTX from baseline to 6 months was found to provide the best compromise between sensitivity (86%) and specificity (48%) to predict the 2-year BMD change at the spine.[26] Similarly, a cut-off of −20% for serum osteocalcin at the same time point resulted in a sensitivity of 79% and a specificity of 53%. The corresponding positive predictive values for NTX and osteocalcin were 92% for both, at the expense of low negative predictive values (33% and 37% respectively). NTX and osteocalcin predicted the BMD response at the hip, forearm and whole body with comparable characteristics.[27] Using a similar approach in 152 women in whom the BMD responses of the spine, hip and forearm were assessed after 3 years of oral HRT at various doses, the same group of investigators proposed two different cut-offs for serum and urinary CTX, serum osteocalcin and bone ALP, favouring either sensitivity or specificity. We have performed a similar analysis in 569 women treated for 2 years with various doses of transdermal oestradiol, as part of two placebo-controlled trials.[20] We determined the sensitivity, probability of positive BMD response and corresponding cut-off values (percentage changes from baseline) of bone markers at 3 and 6 months of treatment with a specificity set at a level of 0.9. This criterion implies that less than 10% of women classified with markers as responders, i.e. as having a subsequent increase in BMD at 2 years ≥2.26% (twice the short-term intra-individual CV), would be false positive. All markers provided a high probability of positive BMD response ranging from 0.82 to 0.91 and a sensitivity higher for resorption markers—serum and urinary CTX (60–68%)—than for formation markers—serum osteocalcin and bone ALP (51–51% at 6 months). With this approach, the cut-offs were −60% and −68% for serum and urinary CTX respectively, and −51% and −49% for osteocalcin and bone ALP respectively. In a group of 307 osteoporotic women treated with alendronate, Garnero et al[26] found, by using a logistic regression analysis, that bone ALP levels and percentage bone ALP change at 6 months were independent predictors of the long-term positive BMD response and that a model combining the percentage decrease and the absolute value of bone ALP provided significantly greater AUC at various times of treatment (6, 12 and 24 months) than either one alone.

In summary, the retrospective analysis of placebo-controlled trials of HRT and alendronate suggest that bone markers can be used for predicting the BMD response. These data have been obtained with measurements on a single sample before treatment and a single sample under treatment. The variability of bone markers can be reduced by averaging values of two samples at baseline, but the trade-off is inconvenience and increased cost. Responders and non-responders should be defined by spinal BMD changes above and below the minimum significant change, in order to avoid misclassification of patients due to the BMD precision error. Because compliance of patients in controlled trials is higher than in clinical practice, placebo-treated patients should be included in such analyses, as they will represent, in clinical practice, patients who are non-compliant. Patients treated with suboptimal doses can also be included, reflecting partially compliant patients. A less than 40% decrease in serum CTX, urinary CTX or urinary NTX is likely to detect most non-responders, while a greater than 60–65% decrease in these markers at either 3 to 6 months of treatment will detect most responders. The corresponding cut-offs for bone ALP and osteocalcin at 6 months are approximately 20% and 40–50% respectively. When the marker change is equivocal, a third measurement a few months later is likely to decrease the number of false-positive and false-negative cases. In order to secure these thresholds, it would be desirable to validate them in other trials.

THE ASSOCIATION BETWEEN CHANGES IN BONE MARKERS AND FRACTURE RISK UNDER TREATMENT

There are several lines of evidence suggesting that the magnitude of the BMD increase under antiresorptive therapy, especially with raloxifene, is only weakly correlated with the occurrence of vertebral fractures, and that changes in other determinants of skeletal strength may play an important role in the mechanism of action of treatment (reviewed in Delmas[35]). Thus, BMD changes may not be an adequate surrogate endpoint to analyse the ability of bone markers to predict fracture risk. There have been few attempts to correlate bone marker changes with fracture risk under antiresorptive therapy. In a retrospective analysis of a small placebo-controlled trial of HRT, Riggs et al[36] suggested that changes in bone turnover—assessed by bone histomorphometry—predict changes in vertebral fracture risk as well as changes in BMD. The short-term changes in serum osteocalcin under raloxifene therapy were found to be associated with the subsequent risk of vertebral fractures in a large subgroup of osteoporotic women enrolled in the MORE study, while changes in BMD were not predictive.[37] Such analyses should be performed in ongoing and recently completed large clinical trials performed in osteoporotic women treated with bisphosphonates, HRT and SERM, with fractures as a primary endpoint. Such studies should define the bone marker response (percentage change and/or absolute values) associated with a low probability of fracture, and should assess whether a subnormal bone turnover level is associated with a high rate of fracture under treatment.

In conclusion, changes in new markers of formation/resorption under antiresorptive therapy for osteoporosis have been adequately documented in many clinical trials. The fact that they decrease rapidly and reach a drug- and dose-dependent plateau within a few months suggests that they could be used to predict the longer-term response to therapy. Statistical models have been recently developed, indicating that the percentage decrease in some bone markers after 3–6 months of HRT or alendronate can be used to predict the 2-year response in BMD with adequate sensitivity and specificity. Validating these models in other cohorts of patients would strengthen the value of these cut-offs in clinical practice. The same approach should be applied to correlate short-term changes in bone markers with the probability of fracture under treatment, as it is the most important endpoint of the treatment of osteoporosis.

REFERENCES

1. Genant HK, Black JE, Steiger P et al, Appropriate use of bone densitometry. *Radiology* 1989; 170: 817–22.
2. Glüer CC, Black G, Lu Y et al, Accurate assessment of precision errors: how to measure the reproducibility of bone densitometry techniques. *Osteoporos Int* 1995; 5: 262–70.
3. Hosking D, Cilvers CED, Christiansen C et al, Prevention of bone loss with alendronate in postmenopausal women under 60 years of age. *N Engl J Med* 1998; 338: 485–92.
4. Liberman YA, Weiss SR, Broll J et al, Effects of oral alendronate on bone mineral density and the incidence of fractures in postmenopausal osteoporosis. *N Engl J Med* 1995; 333: 1437–43.
5. Delmas PD, Bjarnason NH, Mitlak BH et al, Effects of raloxifene on bone mineral density, serum cholesterol concentrations, and uterine endometrium in postmenopausl women. *N Engl J Med* 1997; 337: 1641–7.
6. Overgaard K, Riis BJ, Christiansen C et al, Effect of salcatonin given intranasally on early postmenopausal bone loss. *Br J Med* 1989; 299: 477–9.
7. Delmas PD, Stenner D, Wahner HW et al, Serum bone gla-protein increases with aging in normal women: implications for the mechanism of age-related bone loss. *J Clin Invest* 1983; 71: 1316–21.
8. Johansen JS, Riis BJ, Delmas PD et al, Plasma BGP: an indicator of spontaneous bone loss and effect of estrogen treatment in postmenopausal women. *Eur J Clin Invest* 1988; 18: 191–5.
9. Uebelhart D, Schlemmer A, Johansen J et al, Effect of menopause and hormone replacement therapy on the urinary excretion of pyridinium crosslinks. *J Clin Endocrinol Metab* 1991; 72: 367–73.
10. Prestwood F, Bilbeam CC, Burleson JA et al, The short term effects of conjugated estrogen on bone turnover in older women. *J Clin Endocrinol Metab* 1994; 79: 366–71.
11. Riis BJ, Overgaard K, Christiansen C, Biochemical markers of bone turnover to monitor the bone mass response to postmenopausal hormone replacement therapy. *Osteoporos Int* 1995; 5: 276–80.
12. Delmas PD, Garnero P, Utility of biochemical markers of bone turnover in osteoporosis. In: Marcus R, Feldman D, Kelsey J, eds. *Osteoporosis*. New York: Academic Press, 1996: 1075–85.
13. Chesnut CH, Bell NH, Clark GS et al, Hormone replacement therapy in postmenopausal women: urinary N-telopeptide of type I collagen monitors therapeutic effect and predicts response of bone mineral density. *Am J Med* 1997; 102: 29–37.
14. Rosen CJ, Chesnut CH, Mallinak NJS, The predictive value of biochemical markers of bone turnover for bone mineral density in early postmenopausal women treated with hormone replacement therapy or calcium supplementation. *J Clin Endocrinol Metab* 1997; 82: 1904–10.
15. Hannon R, Blumsohn A, Naylor K et al, Response of biochemical markers of bone turnover to hormone replacement therapy: impact of biological variability. *J Bone Miner Res* 1998; 13: 1124–33.
16. Delmas PD, Pornel B, Felsenberg D et al, A dose ranging trial of a matrix transdermal 170-estradiol for the prevention of bone loss in early postmenopausal women. *Bone* 1999; 24: 517–23.
17. Cooper C, Stakkestad JA, Radowicki S et al, Matrix delivery transdermal 17B-estradiol for the prevention of bone loss in postmenopausal women. *Osteoporos Int* 1999; 9: 358–66.
18. Marcus R, Holloway L, Wells B et al, The relationship of biochemical markers of bone turnover to bone density changes in postmenopausal women: results from the postmenopausal estrogen/progestin interventions (PEPI) trial. *J Bone Miner Res* 1999; 14: 1583–95.
19. Garnero P, Tsouderos Y, Marton I et al, Effects of intranasal 17B-estradiol on bone turnover and serum insulin-like growth factor I in postmenopausal women. *J Clin Endocrinol Metab* 1999; 84: 2390–7.
20. Delmas PD, Hardy P, Garnero P et al, Monitoring individual response to hormone replacement therapy with bone markers. *Bone* 2000; 26: 553–60.
21. Bjarnason NH, Christiansen C, An early response in biochemical markers predicts long-term response in bone mass during HRT in early postmenopausal women. *Bone* (in press).
22. Stepan J, Pospichal J, Schreiber V et al, The application of plasma tartrate-resistant acid phosphatase to assess changes in bone resorption in response to artificial menopause and its treatment with estrogen or norethisterone. *Calcif Tissue Int* 1989; 45: 273–80.
23. Garnero P, Shih WJ, Gineyts E et al, Comparison of new biochemical markers of bone turnover in late postmenopausal osteoporotic women in response to alendronate treatment. *J Clin*

Endocrinol Metab 1994; 79: 1693–700.
24. Greenspan SL, Parker RA, Ferguson L et al, Early changes in biochemical markers of bone turnover predict the long-term response to alendronate therapy in representative elderly women: a randomised clinical trial. *J Bone Miner Res* 1998; 13: 1431–8.
25. Ravn P, Bidstrup M, Wasnisch RD et al, Alendronate and estrogen–progestin in the long-term prevention of bone loss: four-year results from the early postmenopausal intervention cohort study. *Ann Intern Med* 1999; 131: 935–42.
26. Garnero P, Darte C, Delmas PD, A model to monitor the efficacy of alendronate treatment in women with osteoporosis using a biochemical marker of bone turnover. *Bone* 1999; 24: 603–9.
27. Ravn P, Hosking D, Thompson D et al, Monitoring of alendronate treatment and prediction of effect on bone mass by biochemical markers in the early postmenopausal intervention cohort study. *J Clin Endocrinol Metab* 1999; 84: 2363–8.
28. Machado BCA, Hannon R, Eastell R, Monitoring alendronate therapy for osteoporosis. *J Bone Miner Res* 1999; 14: 602–8.
29. Ravn P, Clemmesen B, Riis BJ et al, The effect on bone mass and bone markers of different doses of ibandronate: a new bisphosphonate for prevention and treatment of postmenopausal osteoporosis: a 1-year randomised, double-blind, placebo-controlled dose-finding study. *Bone* 1996; 19: 527–33.
30. Ravn P, Christensen JO, Baumann M et al, Changes in biochemical markers and bone mass after withdrawal of ibandronate treatment: prediction of bone mass changes during treatment. *Bone* 1998; 22: 559–64.
31. Rossini M, Gatti D, Zamberlan N et al, Long-term effects of a treatment course with oral alendronate of postmenopausal osteoporosis. *J Bone Miner Res* 1994; 9: 1833–7.
32. Civitelli R, Gonelli S, Sacchei F et al, Bone turnover in postmenopausal osteoporosis. Effect of calcitonin treatment. *J Clin Invest* 1999; 82: 1268–74.
33. Gonnelli S, Cepollaro C, Pondrelli C et al, The usefulness of bone turnover in predicting the response to transdermal estrogen therapy in postmenopausal osteoporosis. *J Bone Miner Res* 1997; 12: 624–31.
34. Gonnelli S, Cepollaro C, Pondrelli C et al, Bone turnover and the response to alendronate treatment in postmenopausal osteoporosis. *Calcif Tissue Int* 1999; 65: 359–64.
35. Delmas PD, How does antiresorptive therapy decrease the risk of fracture in women with osteoporosis? *Bone* 2000; 27: 1–3.
36. Riggs BL, Melton LJ III, O'Fallon WM, Drug therapy for vertebral fractures in osteoporosis: evidence that decreases in bone turnover and increases in bone mass both determine antifracture efficacy. *Bone* 1996; 18: 197S–201S.
37. Bjarnasson NH, Christiansen C, Sarkar S et al, Six months change in biochemical markers predict 3-year response in vertebral fracture rate in postmenopausal, osteoporotic women: results from the More Study. *J Bone Miner Res* 1999; 14(S1): S157.

15

The use of biochemical markers of bone turnover to monitor response to therapy: statistics and logic

Aubrey Blumsohn

Summary • Introduction • What is the aim of monitoring? • Generation of components of variation • Reproducibility of markers of bone turnover • Comparison between biological variability and response to therapy • Three types of monitoring • Computer simulations • Conclusions and recommendations • References

SUMMARY

The response of markers of bone turnover to antiresorptive therapy has been described in numerous publications. The aim of this chapter is to introduce some issues relating to the statistics and logic of monitoring using these analytes. There is disagreement about the effect of biological variability on the clinical utility of these tests in individual patients. Part of this disagreement is as a result of incorrect statistical analysis, application of inappropriate statistical models, and failure to conceptualize the underlying clinical problem. When monitoring antiresorptive therapy, it is the size of change that is informative, rather than detection of any significant change. We want to know whether the response to therapy is following expected norms. Such norms are not the response of a placebo group. The true heterogeneity of response to antiresorptive therapy is likely to be extremely small. On statistical grounds alone, it would also seem unlikely that change in bone turnover (based on two measurements) would be capable of predicting bone gain in response to therapy in compliant patients. However, markers of bone turnover are likely to be useful to answer other questions, and to detect patients who are non-compliant. Serum-based markers may be most appropriate in this context.

INTRODUCTION

Markers of bone turnover have an established role in therapeutic studies involving skeletally active drugs. The ability of these measurements to provide useful information about individual patients receiving therapy is less certain (*useful: 'able to be used advantageously'*[1]). Recently developed methods for the measurement of telopeptide fragments of type 1 collagen in serum[2] have greatly enhanced the potential for clinical monitoring. However, given the large number of women taking bisphosphonates or oestrogen replacement, the effectiveness of such monitoring clearly needs to be established. The response of these analytes to therapy has been described in numerous publications. What has received much less critical attention is the conceptualization of the underlying problem.

Concerns have been expressed about the effect of measurement variability on the potential utility of these analytes. In contrast, a number of authors have suggested that, since the 'least significant change' between two measurements is greater than the actual change observed with therapy (for some analytes and some therapies), methodological performance is somehow satisfactory. The aim of this chapter is to introduce some issues relating to the statistics of monitoring with these analytes.

WHAT IS THE AIM OF MONITORING?

Perhaps the most basic question related to monitoring of the response to any intervention is simply 'what are we trying to monitor?' This question introduces several difficulties, both practical and logical. To assess the clinical validity of monitoring with these analytes, we need a clear understanding of exactly what information the measurements are intended to provide. This understanding is an essential prerequisite to statistical evaluation. Whatever the specific goal, the main objective has to be to provide the clinician with information and associated rules of interpretation to allow clinically advantageous decisions to be made.

Monitoring is a package deal. This package consists of a particular test, a particular clinical scenario, a particular schedule of monitoring, and the exact clinical question. The package also includes an algorithm, in other words explicit or general rules governing interpretation of the results, and possible clinical actions. There are also costs and risks which have to be offset against any possible benefit. Clearly, a highly useful test may be of little value if the algorithm is inappropriate. If we are interested in reproducibility, what is important is not the reproducibility of the test itself, but the reproducible operation of this algorithm.

A possible example of a simple rule might be 'if a particular marker X decreases by less than 30% at time point *Y* in a patient taking a particular drug, consider non-compliance with therapy'. Possible associated actions might be 'confront the patient regarding possible non-compliance' or 'increase the dose to *Z*'. It is difficult to evaluate, or indeed use, any test if no attempt is made to specify the goal, or to specify exactly how results should be used.

GENERATION OF COMPONENTS OF VARIATION

The total variance of the measured concentration of any analyte in blood or urine can be attributed to four broad components: (1) within-person biological variability (CV_i); (2) pre-analytical variability; (3) methodological variability (CV_a); and (4) between-person variability (CV_g). Variance components should be derived from properly designed, blinded, replicate studies subjected to appropriate statistical analysis.

Correct and incorrect statistical analysis

Comparison of reported coefficients of variation (CV) for markers of bone turnover is complicated by the fact that different (and frequently incorrect) methods for calculating variance components are used. Most reports fail to state how calculations were performed. Variance components are best estimated by nested ANOVA.[3] One error in many publications is to calculate the mean (or median) intra-individual CV based on separate calculations of CV for each subject,[4] i.e.

$$\frac{\sum_{k=1}^{n} CV_k}{n}$$

where CV_k is the CV for an individual subject and n the number of subjects studied. The CV calculated by this method is subject to substantial negative bias. Similarly, calculation of the standard error of the CVs from each subject is not a basis for calculating the confidence interval of the intra-individual CV. Logarithmic transformation before ANOVA is usually required to ensure: (1) that within-subject variability has a normal distribution; (2) to stabilize variances; and (3) to avoid inappropriate exclusion of outliers.[5]

Derived indices

The identification of significant serial changes is complicated by within-subject and analytical variability. The term 'least significant change' (LSC, or critical difference) denotes the minimum difference between two successive results in an individual that can be considered to reflect a real change. More reproducible measurements are generally better for monitoring, but the LSC cannot be interpreted without also considering the magnitude of change expected for a particular analyte. LSC (expressed as percentage of the mean value of the two measurements) can be calculated.[6]

At the 95% confidence level Z is 1.96 for a two-sided difference. Where data are logarithmically transformed, the LSC can be antilogged to give the critical limits on a normal scale. Of more relevance is the 95% confidence interval for the true change, given an observed change:

$$\mathrm{CI} = \%\Delta \pm 2.77\sqrt{\mathrm{CV}_i^2 + \mathrm{CV}_a^2}$$

REPRODUCIBILITY OF MARKERS OF BONE TURNOVER

A large number of studies have shown that markers of bone turnover are highly variable within 1 day[7] and between different days.[8–14] A few studies have shown much lower estimates of variability than most other studies. These differences are difficult to evaluate because: (1) many studies fail to state how variance components were calculated, and where stated the calculations are incorrect in many cases; and (2) the confidence intervals of variance components may be large, and these are usually not determined or are calculated incorrectly. I do not propose to review these data here. However, combined within-subject and analytical variability is probably of the order of 10% (CV) for serum markers of formation or resorption. Most studies have found that reproducibility is substantially higher for serum markers than for urinary markers. This probably relates to the inherent difficulties associated with measuring analytes in urine. CV_i is around 25% for total urinary deoxypyridinoline (DPD). CV_i appears to be greater for telopeptide fragments of type I collagen in urine (CV 25–40%) and less for free pyridinolines (CV about 20%).

These are minimal estimates of variability when samples are collected at the same time of day under clinical research conditions. The use of untimed 'spot' urine collections would greatly increase CV_i.[7] CV_i is likely to be substantially greater in routine clinical practice.[14]

COMPARISON BETWEEN BIOLOGICAL VARIABILITY AND RESPONSE TO THERAPY

In general, markers which show the greatest response to therapy tend to show the most random biological variability as well.[12] The LSC (at $P = 0.05$) based on biological variability tends to be of similar magnitude to the actual response to therapy, at least for urinary markers of bone resorption and serum markers of bone formation. This means that, for most markers, a statistically significant therapeutic change will be detected for some compliant subjects, but not for others. It might be possible to calculate a 'signal-to-noise ratio' for each marker, taking the signal as the mean response to therapy, and the noise as some measure of variability. Although this may be of some interest, the appropriate signal is the true heterogeneity of response to therapy, and not the mean magnitude of response (see below).

THREE TYPES OF MONITORING

Measurement variability is only of relevance when interpreted in the context of the way in which measurements are likely to be used. It is important to distinguish between at least three fundamentally different types of monitoring: (1) where change is not expected; (2) where change is expected; and (3) where change does not matter. The statistical approach to each type differs, as does the performance requirement of the associated method.

Unexpected change

The first type of monitoring is aimed at detection of a perturbation from steady state. In this situation, a patient is assumed to be in steady state with no more than random variation around some set point. Change is not expected, but the clinician would like to know if any change does occur. Change may represent an acute shift to a different steady state, or a developing trend. A typical example is monitoring aimed at detecting the development of metastatic disease in a patient with a malignancy. In statistical terms, we are concerned with the ability of measurement to detect change. This depends on measurement variability and the size of change we would like to detect. The concept of least significant change is appropriate here when we are dealing with paired measurements. Given a typical CV_i of 25% for total DPD creatinine, a change of at least 70% would have to occur in an individual in order for us to conclude that any real change had occurred (at $P = 0.05$). One-sided confidence intervals may be appropriate. When more than two measurements are made, some form of time series analysis may be appropriate.[15] This type of monitoring is well understood, and markers of bone turnover have an important potential role in this context. Serum-based measurements are likely to prove more useful.

Expected change

A second, and completely different, kind of monitoring relates to the situation where change is expected. The best and most studied example is monitoring a paediatric growth.[16] When we monitor growth, we are not asking whether a child is growing. Indeed, given a sufficiently precise measuring instrument, statistically significant growth will always be detected over a reasonable timescale. Any observed change represents a mixture of true change and measurement noise. What the clinician wishes to know is whether the change is following expected norms. In other words, we want to be able to distinguish between individuals in terms of their change. The ability to do this depends on measurement uncertainty as well as the true heterogeneity of change in individuals of that age.

Similarly, when monitoring antiresorptive therapy, it is the size of change that is informative, rather than simple information about whether any significant change has occurred. We want to know whether the response to therapy is following expected norms. Such norms are not the response of a placebo group unless the sole aim of monitoring is simply to detect ingestion of any amount of a pharmacologically active compound. The appropriate norm is the true distribution of response to therapy at any given time point, after taking account of measurement noise. Change could be expressed as absolute change, percentage change or, indeed, change on any other scale. The concept of least significant change is of minimal relevance in this context.

Suppose the observed decrease in total DPD excretion in an individual patient was 30% following oestrogen therapy. Given an optimistic estimate of $CV_i = 20\%$ and $CV_a = 5\%$, the 95% CI of the observed change in that individual would be from +27.1% (increase) to −87.1% (decrease). Clearly, the prospects for distinguishing a subject with a response of 20% from a subject with a response of 40% are not good.

It might be possible to distinguish between the responses of subjects using measurements with lower CV_i. However, critical to the prospects of being able to achieve this is the magnitude of the true heterogeneity of response. Given the observed heterogeneity of response to antiresorptive therapy, and the known measurement variability, it is possible to calculate the distribution of the true (unobservable) responses to therapy. It is possible to use data from the literature to calculate this. Based on analysis of most studies showing response to therapy, the heterogeneity of the true underlying response to therapy in compliant subjects in clinical trials is likely to be extremely small, and certainly much smaller than measurement noise. Indeed, in most studies, the observed range of responses is similar to, or even smaller than, that predicted from measurement noise alone (see data from Gertz et al[10] as an example).

The timescale of clinical studies is also important. The initial increase in bone mass is likely to reflect the remodelling transient,[17] and may not reflect the long-term effect of therapy. The magnitude of the bone density change associated with this remodelling transient is likely to be larger when the change in bone turnover is large.

Change not important

The third kind of monitoring is where the aim is to direct a particular measurement towards a target. In this scenario, change as such is unimportant. What is important here is our ability to know that we have hit the target. This will depend on the size of the target, and the uncertainty associated with our navigation instruments. The target is often a range of arbitrarily chosen thresholds, which may represent particular percentiles of the population distribution.

The targeting of specific '*t*-scores' as a goal in patients receiving antiresorptive agents is an example of this type of monitoring.[18] This scenario is not fundamentally different from the use of markers of bone turnover to distinguish between patients within the reference range for purposes of predicting bone loss or fractures. It is well recognized that when single urinary measurements are used to classify patients with respect to their turnover status, a considerable number of patients will be classified differently on different occasions, and single measurements cannot be used as the basis for making clinical decisions. Indeed, if the aim is to classify patients into tertiles of bone turnover, less than 60% are classified into the same tertile on two occasions using urinary markers.[8] The problem with any classification is that it is greatly influenced by the way in which we draw our cut-off limits. We have to compare the number correctly classified with the number who would be so classified by chance. In the case of tertile classification, the expected number of correctly reclassified subjects on a random basis is expected to be 33%. Serum markers are likely to be more appropriate in this context.

COMPUTER SIMULATIONS

The overall utility of monitoring would best be assessed in the context of long-term clinical trials, in which patients are randomized to receive or not to receive monitoring. Such trials will almost certainly never be conducted. An alternative is to use computer simulations to explore the possible outcome of such a trial were it to be conducted. Computer modelling is also useful to explore the possible outcomes of clinical research studies.

The simple simulation demonstrated here is a model of a randomized trial. The aim of the trial is to assess the likely ability of a change in a particular marker of bone turnover (X) to predict the initial short-term (2 years) bone gain following onset of therapy. Parameter values were chosen so as to err on the side of maximizing the potential utility of the marker. It was assumed that the mean response of the particular marker to therapy was 50%. The true underlying heterogeneity of the response of X was chosen to be substantially larger than predicted from the literature. The true underlying hetero-

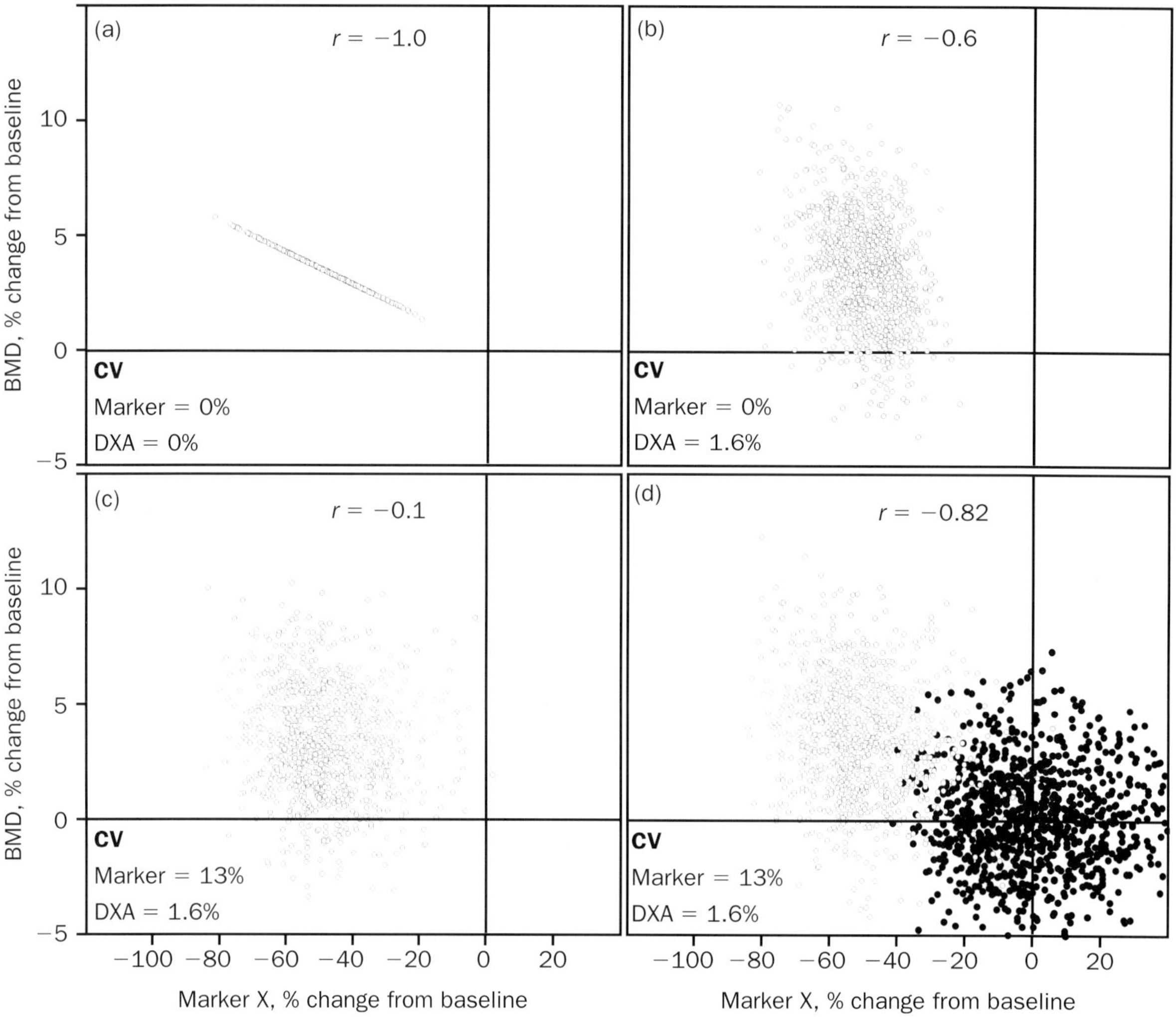

Fig. 15.1a–d Simulation results for the model discussed in the text. Figures a–c represent the response of the treated group (n = 1000). Figure d is the response for placebo and treated groups combined.

geneity of the bone mineral density (BMD) response and mean magnitude of response were derived from the literature. Finally, the true underlying ability of the marker to predict bone gain was assumed to be perfect ($R = 1.0$). Fig. 15.1a shows the relationship between change in BMD and the change in the marker in the treated group when CV_{DXA} and CV_X are set at zero. This line represents the underlying model. Fig. 15.1b shows the relationship between change in BMD for the treated group and the change in the marker when CV_{DXA} is set at 1.6% and CV_X is set at zero. The scatter of observed results ($R = 0.6$) represents the maximum correlation that would be expected given expected DXA imprecision and an infinitely precise marker (for an optimistic model). Fig. 15.1c shows the relationship between change in BMD and the change in the marker when CV_{DXA} is set at 1.6% and CV_X is set at 13%.

The scatter of observed results ($R = 0.1$) represents the maximum correlation that would be expected, given expected DXA imprecision and a marker with the low biological variability expected when using a serum marker of bone resorption. Using a marker with a $CV_X = 25\%$, the correlation falls to zero. It is certainly possible to obtain a much better correlation by including the placebo group in the analysis[18] (Fig. 15.1d), by including subjects taking different drugs or different doses,[19] or by including subjects known to be non-compliant on the basis of simple questioning.

CONCLUSIONS AND RECOMMENDATIONS

There is little disagreement regarding the greater bone specificity of several new biochemical markers of bone turnover. In particular, new serum markers of bone resorption may have much to offer in a monitoring context. The design of most studies is, however, focused on a population-averaged, rather than a patient-specific, view.

When monitoring antiresorptive therapy, it is the size of change that is informative, rather than detection of any significant change. The true heterogeneity of response to therapy in compliant patients is likely to be extremely small. On statistical grounds alone, it would also seem unlikely that change in bone turnover would be capable of predicting bone gain in response to therapy in compliant patients. Markers of bone turnover are likely to be useful to answer other clinical questions, as well as to detect non-compliance. New serum-based markers are likely to be most appropriate in this context.

REFERENCES

1. *Collins English Dictionary*. London: Collins, 1986.
2. Rosenquist C, Fledelius C, Christgau S et al, Serum CrossLaps One Step ELISA. First application of monoclonal antibodies for measurement in serum of bone-related degradation products from C-terminal telopeptides of type I collagen. *Clin Chem* 1998; 44: 2281–9.
3. Searle SR, Variance components in the unbalanced 2-way nested classification. *Ann Math Stat* 1961; 32: 1161–6.
4. Kress BC, Mizrahi IA, Armour KW et al, Use of bone alkaline phosphatase to monitor alendronate therapy in individual postmenopausal osteoporotic women. *Clin Chem* 1999; 45: 1009–17.
5. Keene ON, Log transformation is special. *Stat Med* 1995; 14: 811–19.
6. Fraser CG, Harris EK, Generation and application of data on biological variation in clinical chemistry. *Crit Rev Clin Lab Sci* 1989; 27: 409–37.
7. Blumsohn A, Herrington K, Hannon RA et al, The effect of calcium supplementation on the circadian rhythm of bone resorption. *J Clin Endocrinol Metab* 1994; 79: 730–5.
8. Blumsohn A, Eastell R, The performance and utility of biochemical markers of bone turnover: do we know enough to use them in clinical practice? *Ann Clin Biochem* 1997; 34: 449–59.
9. Blumsohn A, Hannon RA, Eastell R, Long-term retest-reliability of biochemical markers of bone turnover in healthy postmenopausal women. *J Bone Miner Res* 1995; 10: S182.
10. Gertz BJ, Shao P, Hanson DA et al, Monitoring bone resorption in early postmenopausal women by an immunoassay for cross-linked collagen peptides in urine. *J Bone Miner Res* 1994; 9: 135–42.
11. Panteghini M, Pagani F, Biological variation in urinary excretion of pyridinium crosslinks: recommendations for the optimum specimen. *Ann Clin Biochem* 1996; 33: 36–42.
12. Hannon R, Blumsohn A, Naylor K, Eastell R, Response of biochemical markers of bone turnover to hormone replacement therapy: impact of biological variability. *J Bone Miner Res* 1998; 13: 1124–33.
13. Beck Jensen JE, Kollerup G, Sorensen HA et al, Intraindividual variability in bone markers in the urine. *Scand J Clin Lab Invest* 1997; 57(suppl 227): 29–34.
14. Hart S, Bainbridge P, Hannon R et al, Short term variability of biochemical markers of bone turnover in clinical practice. *J Bone Miner Res* 1999; 14: 1042.
15. Gordon K, Smith AFM, Modeling and monitoring biomedical time series. *J Am Stat Assoc* 1990; 85: 328–37.
16. Himes JH, Minimum time intervals for serial

measurements of growth in recumbent length or stature of individual children. *Acta Paediatr* 1999; 88: 120–5.
17. Heaney RP, The bone-remodeling transient: implications for the interpretation of clinical studies of bone mass change. *J Bone Miner Res* 1994; 1515–19.
18. Garnero P, Shih WJ, Gineyts E et al, Comparison of new biochemical markers of bone turnover in late postmenopausal osteoporotic women in response to alendronate treatment. *J Clin Endocrinol Metab* 1994; 79: 1693–700.
19. Ravn P, Christensen JO, Baumann M et al, Changes in biochemical markers and bone mass after withdrawal of ibandronate treatment: prediction of bone mass changes during treatment. *Bone* 1998; 22: 559–64.

16

Prediction of response in bone mass by biochemical markers of bone turnover during antiresorptive therapy for prevention of osteoporosis

Pernille Ravn

Summary • Introduction • Monitoring of treatment response in clinical trials • Prediction of treatment response in individual patients • References

SUMMARY

Biochemical markers can be used for valid prediction of long-term response in bone mass during antiresorptive therapy. The advantage of the biochemical markers is a pronounced response immediately after start of therapy, which allows for an early estimation of treatment effect on bone mass. In contrast, about 1 year of treatment is necessary for a significant detection of effect by bone densitometry, because the average yearly response in bone mass is within the precision error of the equipment. The association between change in biochemical markers and change in bone mass is fully expressed when the biochemical marker is suppressed to a stable level during treatment. As steady state is obtained within 3 months after start of therapy in the bone resorption markers, these markers are superior in this context. Serum and urinary C- and N-terminal telopeptides of type I collagen (CTX, NTX) are among the most sensitive biochemical markers presently available. Serum CTX combines a pronounced response to antiresorptive therapy and a low long-term spontaneous variability. Both are important characteristics for optimal accuracy of prediction of long-term bone mass changes. In clinical trials, the group-wise magnitude of suppression in the biochemical marker is strongly related to the group-wise size of the long-term response in bone mass, which allows for early, numerical anticipation of long-term treatment effect on bone mass. In individual patients, the biochemical marker should be used as a threshold analysis, where a decrease below a cut-off value indicates response to treatment, and lack of suppression indicates non-response to treatment. Sensitive, recently developed biochemical markers like serum CTX have a sensitivity and specificity of up to 90–95%.

INTRODUCTION

Biochemical markers can be used for group-wise and individual prediction of bone mass response during antiresorptive therapy.[1–13] In clinical trials, group-wise prediction is useful because the exact size of the bone mass response according to the dose of antiresorptive agent can be estimated based on degree of suppression in the biochemical markers within the respective treatment groups.[3,4,8,9,11,13] For individual prediction, the dispersion of the association between change in biochemical markers and change in bone mass becomes important. To address the clinical question about response to treatment in individual patients, biochemical markers must thus be used with cut-off values to indicate response or non-response to treatment, whereas the exact size of bone mass response cannot be accurately estimated in this context.[1,4,6,10–14] An indication of response (stable or increasing bone mass during treatment) or non-response (significant bone loss during treatment) is, however, mostly an adequate estimate in terms of prevention of postmenopausal osteoporosis, because the therapeutic aim is simply to stabilize bone mass.[1,4,6,10,11]

MONITORING OF TREATMENT RESPONSE IN CLINICAL TRIALS

The biochemical markers of bone turnover reveal a dose-related decrease during antiresorptive therapy.[3,4,10,11,15–19] In clinical trials, the biochemical markers allow for detection of significant differences between treatment groups by 3 months after start of therapy.[5,15,17] The size of the short-term decrease in the biochemical markers is associated with the long-term response in bone mass.[1–13] For example, alendronate 10 mg/day caused a decrease in serum and urinary C-telopeptides of type I collagen (sCTX and uCTX) of 70–80% in the dose-finding study of alendronate treatment for osteoporosis prevention (Fig. 16.1).[10] Similar decreases in these markers were found during treatment with ibandronate 2.5–5 mg/day (Fig. 16.2).[4,18] These two treatment regimens were also found to cause comparable increases in bone mineral density (BMD) indicating an equivalent effect on bone.[16,18]

The correlation between change in the biochemical markers and change in BMD during antiresorptive therapy is usually reported to be in the range $R = -0.4$ to $R = -0.8$,[1–13] but has been found to be lower in some studies.[14] By grouping the study participants into tertiles of change in the biochemical markers, it is clearly shown that the more the suppression of the biochemical markers, the higher the response in BMD (Fig. 16.3).[11] Fig. 16.3 shows that a decrease of 50–70% in uCTX or urinary N-telopeptide cross-links of type I collagen (NTX) at month 6 corresponded to a 2-year increase in spine and hip BMD of about 2–3% during treatment with alendronate 5 mg/day (middle tertiles). These findings corresponded to the average longitudinal changes in uCTX, NTX and BMD reported in this group.[19,20] The group-wise size of suppression in the biochemical markers during antiresorptive therapy thus reflects the size of the long-term response in BMD. This relationship is, furthermore, remarkably similar during treatment with different antiresorptive agents (Fig. 16.4). The association is of particular use in clinical trials, where the information can be used to give an early indication of anticipated long-term effect on bone mass, and especially to indicate whether applied doses are insufficient in terms of treatment effect. In phase I trials, the biochemical markers can thus be used for early screening of treatment effect on bone. Biochemical markers should also be included in phase II trials for interim analyses, with the purpose of dose selection for phase III trials. Such a strategy is likely to restrict human and technical resources and time used for product development. Both markers of bone formation and markers of bone resorption can be used. Optimal prediction is, however, first achieved when the biochemical marker is suppressed to a new steady state, i.e. after 3–6 months in the case of bone resorption markers and after 6–12 months in the case of bone formation markers (Fig. 16.5; Table 16.1).[1,10]

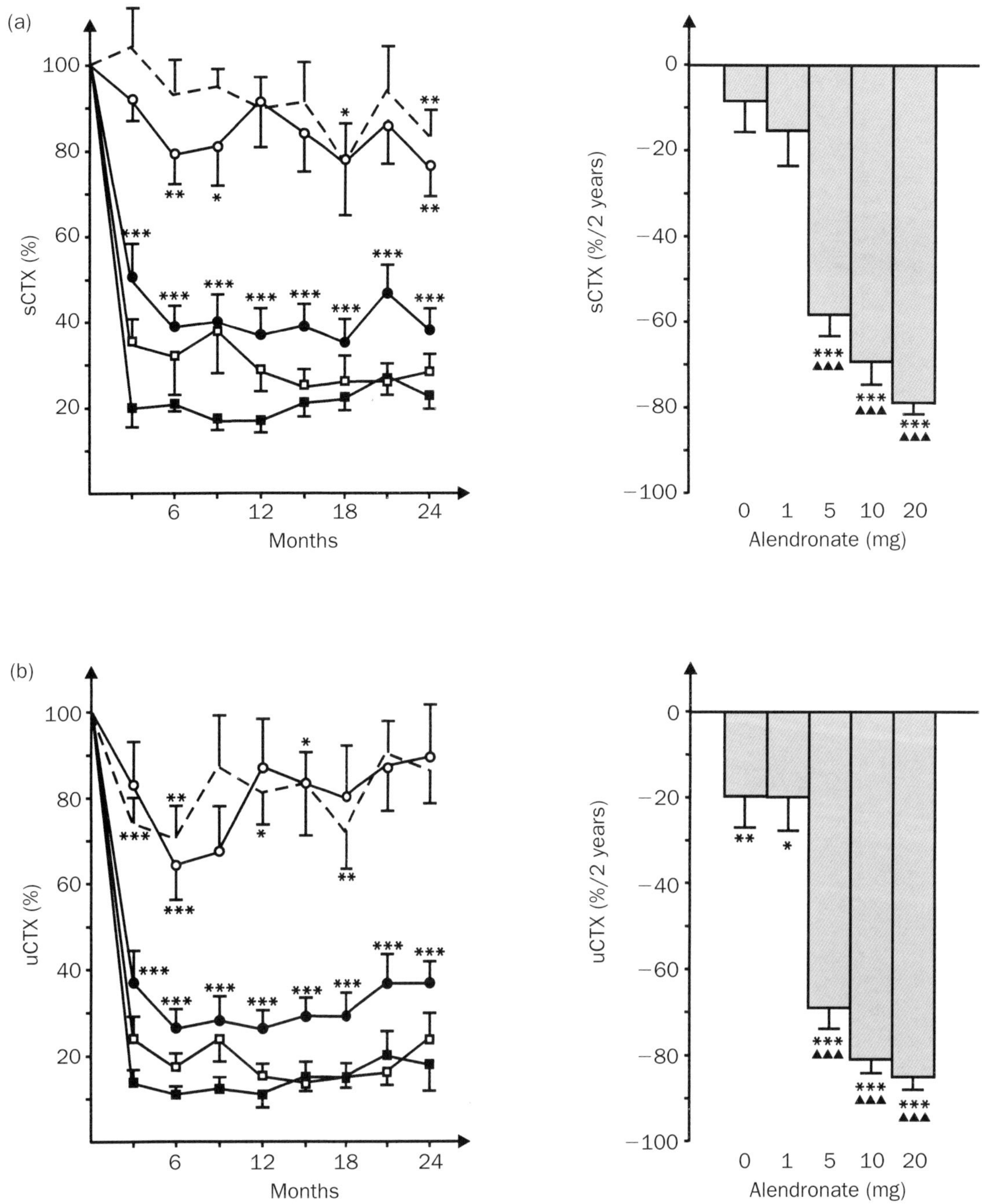

Fig. 16.1a, b (a) Serial measurements of serum and urinary C-telopeptides of type I collagen (sCTX and uCTX) expressed as percentage of baseline values and given as the mean percentage value (SEM) in groups of women treated with alendronate (1–20 mg/day) or placebo. (b) Net response in sCTX and uCTX (month 3–24).[10]
* $p < 0.05$ (compared with baseline); ** $p < 0.01$ (compared with baseline); *** $p < 0.001$ (compared with baseline); ▲▲▲ $p < 0.001$ (compared with placebo).

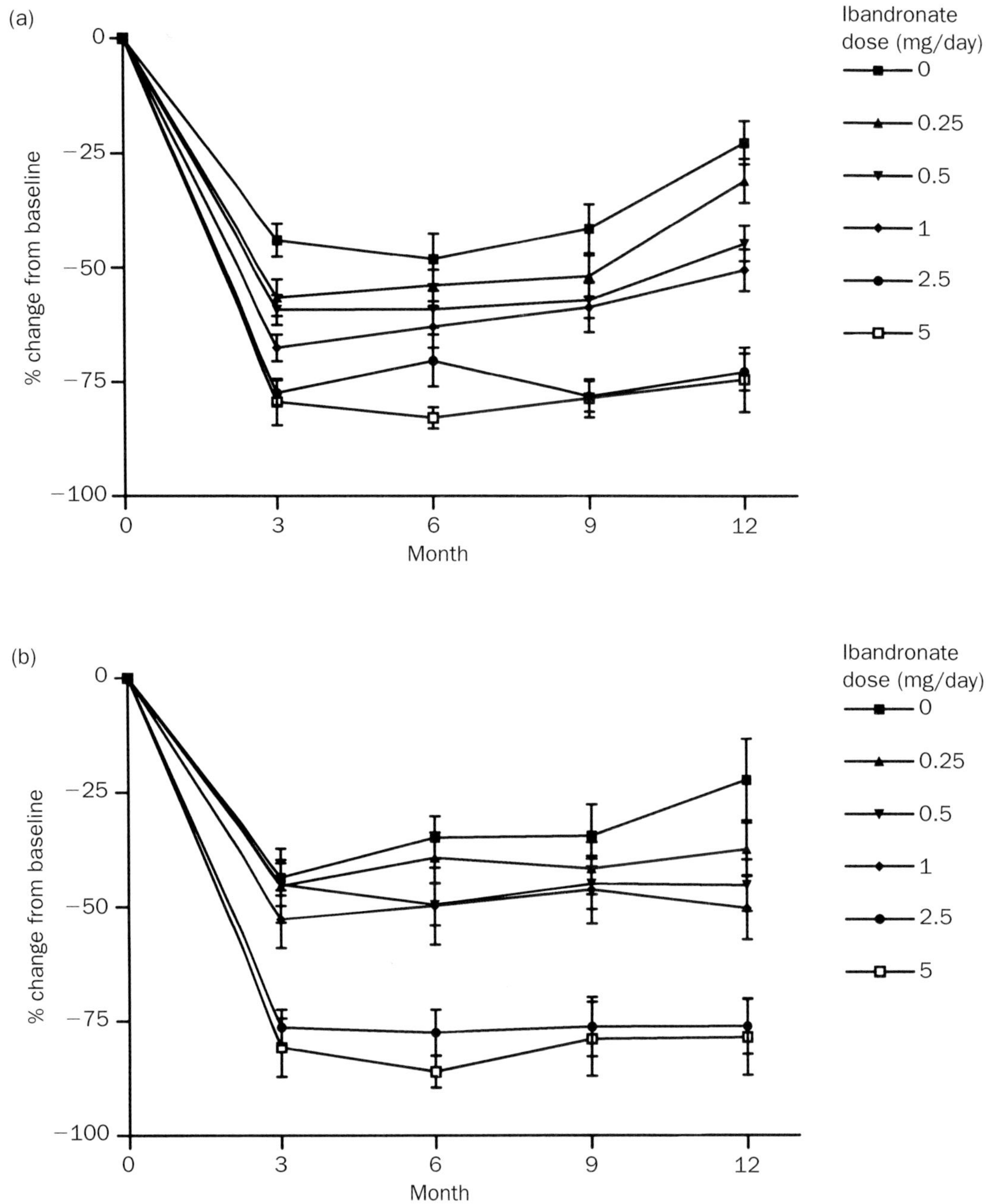

Fig. 16.2a, b Serial measurements of sCTX (a) and uCTX (b), expressed as percentage of baseline values and given as the mean percentage value (SEM) in groups of women treated with ibandronate (0.25–5 mg/day) or placebo.[4]

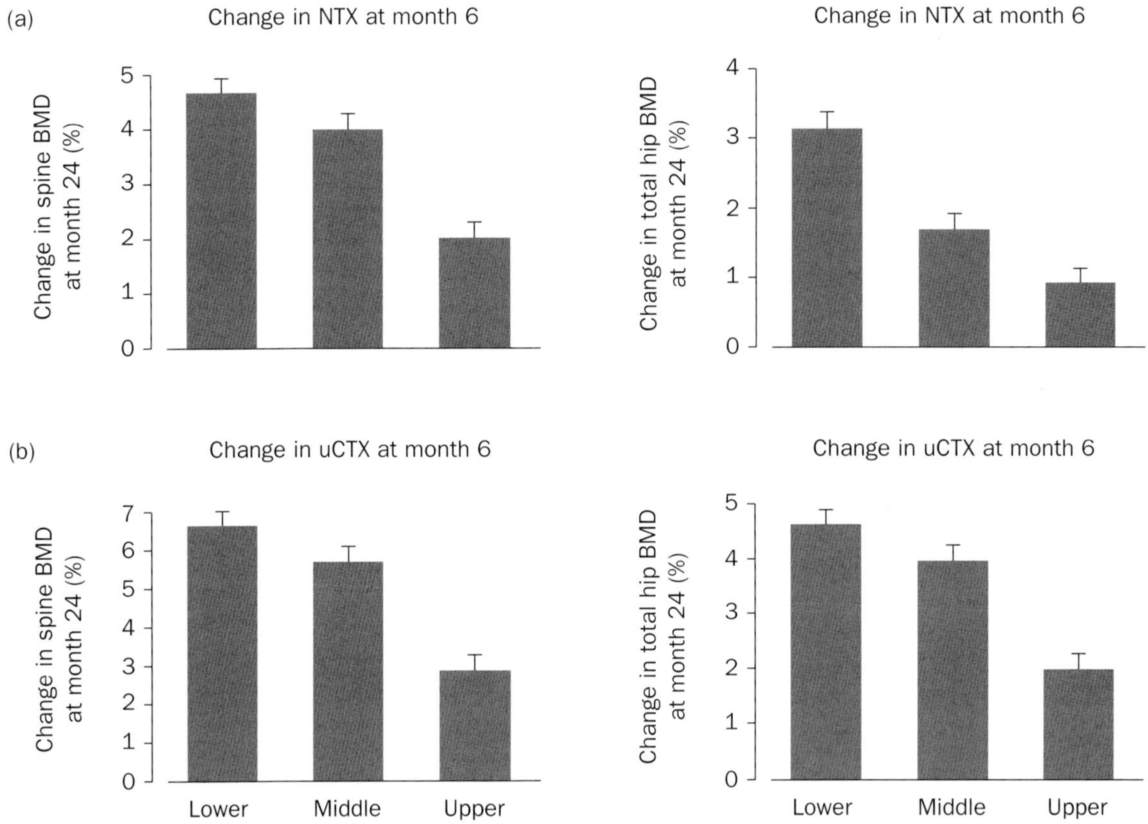

Fig. 16.3a, b Percentage change from baseline at month 24 in spine and hip BMD by tertiles of percentage change from baseline at month 6 in (a) uCTX and (b) urinary N-telopeptide crosslinks of type I collagen (NTX) in women treated with alendronate 5 mg/day. Ranges of decrease in the tertiles were: lower (below 70%), middle (70–50%), and upper (above 50%).[11]

PREDICTION OF TREATMENT RESPONSE IN INDIVIDUAL PATIENTS

Measurement of BMD is a widely used method to assess outcome of osteoporosis intervention.[21] The response to treatment seen in BMD over 1 year is, however, in the order of the precision error of the BMD equipment.[21] Accordingly, more than 1 year has to pass before a BMD measurement reliably detects a response to treatment. This is a serious problem for individual monitoring of treatment response, for several reasons. Many women on hormone replacement therapy (HRT) choose to discontinue treatment much earlier because of dislike of side-effects, fear of cancer, or general concerns with regard to long-term medical intervention. Bisphosphonates are poorly absorbed from the gut. Oral therapy thus demands strict and rather complex administration routines, which challenge long-term adherence to therapy. Patients might also be unaware of inappropriate tablet intake and being non-compliant as a result. In all cases, non-compliant patients should be detected as early as possible. Early detection of a potential positive effect of the

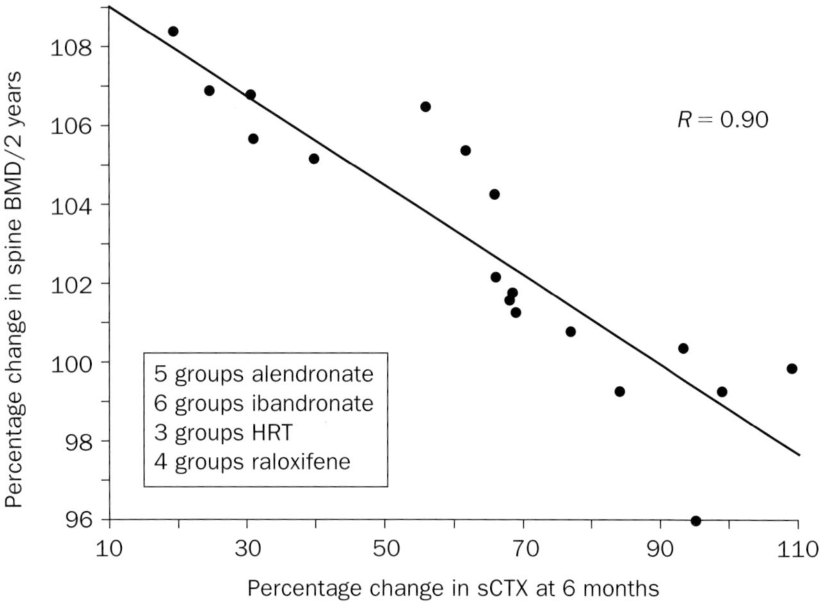

Fig. 16.4 Correlation between mean change in sCTX and mean change in BMD grouped by dose of different antiresorptive agents (unpublished data).

treatment is, moreover, likely to increase adherence to therapy over the longer term, probably even more so in cases of expensive treatments, complex administration schedules, and treatments likely to induce side-effects.

As indicated above, antiresorptive therapy induces a decrease in the biochemical markers in the order of 30–80% within 3–6 months.[10,11,16–19] However, the extent of change necessary to demonstrate antiresorptive efficacy needs to be defined based on both long-term variability and on anticipated change during antiresorptive therapy.[22,23] Biochemical markers revealing a response beyond the minimum significant change (MSC), defined as twice the long-term coefficient of variation, are therefore advantageous.[22,23] Examples of MSCs are 30% for sCTX, 55% for NTX, and 3–4% for spine BMD,[21–23] which indicate the potential benefit of the biochemical markers over BMD as early predictors of individual treatment response.

The choice of biochemical markers has increased considerably over the last decade. Serum and urinary CTX and NTX are, however, among the most sensitive markers presently available.[1,4,5,10,22,23] These markers seem to be more strongly associated with BMD response during antiresorptive therapy than the previous 'gold standard' biochemical marker of bone resorption, urinary deoxypyridinoline (DPD) (Fig. 16.5).[5,10,22,23] The association between change in bone resorption markers and long-term response in BMD increases with increasing time after start of antiresorptive therapy, but is almost fully expressed already after 3 months (Table 16.1).[1,10] In contrast, the association

Fig. 16.5a, b ROC plots showing the predictive accuracy of seven different biochemical markers in women treated with alendronate (1–20 mg/day) or placebo: N-terminal mid-fragment osteocalcin measured by ELISA (OC [1–43] (ELISA)) and RIA (OC [1–43] (RIA)), bone-specific alkaline phosphatase (bone ALP), sCTX, uCTX, NTX, and deoxypyridinoline (DPD). Values of percentage change from baseline at month 12 in the biochemical markers were used. Prevention of spinal bone loss (0%/year) was used as reference. The areas under the curves (AUC) are also given.[10]

(a)

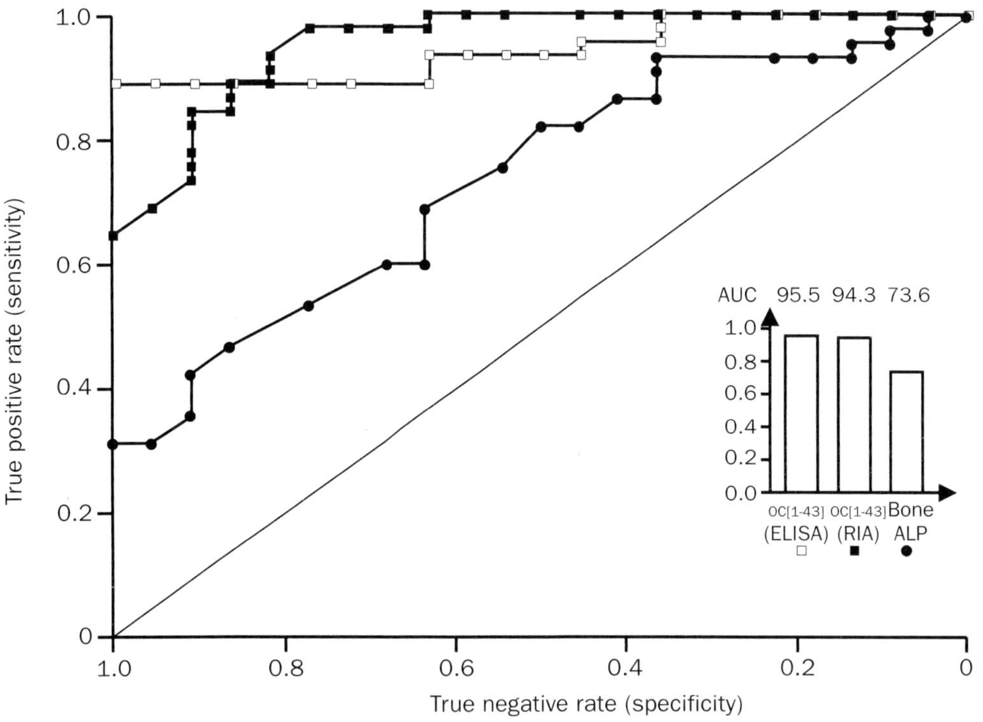
1.0
0.8
0.6
0.4
0.2
0
True positive rate (sensitivity)
1.0 0.8 0.6 0.4 0.2 0
True negative rate (specificity)
AUC 95.5 94.3 73.6
1.0
0.8
0.6
0.4
0.2
0.0
OC[1-43] (ELISA)
OC[1-43] (RIA)
Bone ALP

(b)

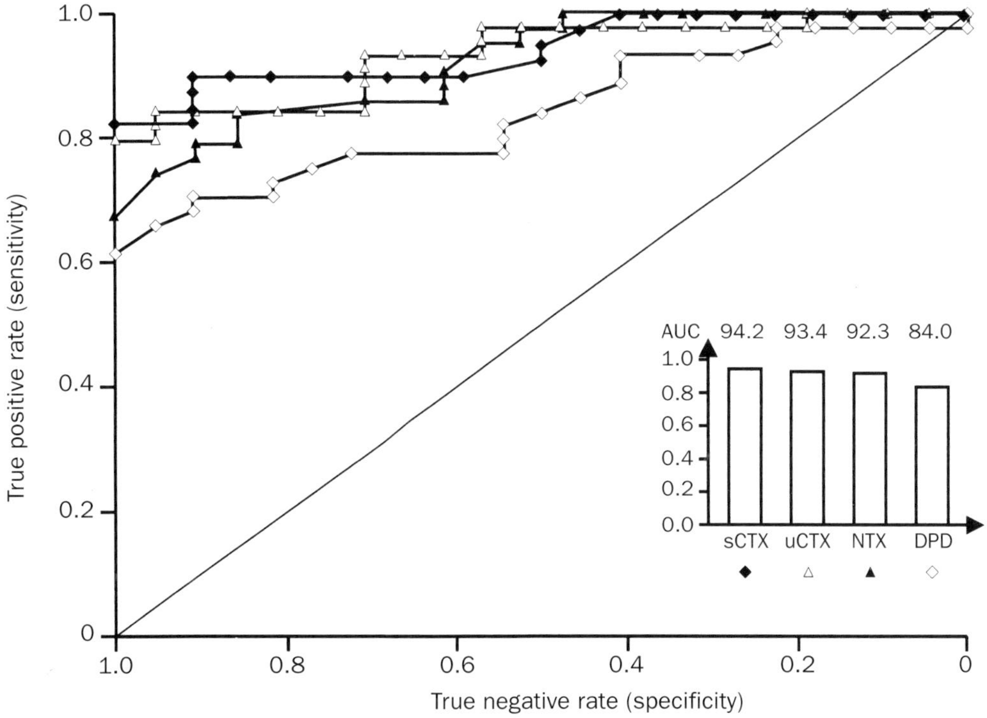
1.0
0.8
0.6
0.4
0.2
0
True positive rate (sensitivity)
1.0 0.8 0.6 0.4 0.2 0
True negative rate (specificity)
AUC 94.2 93.4 92.3 84.0
1.0
0.8
0.6
0.4
0.2
0.0
sCTX uCTX NTX DPD

Table 16.1 Correlations between change in biochemical markers at different time points and change in regional BMD at month 24 in women treated with alendronate (1–20 mg/day) or placebo.[10]

		sCTX	uCTX	NTX	OC [1–43] (ELISA)	OC [1–43] (RIA)
Spine BMD	12 months	−0.76***	−0.71***	−0.62***	−0.78***	−0.70***
	6 months	−0.77***	−0.68***	−0.45***	−0.71***	−0.59***
	3 months	−0.73***	−0.62***	−0.61***	−0.59***	−0.58***
Hip BMD	12 months	−0.53***	−0.50***	−0.39***	−0.54***	−0.36**
	6 months	−0.49***	−0.47***	−0.27*	−0.50***	−0.39***
	3 months	−0.50***	−0.40***	−0.38***	−0.40***	−0.36**

*$P < 0.05$, **$P < 0.01$, ***$P < 0.001$.
OC [1–43], N-mid fragment of osteocalcin.

between change in bone formation markers and BMD during antiresorptive therapy first reaches optimal values 3–9 months later (Table 16.1).[1,10] This reflects the fact that antiresorptive therapies primarily reduce the level of bone resorption and secondarily the level of bone formation as a result of coupling of these two bone turnover processes.

On-treatment values of the biochemical markers can be expressed as percentage changes from baseline or as absolute values. The predictive accuracy of the biochemical markers is higher if values relative to baseline are used (Fig. 16.6).[1] The increase in predictive accuracy by use of values relative to baseline is, however, relatively small (Fig. 16.6).[1] Although values relative to baseline are superior over absolute values, an approach using absolute values may nevertheless be considered attractive from a clinical point of view. However, considerable effort is needed to standardize the analyses between different laboratories, and the costs of standardization are likely to exceed the costs of including a baseline measurement for each individual. Measurements of baseline biochemical markers furthermore provide additional information for the assessment of risk of osteoporosis likely to be performed in a clinical situation before start of treatment.[24] Use of values relative to baseline also compensates for potential differences in local sampling procedures and allows for extrapolation of results across laboratories.

Based on these considerations, 3- or 6-month values relative to baseline of serum or urinary CTX or NTX appear to be the most valid choices for early individual prediction of long-term response to antiresorptive therapy.

As examples, Fig. 16.7 shows the correlation between change from baseline at month 6 in sCTX, uCTX and NTX, and change in spine BMD after 2 years of treatment with alendronate for osteoporosis prevention.[10] Because prevention of bone loss was considered to be the therapeutic aim in this study of postmenopausal women with normal bone mass at baseline,[16] 0% was chosen as the cut-off value for BMD.[10] By use of the same cut-off value for all bone resorption markers, a 50% decrease from baseline, sensitivities ranged from 80% to 90% and specificities from 70% to 100%. In a study of oral oestradiol, 1–2 mg/day combined

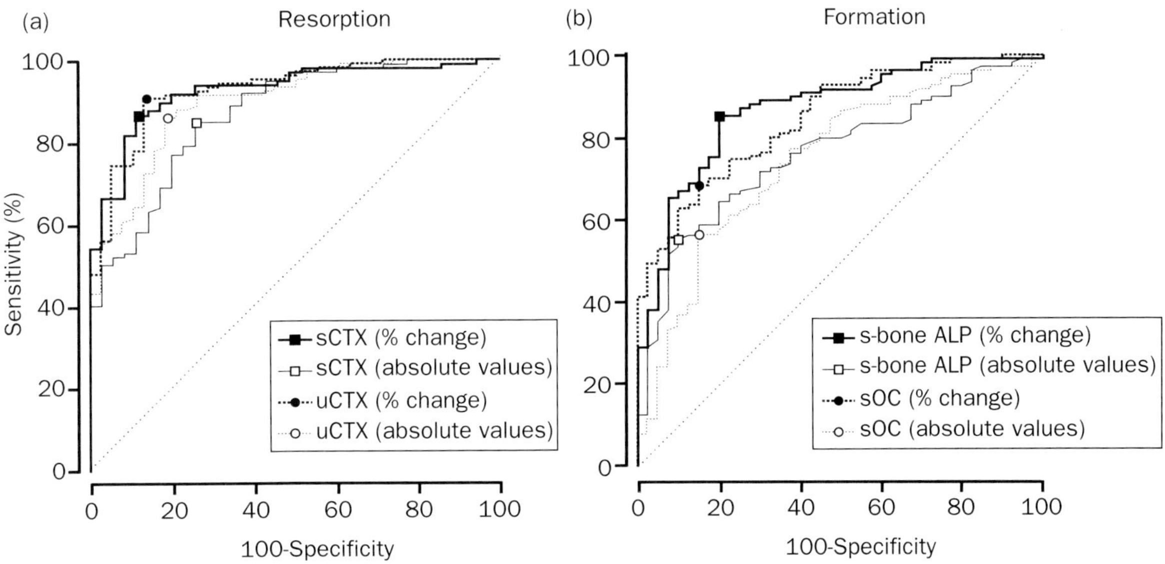

Fig. 16.6a, b ROCS plots showing the predictive accuracy of absolute and change from baseline values of biochemical markers of (a) bone resorption (sCTX and uCTX) and (b) bone formation (bone-specific alkaline phosphatase (bone ALP) and osteocalcin (OC)) after 6 months of intervention with oestradiol (1–2 mg/day). Prevention of spinal bone loss (0%/year) was used as reference.[1]

with gestagen 25–50 μg/day for osteoporosis prevention,[25] the predictive accuracy of sCTX and uCTX were superior to that of serum bone-specific alkaline phosphatase and osteocalcin.[1] At cut-off values of a 40% and 30% decrease for sCTX and uCTX, respectively, sensitivities and specificities were about 90% (Fig. 16.8).[1] Both studies revealed high positive predictive values, which indicated a 90–100% probability that prevention of bone loss would be achieved if the bone resorption marker decreased below the cut-off value at month 6. The lower negative predictive values of about 70–75% indicated that the biochemical markers were less able to predict bone loss or non-response to treatment. Nevertheless, most women could be classified correctly as long-term BMD responders simply by measuring a biochemical marker at baseline and after 6 months (upper left squares in Figs 16.7 and 16.8). In a clinical situation, these women may be considered to be adequately treated, with no need of further follow-up. A conservative strategy for monitoring of individual patients could then be to restrict BMD follow-up measurements to women who do not reveal a decrease below the cut-point in the biochemical marker at month 6 during antiresorptive therapy (upper and lower right squares in Figs 16.7 and 16.8). These women are either inadequately treated or represent false negatives. If compliance is assured, a repeated BMD measurement will reveal true non-responders. The result could then indicate the need for a potential increase of dose or change of therapy. With this strategy, only a small number of patients are missed as non-responders (lower left squares in Figs 16.7 and 16.8). At the same time, costs for follow-up measurement are restricted considerably. Alternatively, a recent report has indicated improved accuracy of prediction of response in bone mass by combining the level of a biochemical marker (alkaline

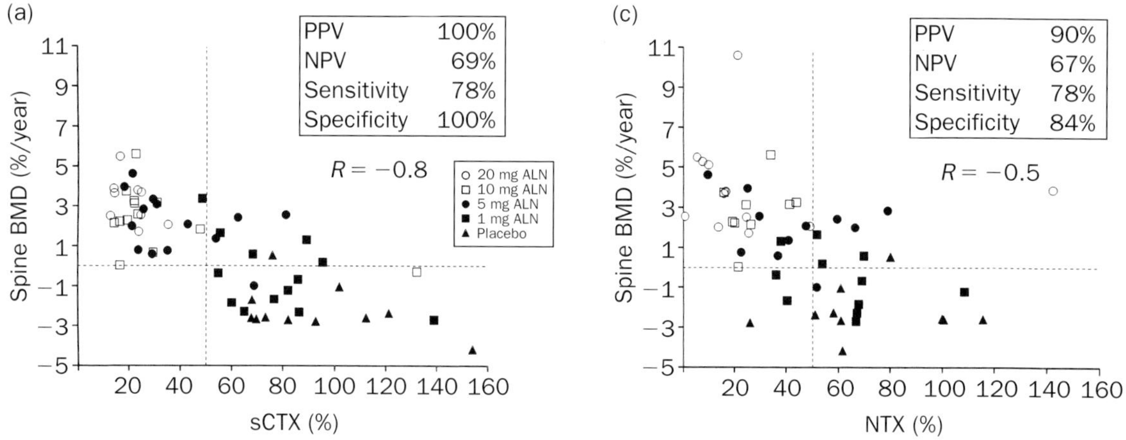

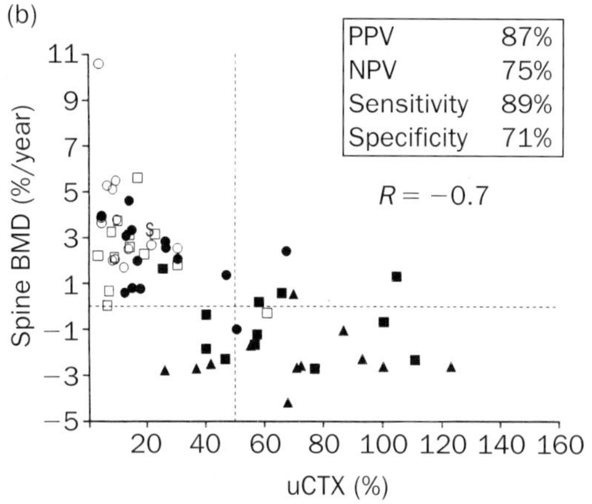

Fig. 16.7 Plots showing predictive accuracy of three biochemical markers: (a) sCTX; (b) uCTX; (c) NTX in women treated with alendronate (ALN) (1–20 mg/day) or placebo. Values of percentage change from baseline at month 6 were used. Prevention of spinal bone loss (0%/year) was used as reference.[10] PPV, positive predictive value; NPV, negative predictive value.

phosphatase) with percentage change from baseline during antiresorptive therapy.[6] Both values were found to contribute independently to the prediction of long-term response in BMD, and increased the specificity of prediction above 90%. The results stress the influence of individual level of bone turnover on the size of response to antiresorptive therapy, which is consistent with studies showing a trend towards a more pronounced response in bone mass in women with a high baseline bone turnover.[5,7,10,11,13]

In conclusion, biochemical markers can be used for valid prediction of response in bone mass during antiresorptive therapy. CTX and NTX appear to be superior to other biochemical markers as predictors of individual response to antiresorptive therapy. Serum CTX combines a pronounced response to treatment with a low MSC, and values should be expressed as change from baseline at month 3 or 6 for optimal individual prediction of BMD response to treatment. The optimal cut-off value is probably a decrease from baseline around 40%, but cut-off values and potential combinations of baseline and on-treatment levels of the biochemical markers remain to be further investigated to define optimal strategies of prediction of BMD response during antiresorptive therapy.

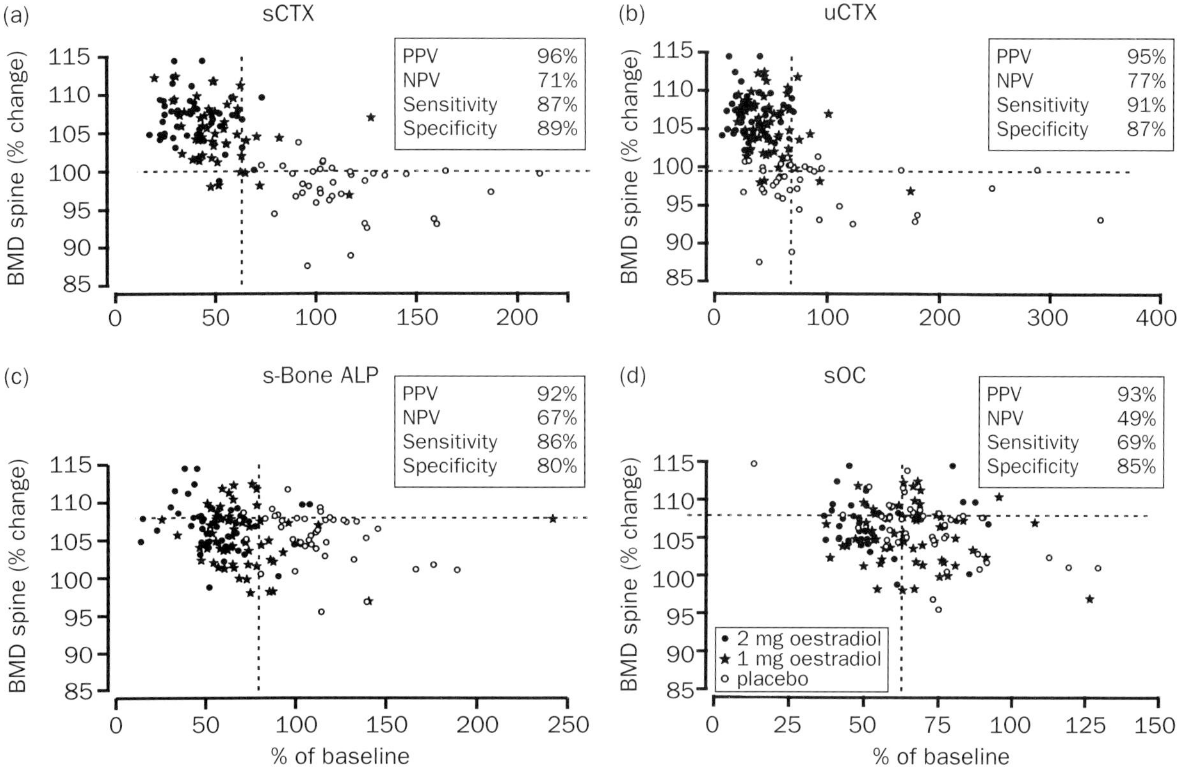

Fig. 16.8 Plots showing predictive accuracy of four biochemical markers: (a) sCTX; (b) uCTX; (c) bone ALP and (d) osteocalcin (OC) in women treated with oestradiol (1–2 mg/day) or placebo. Values of percentage change from baseline at month 6 were used. Prevention of spinal bone loss (0%/year) was used as reference. Cut-off values were: −32% (sCTX), −37% (uCTX), −21% (bone ALP), −33% (OC), 0% (hip BMD).[7] PPV, positive predictive value; NPV, negative predictive value.

REFERENCES

1. Bjarnason NH, Christiansen C, Early response in biochemical markers predicts long-term response in bone mass during hormone replacement therapy in early postmenopausal women. *Bone* 2000; 26(6): 561–9.
2. Bjarnason NH, Bjarnason K, Hassager C, Christiansen C, The response in spinal bone mass to tibolone treatment is related to bone turnover in elderly women. *Bone* 1997; 20(2): 151–5.
3. Chesnut CH 3rd, Bell NH, Clark GS et al, Hormone replacement therapy in postmenopausal women: urinary N-telopeptide of type I collagen monitors therapeutic effect and predicts response of bone mineral density. *Am J Med* 1997; 102(1): 29–37.
4. Christgau S, Rosenquist C, Alexandersen P et al, Clinical evaluation of the Serum CrossLaps One Step ELISA, a new assay measuring the serum concentration of bone-derived degradation products of type I collagen C-telopeptides. *Clin Chem* 1998; 44(11): 2290–300.
5. Garnero P, Shih WJ, Gineyts E et al, Comparison of new biochemical markers of bone turnover in late postmenopausal osteoporotic women in response to alendronate treatment. *J Clin Endocrinol Metab* 1994; 79(6): 1693–700.
6. Garnero P, Darte C, Delmas PD, A model to monitor the efficacy of alendronate treatment in women with osteoporosis using a biochemical marker of bone turnover. *Bone* 1999; 24(6): 603–9.
7. Gonnelli S, Cepollaro C, Pondrelli C et al, The usefulness of bone turnover in predicting the

response to transdermal estrogen therapy in postmenopausal osteoporosis. *J Bone Miner Res* 1997; 12(4): 624–31.

8. Greenspan SL, Parker RA, Ferguson L et al, Early changes in biochemical markers of bone turnover predict the long-term response to alendronate therapy in representative elderly women: a randomized clinical trial. *J Bone Miner Res* 1998; 13(9): 1431–8.
9. Ravn P, Christensen JO, Baumann M, Clemmesen B, Changes in biochemical markers and bone mass after withdrawal of ibandronate treatment: prediction of bone mass changes during treatment. *Bone* 1998; 22(5): 559–64.
10. Ravn P, Clemmesen B, Christiansen C, Biochemical markers can predict the response in bone mass during alendronate treatment in early postmenopausal women. Alendronate Osteoporosis Prevention Study Group. *Bone* 1999; 24(3): 237–44.
11. Ravn P, Hosking D, Thompson D et al, Monitoring of alendronate treatment and prediction of effect on bone mass by biochemical markers in the Early Postmenopausal Intervention Cohort Study. *J Clin Endocrinol Metab* 1999; 84(7): 2363–8.
12. Riis BJ, Overgaard K, Christiansen C, Biochemical markers of bone turnover to monitor the bone response to postmenopausal hormone replacement therapy. *Osteoporos Int* 1995; 5(4): 276–80.
13. Rosen CJ, Chesnut CH 3rd, Mallinak NJJ, The predictive value of biochemical markers of bone turnover for bone mineral density in early postmenopausal women treated with hormone replacement or calcium supplementation. *Clin Endocrinol Metab* 1997; 82(6): 1904–10.
14. Marcus R, Holloway L, Wells B et al, The relationship of biochemical markers of bone turnover to bone density changes in postmenopausal women: results from the Postmenopausal Estrogen/Progestin Interventions (PEPI) trial. *J Bone Miner Res* 1999; 14(9): 1583–95.
15. Harris ST, Gertz BJ, Genant HK et al, The effect of short term treatment with alendronate on vertebral density and biochemical markers of bone remodeling in early postmenopausal women. *J Clin Endocrinol Metab* 1993; 76(6): 1399–406.
16. McClung M, Clemmesen B, Daifotis A et al, Alendronate prevents postmenopausal bone loss in women without osteoporosis. A double-blind, randomized, controlled trial. Alendronate Osteoporosis Prevention Study Group. *Ann Intern Med* 1998; 128(4): 253–61.
17. Prestwood KM, Pilbeam CC, Burleson JA et al, The short-term effects of conjugated estrogen on bone turnover in older women. *J Clin Endocrinol Metab* 1994; 79(2): 366–71.
18. Ravn P, Clemmesen B, Riis BJ, Christiansen C, The effect on bone mass and bone markers of different doses of ibandronate: a new bisphosphonate for prevention and treatment of postmenopausal osteoporosis: a 1-year, randomized, double-blind, placebo-controlled dose-finding study. *Bone* 1996; 19(5): 527–33.
19. Ravn P, Bidstrup M, Wasnich RD et al, Alendronate and estrogen–progestin in the long-term prevention of bone loss: four-year results from the Early Postmenopausal Intervention Cohort Study. A randomized, controlled trial. Early Postmenopausal Intervention Cohort Study Group. *Ann Intern Med* 1999; 131: 935–42.
20. Hosking D, Chilvers CED, Christiansen C et al, Prevention of bone loss with alendronate in postmenopausal women under 60 years of age. *N Engl J Med* 1998; 338(8): 485–92.
21. Genant HK, Engelke K, Fuerst T et al, Noninvasive assessment of bone mineral and structure: state of the art. *J Bone Miner Res* 1996; 11(6): 707–30.
22. Rosen HN, Moses AC, Garber J et al, Serum CTX: a new marker of bone resorption that shows treatment effect more often than other markers because of low coefficient of variability and large changes with bisphosphonate therapy. *Calcif Tissue Int* 2000; 66(2): 100–3.
23. Rosen HN, Moses AC, Garber J, Ross DS, Lee SL, Greenspan SL, Utility of biochemical markers of bone turnover in the follow-up of patients treated with bisphosphonates. *Calcif Tissue Int* 1998; 63(5): 363–8.
24. Garnero P, Hausherr E, Chapuy MC et al, Markers of bone resorption predict hip fracture in elderly women: The EPIDOS Prospective Study. *J Bone Miner Res* 1996; 11(10): 1531–8.
25. Bjarnason NH, Byrjalsen I, Hassager C, Haarbo J, Christiansen C, Low dose estradiol in combination with gestodene is fully preventive in early postmenopausal women. *Am J Obstet Gynecol* 2000; 183(3): 550–60.

17

Biochemical markers of bone turnover as predictors of bone loss and response to therapy

Clifford J Rosen

Summary • Introduction • Factors which affect the predictive value of BTM for bone loss or response to treatment • Studies demonstrating that baseline BTM can predict bone loss in untreated individuals • Analysis of studies investigating whether baseline BTM can predict bone density changes in patients treated with antiresorptives • Analysis of baseline BTM to predict response to anabolic therapy • Conclusions • References

SUMMARY

Although biochemical markers of bone turnover (BTM) are now readily available at the point of care, there is still significant debate about their clinical utility. In this chapter, the issue of baseline BTMs as predictors of bone loss or response to therapy will be examined. In general, there is significant heterogeneity in studies which have addressed the question of whether baseline BTM can predict subsequent bone mass responsiveness. Probably the strongest evidence in support of their use comes from studies of both young and older postmenopausal women, in which baseline markers were used to define rate of bone loss without treatment. On the other hand, even in early postmenopausal women, baseline BTMs are poor indicators of treatment response to the bisphosphonates and are only fair predictors of change in BMD in response to hormone replacement therapy. Several factors can be identified which lead to significant heterogeneity in respect to the results These include: (1) time to follow-up; (2) magnitude of change in BMD; (3) variability in BTMs; and (4) the cohort under investigation. Further studies and systematic meta-analyses may provide critical information to guide providers in respect of the utility of baseline BTMs in clinical practice.

INTRODUCTION

Over the last half-decade, tremendous strides have been made in defining markers of bone turnover, and assessing their utility in selected populations. One of the earliest and most relevant uses of these markers was to predict bone loss.[1] However, after numerous studies, the role of baseline markers of bone turnover in predicting bone loss or gain following treatment remains nebulous. There is no doubt that rapid bone loss occurs both in early and late postmenopausal women and can lead to skeletal fragility and an increased risk of fracture. Although oestrogen deficiency is a principle cause of bone loss in the immediate perimenopausal period, other factors, including poor calcium intake, vitamin D deficiency and secondary hyperparathyroidism, probably contribute to marked acceleration of bone resorption in the 8th and 9th decades of life. Despite the fact that the pathogenesis of osteoporotic fractures is complex, low bone mass and high bone turnover are often considered the two major risk factors for fractures in postmenopausal women. Hence, an important goal for clinicians managing these women should be to identify individuals who are at greatest risk of losing bone rapidly. The definition of rapid bone loss is arbitrary, although most would agree that >2.0% loss over 1 year would be considered significant, especially if this occurred in the lumbar spine. Since several studies have established that rapid bone loss in most women is often sustained beyond 24 months, the case for intervention becomes even more compelling. Moreover, there is a suggestion that those individuals with the lowest bone mineral density (BMD) and greatest rate of bone turnover exhibit the greatest response to certain antiresorptive drugs in respect to change in BMD. Hence, a compelling case can be made that BTMs could be extremely useful in the management of postmenopausal osteoporosis. In this chapter, I will examine the published evidence which could define the role of baseline BTM in clinical practice, with particular attention to their predictive value for bone loss or gain, and the reasons for heterogeneity across studies.

FACTORS WHICH AFFECT THE PREDICTIVE VALUE OF BTM FOR BONE LOSS OR RESPONSE TO TREATMENT

Several factors contribute to heterogeneity across studies of bone turnover and changes in bone density which lead to a relative lack of confidence in the ability of baseline BTM to predict change in bone mass and fracture risk. First and foremost, estimation of change in BMD is a critical factor. In most research and clinical centres, measurement of BMD by dual energy X-ray absorptiometry (DXA) is both accurate and precise. Because of these properties, DXA has become the gold standard with which other tools are compared with respect to relative risk of fracture and bone loss. In principle, the low precision error for DXA provides the clinician with some degree of confidence about estimation of bone loss related to oestrogen deficiency or other secondary causes of bone loss. However, changes in bone mass must be of significant magnitude to enhance confidence that other risk factors or baseline measurements are indeed predictive. Hence, studies which examine rapid bone losers, especially in large cohorts over extended time periods, are more likely to be illustrative than short-term studies in populations with minimal bone loss.

Trial duration is a second factor that must be considered when examining the published evidence about BTMs. Short-term studies of 1 year or less are likely to be influenced by noise in the measurement as well as uncertainty concerning changes in bone mass. For example, in the first year of the Fracture Intervention Trial, 1.4% of the subjects on alendronate lost more than 4% in total hip BMD, and 2.5% of the placebo group had a greater than 8% increase in BMD.[2] However, during year 2, those who lost bone on alendronate had a 92% chance of gaining bone during the second year, while those who gained on placebo during year 1 lost an average of 1% during the second year.[3] This 'regression to the mean' phenomenon lessens confidence in the predictive value of any baseline marker in respect of change in BMD over 1 year. Similarly, although mean values of bone loss (or gain) in a trial are relevant, percentage loss to follow-up must be considered, especially since those losing bone are more likely to drop out of the trials and therefore not be considered in the final analysis.

The cohorts under investigation are important variables which require careful scrutiny when considering the predictive value of baseline BTMs. Certainly, heterogeneous populations would increase the likelihood of a type I error, especially small cohorts followed over a short time course. Hence, variable characteristics of a population, ranging from demographics, to age, to exposure to medications, are important to consider.

Finally, and probably most importantly, variability in the baseline BTM could account for huge effects (or lack thereof). Since the coefficient of variation for one BTM can be as high as 30%, and percentage change is not considered in this analysis, a high noise-to-signal ratio reduces confidence in the predictive measure.[4] At least two baseline measurements, or repeated measures among individuals over the course of a placebo arm, reduce the noise of the assay, and may strengthen the association between a baseline measurement and bone loss (or gain).

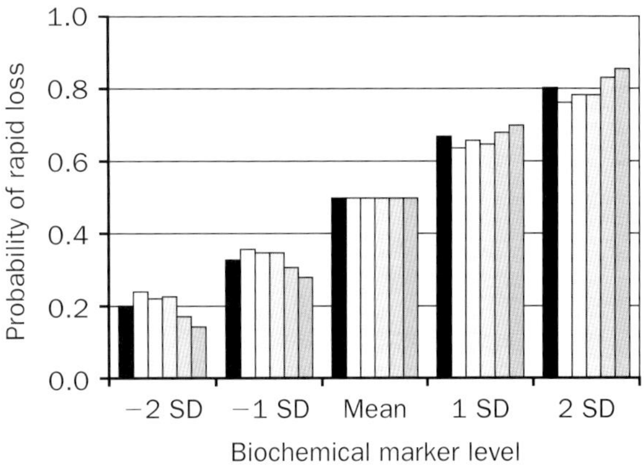

Fig. 17.1 Probability of rapid bone loss as a function of biochemical marker level. At each level, from left to right, the markers were bone ALP by Metra kit (■; mean, SD = 22.8, 9.4 U/l), OC by Metra kit (; 6.9, 2.6 ng/ml), DPD by Metra kit (; 5.2, 1.8 nm/mm), PYD by Metra kit (; 31.5, 10.4 nm/mm), bone ALP by Hybritech kit (; 14.2, 6.3 μg/l), and CTX by Osteometer kit (; 179.1, 119.4 μg/mmol).

STUDIES DEMONSTRATING THAT BASELINE BTM CAN PREDICT BONE LOSS IN UNTREATED INDIVIDUALS

The strongest evidence relating BTM to changes in bone mass comes from longitudinal studies of cohorts who have not undergone treatment for osteoporosis. These studies suggest that a baseline BTM can reliably predict women who will subsequently lose bone mass. Several important studies fulfil the criteria noted above, including relatively long follow-up, large cohorts, and a significant magnitude of effect. Ross and Knowlton studied a subset of 200 women in the longitudinal component of the large Hawaii osteoporosis study (HOS), which included over 1100 postmenopausal women.[5] They subdivided those women who had baseline markers and at least eight calcaneal BMD measurements over 13 years into one group with rapid bone loss (>2.2%/year) and those with the slowest loss (<0.4%/year). Each of the formation (osteocalcin and bone-specific alkaline phosphatase (bone ALP)) and resorption (pyridinoline and free deoxypyridinoline) markers was strongly associated with rapid loss; that is, the odds of rapid loss increased by 1.8–2.0 times for each SD increase in the marker (Fig. 17.1; Table 17.1).[5] For bone ALP, a 2 SD above the mean value was associated with an 80% probability of rapid bone loss. Other markers had similar probabilities for a 2 SD baseline value. In another cohort, women from the Study of Osteoporotic Fractures (410 subjects, aged 67 and older) were followed for 4 years with baseline and subsequent femoral BMD. Baseline markers of bone resorption were significant predictors of bone loss rate; in addition, the average loss of bone was twice as great for those women in the highest quartile compared to those in the lowest quartile.[6] It should be noted that, in both these studies, women were older, long past menopause, and had sizeable bone loss over the time of the study.

Observational and randomized trials in younger postmenopausal women provide interesting and somewhat corroborative data to the

Table 17.1 Odds of BMD loss by increase of 1 SD in marker value[a]

Marker	Odds ratio	95% CI
NTX	2.1	1.3, 3.4
DPD	1.5	1.0, 2.4
OC	1.6	1.0, 2.4
BAP	1.1	0.8, 1.7

[a]Modified from Greenspan et al[16] with permission.

cohort studies of older women with respect to the predictive value of baseline BTMs. Gorai et al studied lumbar BMD in 82 premenopausal (mean age 45) and 325 postmenopausal (mean age 60) Japanese women over 4 years and examined markers of bone resorption and formation, as well as cytokine levels. There was no relationship between baseline BTM and bone loss in the premenopausal women.[7] However, in the postmenopausal group, there were major differences in the rate of bone loss at year 1 when comparing women with a baseline BTM in the lowest quartile of N-terminal cross-linking telopeptide of type I collagen (NTX) with those in the highest quartile (Q1 = 0.28% versus Q4 = −2.4%; $P < 0.001$). This relationship held up in subsequent years (i.e. year 4), with the Q1 baseline BTM women exhibiting minimal bone loss (−1%) compared to Q4 women (−4.1%). With a cut-off for NTX of 37 BCE/mmol creatinine (Cr), the sensitivity and specificity of NTX in postmenopausal women were 65% and 66% respectively.[7]

Similar data on NTX were noted in a randomized placebo-controlled trial conducted by Rosen et al over 1 year in 250 younger postmenopausal women (mean age 55 years). Those receiving just calcium (no hormone replacement therapy (HRT)) after 1 year lost nearly 2% of their spine BMD. The authors averaged four BTM values at baseline and during the course of the study and found that the odds ratio of spine BMD loss per 1 SD increase in NTX was 2.1.[8] A similarly significant odds ratio was also noted for deoxypyridinoline (DPD) and osteocalcin, although not for bone ALP. Similar findings were noted in a longitudinal study of untreated pre- and postmenopausal women followed for 3 years by Cosman et al.[9] For every 1 SD increase in formation or resorption, bone loss from the spine and femoral neck averaged approximately 1–1.5%/year. Finally, in the EPIC trial, a subset of younger postmenopausal (mean age 53) women from the Danish cohort were analysed ($N = 67$). Baseline urinary cross-links were negatively associated ($R = -0.54$, $P = 0.05$) with rate of bone loss in the placebo group over 2 years.[10]

These studies are all relatively consistent and support the hypothesis that baseline BTM can predict with relative confidence significant bone loss in women who are oestrogen deficient and are not receiving antiresorptive therapy. Hence, there could be utility for such indices in the clinic, assuming that the same criteria noted above apply to the individual patient. These would include: minimizing variance by repeated baseline measures of BTM, studying high-risk individuals who are likely to lose significant bone density, and quality assurance with maintenance of strong precision for the bone mass measurement tool. Clearly, these conditions are not likely to be found at the primary care level or even among subspecialists who are not conducting active research trials.

ANALYSIS OF STUDIES INVESTIGATING WHETHER BASELINE BTM CAN PREDICT BONE DENSITY CHANGES IN PATIENTS TREATED WITH ANTIRESORPTIVES

Since the seminal study by Civitelli et al more than a decade ago, which defined a positive bone density response to calcitonin in women with high serum osteocalcin, investigators have examined the predictive value of several baseline BTM in relation to changes in bone mass, with variable results.[11] Factors which affect the predictive value of these markers include the age of the subject (young premenopausal versus old premenopausal), the type of antiresorptive treatment (HRT versus bisphosphonates versus calcitonin) and the duration of treatment. In newly postmenopausal women randomized to HRT, Rosen et al demonstrated a stepwise increase in the change in lumbar BMD according to quartile of urine NTX at baseline.[8,12] Moreover, women who gained BMD on HRT had a 50% greater baseline NTX than those women who showed a loss of spine BMD. Similarly, Ravn et al examined the predictive value of several baseline BTM in respect to change in BMD with 5–20 mg of alendronate.[10] Serum, but not urinary, cross-links were positively associated with change in both spine and hip BMD (R=0.29 and 0.25 respectively), although the amount of variance for the change in BMD accounted for by these markers was far less than 10%. In another study of younger postmenopausal women in EPIC, McClung et al noted that in 442 women randomized to 5 mg of alendronate with osteopaenia, the highest tertile of NTX at baseline was associated with the greatest increase in spine BMD (3.9%), although the difference from the lowest tertile of NTX (a +3.0% increase in BMD) was not statistically significant.[13]

Studies in elderly women provide more disparate results. In a subset of FIT in which 866 women had BTM at baseline and follow-up BMD measurements, there was no correlation between baseline NTX or other parameters and subsequent spine or hip BMD ($R = -0.01$ to -0.08).[14] In another randomized trial with alendronate in elderly women with low bone mass but not more than one vertebral fracture, Bone et al showed that there was a weak correlation between baseline NTX and change in spine BMD after 2 years of 2.5 mg alendronate.[15] Similarly, this group showed that baseline urinary DPD and bone ALP also weakly predicted change in spine BMD in women treated with 5 mg/day of alendronate. Finally, Greenspan et al studied elderly women with low bone mass, of whom about half fulfilled the criteria for the WHO diagnosis of osteoporosis.[16] Once again, in this cohort of 120 women, baseline NTX was weakly correlated with total body BMD at 18, 24 and 30 months, with spine BMD at 12 months, and with trochanteric BMD at 24 months ($R = 0.28$–0.36, $P < 0.05$). However, as in the other studies, the predictive value of these measurements was quite weak and far less than change in BTM versus change in BMD. In all these studies, baseline BTM did correlate inversely with ultimate change in marker, so that the higher the turnover rate initially, the lower the BTM is at the end of the study. Whether this consistent relationship ($R = -0.7$ to $R = -0.9$) will be of any clinical value remains unclear.

In summary, baseline BTM only weakly predicts the response to therapy with all antiresorptive drugs. Whether these markers may be useful in certain situations, such as women going on HRT, will require further studies. Compared to the predictive value of baseline BTMs for bone loss, the sensitivity and specificity for bone gain are quite low.

ANALYSIS OF BASELINE BTM TO PREDICT RESPONSE TO ANABOLIC THERAPY

Anabolic therapies increase bone mass by stimulating new bone formation. At the present time, the one anabolic agent most likely to reach the market is parathyroid hormone (PTH). There are several observational and uncontrolled studies showing that BTM increase significantly in postmenopausal women and men receiving daily intermittent PTH by subcutaneous injection. Most importantly, there appears to be a time-dependent difference, so that changes in bone formation occur earlier than bone resorption. Moreover, each type of marker increases dramatically during therapy, and these changes provide some insight into the mechanism and magnitude of change in bone density. However, there are too few published randomized trials with PTH, and therefore there is a paucity of data with regard to the true predictive value of BTMs for fracture risk reduction or change in BMD. In addition, there are virtually no data on baseline BTM as a predictor of subsequent response to treatment. However, one randomized placebo-controlled trial in men does provide some insight into the possible utility of baseline BTM to predict skeletal response to PTH.

Kurland et al, conducted an 18-month randomized placebo-controlled trial in 23 men with idiopathic osteoporosis, most of whom also had sustained vertebral fractures.[17] Men with idiopathic osteoporosis are generally heterogeneous with respect to the pathogenesis of their disease, although several groups have reported that the majority of these men have low bone turnover. The Kurland cohort at baseline had normal levels of PTH, 25-hydroxyvitamin D, urinary calcium excretion, procollagen type I C propeptide (PICP), bone ALP, NTX and free DPD. In response to PTH, pyridinoline (PYD) increased by 130% and NTX by 261% after 18 months, while osteocalcin peaked at 200% (i.e. osteocalcin) of baseline at 12 months, and bone ALP peaked at 168%. Similarly, bone density of the spine increased by nearly 14% after 18 months of treatment with PTH. In this trial, both baseline BTMs and change in BMD were analysed with respect to predictive value for change in lumbar BMD. Kurland et al noted several interesting aspects concerning BTM. Using multiple regression analysis, urinary PYD at baseline contributed 28% of the variance in BMD at 18 months, while three other markers (NTX, osteocalcin and PYD) contributed less than 3% of the variance. Furthermore, baseline PYD in the control subjects did not correlate with any markers of bone formation at baseline, while NTX did. However, follow-up PYD did strongly correlate with formation indices at 18 months. Overall, the change in osteocalcin at 3 months and the baseline PYD contributed over 70% of the variance in BMD observed at 18 months. The authors conclude that PTH linked PYD excretion to bone formation, a relationship that was not present prior to therapy, presumably because of low turnover. More importantly, it appears that the greater the turnover at baseline, at least as measured by PYD, the greater the response to PTH with respect to bone density. Clearly, more studies will be needed in larger cohorts before definitive conclusions can be drawn about the predictive value of bone turnover markers for both change in density and change in fracture risk.

CONCLUSIONS

Baseline markers of bone turnover provide insight into the state of the remodelling units in the adult skeleton. In both younger and older postmenopausal women, resorption and formation indices do predict with some degree of confidence the degree of bone loss which may occur without therapeutic intervention. On the other hand, the published data would suggest that baseline BTMs are not helpful in defining the response to therapy among older, more severely affected postmenopausal women. The value of baseline BTM for predicting response to therapy in young postmenopausal women is variable and may depend on the type of antiresorptive agent being utilized. With respect to the predictive value of baseline BTMs in anabolic treatment regimens, there are not enough data to make a definitive conclusion. Conflicting results from several studies of baseline BTM in both treated and untreated individuals almost certainly result from co-variates which influence not only the measurement of markers, but also BMD. Furthermore, the characteristics of the cohort, the duration of treatment and the time to follow-up may all be critical in defining heterogeneity with respect to results from several trials. Future studies will be necessary before definitive guidelines about the use of baseline BTMs in clinical practice can be established.

REFERENCES

1. Hansen MA, Kirsten O, Riis BJ et al, Role of peak bone mass and bone loss in postmenopausal osteoporosis. *BMJ* 1991; 303: 961–4.
2. Black DM, Cummings SR, Karpf DB et al,

Randomized trial of effect of alendronate on the risk of fracture in women with existing fractures. *Lancet* 1996; 348; 1535–41.
3. Cummings SR, Palermo L, Browner W et al, Monitoring osteoporosis therapy with bone densitometry. Misleading changes and regression to the mean. *JAMA* 2000; 283: 1318–21.
4. Machado ABC, Hannon R, Eastell R, Monitoring alendronate therapy in osteoporosis. *J Bone Miner Res* 14: 602–8.
5. Ross PD, Knowlton W, Rapid bone loss is associated with increased levels of biochemical markers. *J Bone Miner Res* 1998; 13: 297–302.
6. Cummings SR, Black D, Ensrud K et al, Urine markers of bone resorption predict hip loss and fractures in older women: the Study of Osteoporotic Fractures. *J Bone Miner Res* 1996; 11: S128.
7. Gorai I, Yoshikata H, Kikuchi R et al, The predictive value of biochemical markers of bone turnover for BMD in postmenopausal Japanese women. *J Bone Miner Res* 1999; S365.
8. Rosen CJ, Chesnut CH III, Mallinak NJS, The predictive value of biochemical markers of bone turnover for bone mineral density in early postmenopausal women treated with hormone replacement or calcium supplementation. *J Clin Endocrinol Metab* 1997; 82: 1904–10.
9. Cosman F, Nieves J, Wilkinson C et al, Bone density change and biochemical indices of skeletal turnover. *Calcif Tissue Int* 58: 236–43.
10. Ravn P, Clemmesen B, Christiansen C for the Alendronate Osteoporosis Prevention Study Group, Biochemical markers can predict the response in bone mass during alendronate treatment in early postmenopausal women. *Bone* 1999; 24: 237–44.
11. Civitelli R, Gonnelli S, Zacher F et al, Bone turnover in postmenopausal osteoporosis: effect of calcitonin treatment. *J Clin Invest* 1988; 82: 1268–74.
12. Chesnut CH, Bell NH, Clark G et al, Hormone replacement therapy in postmenopausal women: urinary N-telopeptide of type I collagen monitors therapeutic effect and predicts response of bone mineral density. *Am J Med* 1997; 102: 29–37.
13. McClung MR, Faulkner KG, Ravn P et al, Inability of baseline biochemical markers to predict bone density changes in early postmenopausal women. *J Bone Miner Res* 1996; 11: S127.
14. Ott SM, Bauer DC, Santora A et al, Ability of biochemical markers to predict 4 year changes in bone density in postmenopausal women. *Bone* 1998; S159.
15. Bone HG, Downs RW, Tucci JR et al, Dose response relationships for alendronate treatment in osteoporotic elderly women. *J Clin Endocrinol Metab* 1997; 82: 265–74.
16. Greenspan SL, Parker RA, Ferguson L et al, Early changes in biochemical markers of bone turnover predict the long term response to alendronate therapy in representative elderly women: a randomized clinical trial. *J Bone Miner Res* 1998; 13: 1431–8.
17. Kurland ES, Cosman F, McMahon DJ et al, PTH as a therapy for idiopathic osteoporosis in men: effects of bone mineral density and bone markers. *Lancet* 2000; 85: 3069–76.

Discussion paper C

Panel: *Aubrey Blumsohn, Pierre D Delmas, Susan L Greenspan, Michael R McClung, Pernille Ravn, Clifford J Rosen, Markus J Seibel and Nelson B Watts*

McClung During the previous discussion on circadian rhythm, a statement was made that there were data somewhere. We have discovered the data to demonstrate. So, let us do that quickly while we get organized and then we'll launch into our discussion of the monitoring issues. [The reader is referred to Discussion paper B.]

Christgau I'd like to thank the organizers for giving me this opportunity to show some data and I would like to build on the comments made by Dr Fraser in the last session, when we learnt that markers of bone resorption both in urine and in serum have a huge, pronounced circadian variability. This, of course, raises a lot of issues for the clinical use of the markers. So, I just want to take this opportunity to show some data that we've obtained in collaboration with Drs Bjarnason and Christiansen in Denmark (*see Discussion Fig. 1*[†]).

We agree with Dr Fraser that there is this circadian variability and this is also shown here in this slide, which shows data from 15 postmenopausal women who were subjected to a randomized crossover study where, in one part of the study, they were fasted before they entered the 24-h study period but were allowed to eat throughout the study period. As you can see, there was circadian variation, as indeed Fraser also showed. However, we observed that when the women fasted throughout the study period, the circadian variability was significantly reduced. This we assessed further in another study. A small study which shows short-term variability over 2 weeks in postmenopausal women who were sampled either after overnight fasting or non-fasting. As you can see, the variability in serum CrossLaps was almost half in the fasting women as compared with the non-fasting women. This, of course, has big implications for the clinical use of the marker and, as you heard from Dr Delmas and Dr Ravn, when you take morning fasting samples, you can actually use the serum CrossLaps marker to monitor the response to various antiresorptive therapies on an individual level. With this last slide I want to show the effects in three different studies with ibandronate, tibolone and HRT. It is apparent that a 6-month response in serum CrossLaps is significant for the majority of the treated women compared with the spine BMD response after 2 years. I hope this can clarify some questions.

[†] See Christgau S, Bitsch-Jensen O, Hanover Biarnason N et al, Serum crosslaps for monitoring the response in individuals undergoing antiresorptive therapy. *Bone* 2000; 26: 505–11.

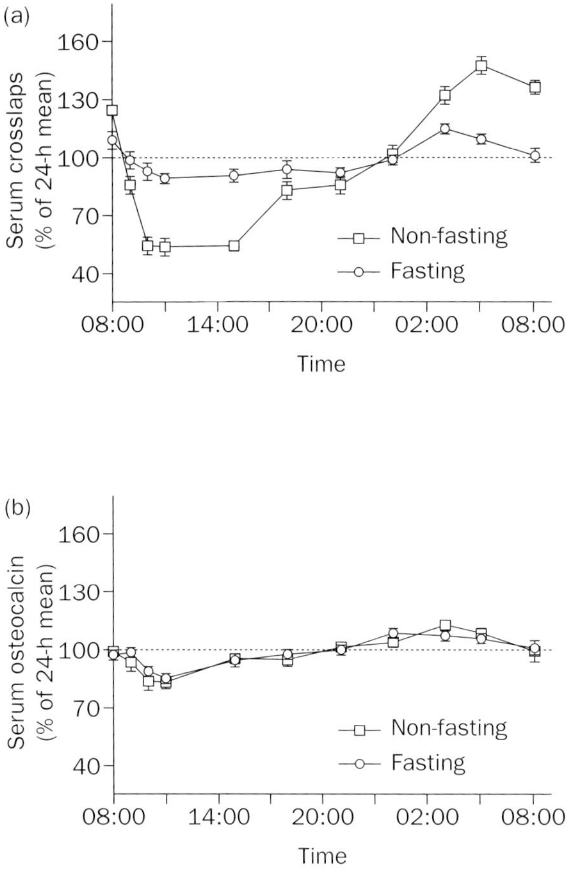

Discussion Fig. 1 (a) Circadian variation in serum CTX concentration. Fifteen postmenopausal women were sampled at 08:00, 09:00, 10:00, and 11:00, and then every 3 h over a 24-h period. In one part of the experiment, the women fasted throughout the study, in the other part they were allowed to eat and drink without restriction. Results are shown as mean ± SEM. (b) Osteocalcin measurements of the same samples as in (a).

McClung All right, thank you for that clarification. Comments, Dr Fraser?

Fraser I'd like to comment on that. Clearly, what you are seeing here when you present your first graph is that the mean has increased, as was demonstrated with Richard Eastell's data and my own data. If you now calculate the values over the 24-h cycle, then you get that change in the relative percentage value that you have seen there. I cannot understand how not eating at 8 o'clock in the morning could have a major effect on the value at 7:00, 7:30, 8:00, 8:30 or 9:00, although we do see that small percentage change, but the distribution stays exactly the same. So, that would be my comment back on the data. When I asked Claus Christiansen this question in Garmisch a couple of weeks ago, he said that I am right. The values are no different at 7:00, 7:30, 8:00, 8:30, and your data are entirely similar to the data that we have got on the fasting samples at 9:00, 10:00 and 11:00. So it is a relative percentage effect when the data are presented as a percentage of the mean.

McClung Do you want to add anything, Pierre (Delmas)?

Delmas Yes. The only relevant issue is the clinical value of the marker, and I think the data that you are going to see this afternoon clearly show that if you take the serum CrossLaps morning fasting, you have a prediction, for a fracture for example. If there is no fasting, there is a high variability and you do not predict fracture. So, if you forget about statistics and just think about the clinical relevance, I think the data that have been shown and published or are in press with the SOF study, the EPIDOS study and the OFELY study are very suggestive that the best way to collect serum for serum CrossLaps is morning fasting.

McClung Let us move onto the topic of the afternoon, the issue of monitoring. I think it is useful to reflect on what we're trying to do with monitoring therapy. Pierre (Delmas) made the point a time or two that the two major things that we are trying to do with monitoring are, in patients on treatment, to identify those who respond (I would actually contend that we are trying to identify those who don't respond, because it is those patients in whom the changes in management are going to be made) and to influence long-term adherence to therapy. While that is a theoretical idea, there are, as Dr Delmas said, no data yet to document that has an effect. I think it is also important to distinguish, as Cliff Rosen and Dr Ravn said, between patients who are not being treated actively, where we are trying to follow and assess bone loss in a control population essentially, and clinical questions that have to do with patients who are on active therapy. The issues involved in those two sets of patients are, I think, quite different. As we have our discussion, it will be important to keep in mind that those are both the objectives and the types of patients that we see in clinical practice. Our group has a number of questions that we would like to pose to the speakers. Let us attend to those, and then we will come back and have time for discussion from the floor as well. Nelson (Watts), let me ask you to begin if you will.

Watts I have questions for Dr Delmas and Dr Ravn. In your analysis of the relationship between change in marker and change in BMD in studies that included both treatment and placebo groups, you have included the placebo group, the argument being that in the real world there are patients who do not take their medication or do not absorb it appropriately. But I think that the real question is whether or not patients who are taking treatment are showing a response, and the inclusion of the placebo patients in that analysis distracts me from the question of what is happening to the patients who are on treatment. If you remove the placebo patients from these analyses, how strong is the relationship?

Ravn As you know, in the Early Postmenopausal Intervention Cohort Study we analysed both the total cohort treated with placebo, 2.5 and 5 mg, and also the groups that were treated with the 5-mg dose. What happens is that if you only look at the 5-mg group, you increase the positive predictive value considerably, but then you still have a problem with a low negative predictive value. So it comes back to the problem of specificity of the test, and I think that what you can use the marker for is to say 'OK, those who are decreased below the cut-off point are responding,' and then, 'What do we do with those who do not decrease?' As I was suggesting, you can then do a follow-up measurement of BMD. Also, I know that Dr Delmas has performed studies where he repeated the bone marker measurement or combined absolute and percentage change, allowing an increase in the specificity. The bottom line is that you get more or less the same results if you only look at the 5-mg group. The problem is that you decrease the specificity of the test slightly. That is our experience from the Early Postmenopausal Intervention Cohort Study.

Delmas Yes, I agree entirely. Actually, in the large HRT study that I showed where we derived this cut-off, only 25% of the patients were on placebo.

McClung You focused on the idea that percentage change rather than an absolute number was the issue. What effect does the baseline value have on percentage changes? Patients who have very high values, for example of a urinary resorption marker, have more opportunity to have a bigger absolute change, and even a bigger percentage change. Is that a factor in this whole issue? Does the baseline value influence how we interpret percentage change?

Delmas I do not think so, because we are not looking at the absolute decrease but at the percentage change.

McClung But is the percentage change the same in the lowest quartile, for example? If you have a group of patients treated with HRT or alendronate and you divide the cohort into tertiles or quartiles according to their baseline value, is the percentage change the same in the lower group versus the higher group?

Delmas Well, it depends on the study. The percentage change tends to be higher in groups that start with a high value, in the same way as the BMD increase tends to be higher than in those who start with a lower BMD. I do not think that it makes a big difference in the analysis because, again, you are not looking at the absolute decrease but a percentage decrease. I think it is going to become more important and more interesting once we have fracture data. One of the questions with the bisphosphonates, for example, is that if you go to very low absolute levels under treatment, do you have an indication that you could get an increase with fracture, for example? But, in terms of BMD, I am not sure that is too important.

McClung Actually, that was similar, I know, to a question that Nelson (Watts) had about baseline values.

Watts Another problem is if the baseline level is low, the precision for measuring changes is reduced. Is there a level of baseline marker below which looking for percentage change is irrelevant?

Blumsohn The bottom line is that, whatever data you choose from the literature, if you back-calculate to remove measurement noise, the response on a percentage scale is almost identical in different patients.

Greenspan I would like to ask the group a question. We have heard a lot about HRT and alendronate in terms of prediction of response and monitoring. If we do monitor a patient, is there a best time to monitor a patient and are there differences with the different agents? For example, when should we monitor HRT versus alendronate versus a SERM?

Delmas That is an easier question... for HRT and alendronate. With SERMs we cannot really correlate a marker change with a change in BMD so we have to wait until we have a correlation with fracture. But probably also with raloxifen, 3 months or 6 months work about the same. Even with bone formation markers at 3 months, you have, in many studies, at least with alendronate, a good prediction of the increase in BMD. The problem is that the signal-to-noise ratio is lower because you have not reached a plateau. So I think, to be 'simple', the best for resorption markers is to do it either at 3 or 6 months, depending on the clinician, and for formation markers probably to wait until 6 months of treatment. I think the cut-offs are different for bisphosphonate and for HRT, but I do not think that the time course is going to be that critical.

McClung But you made the comment that monitoring BMD changes with therapies that have a modest change in bone density, like raloxifen, is more difficult; exactly the same thing holds true for markers. For agents which have a less potent effect on marker reduction, it makes it more difficult to distinguish those who do and do not respond with markers than with bisphosphonates and oestrogens, where the changes are so big that it will be easier with those drugs.

Delmas Again, I cannot comment on calcitonin, but with raloxifen the mean decrease in the urinary CrossLaps is in the order of 30–40%. It is not a marginal decrease.

McClung Fair enough, but it is intermediate between the changes seen with bisphosphonates and even oestrogens.

Delmas Right, but the change should be enough to see, for it is larger than the least significant change for at least a substantial proportion of the patients. Again, what needs to be demonstrated is what correlates with the fracture incident.

Seibel I have a question for Pierre (Delmas) as well as Dr Ravn. It kind of relates to what you have just said. We agree that markers do not really predict bone loss at most sites very well.

Delmas In untreated or treated?

Seibel Mainly in treated patients. We also agree that BMD, and that is what you said in the beginning of your talk, is not a good measure of treatment outcome in many cases; see, for example, raloxifen or the fluorides. With the latter, we have huge increases in BMD but little effect on fractures. The next argument you make is that with the markers we see correlations with change in BMD during treatment, and therefore markers are supposed to be predictors of therapeutic response or at least feasible for monitoring treatment. Now that sounds like the blind leading the lame across a very busy street, it is a circular argument. Unless we have the fracture data, the argument is just not round; it just does not make much sense to me.

Delmas OK. First of all, I think that for clarity, this round table discussion is dealing with the monitoring of treatment. I think that we are going to have another about prediction of bone loss in fracture and treated patients, so let us not get confused with different issues. I actually disagree with your statement, because the relationship between BMD change and fracture is mainly in the grey zone, and that is why I said that I prefer the analysis looking at the true responder, namely an increase in BMD over, for example 3%, and the true non-responders. If you do that, there is an association with fracture reduction. So, what you do by excluding those patients who have an equivocal response is that you exclude patients for whom that change in BMD is within the coefficient of variation. I think it is quite legitimate. I accept the other analysis, because then you do not exclude any patients, but I think that it is not a circular argument. What we do not know is the relationship between the fracture rate and the change in BMD when you have a very small change in BMD, and those patients were excluded from the analysis that I showed.

Seibel OK, but that has implications for other studies. For example, in the raloxifen study with the small increase in BMD and the significant decrease in fracture, the markers will not help us because they are unlikely to predict both BMD and fracture outcome. There is a very small change in markers in that study.

Delmas Again, you are mixing up two different things. The statements that were made this afternoon were made with the use of bone markers to predict the response to HRT and to alendronate. I made it very clear in my presentation that for raloxifen we need to have the full data with the fracture. So, let us not confuse different data. With raloxifen, the only data that we have today are the data presented by Nina Bjarnason that show that if you divide into tertiles you have a significant association between the decrease in the bone markers and the fracture incidence. That is all that we have. This is on a group basis; we do not have any data on the use of markers to follow individual patients on raloxifen.

Greenspan So, what do we think if we now try to bring this back to the patient if we have these studies where we can look at monitoring and response of bone mass with treatment? Can we really make the conclusion now that we would do this in clinical practice to monitor patients with these markers?

Delmas Yes.

Greenspan For all patients on therapy or particular patients?

Delmas Well, again, when I think of the data that were presented, I think it is interesting to see that for the three lectures on this specific topic, the data are very consistent. If you scrutinize the literature looking at cut-offs for HRT and alendronate, the data are very consistent from one study to the other. Now, your question is, which cut-off for which patient according to treatment? This is, of course, a bit more difficult, because you have different treatments and different markers. I agree with the cut-offs that were suggested by Dr Ravn. I think that, again, it depends on whether you want to have a high specificity or high sensitivity, and I think what we need today, what would be nice, is to apply this cut-off to different cohorts. You derive a model from a study and then you need to validate it in a different cohort. What I would like to see is the cut-offs that were suggested applied to another cohort where you have BMD measurement and marker measurement and see if you get the same sensitivity and specificity. You will probably not get the same values, because the cut-off was not modelled on this cohort, but I think it would strengthen the relationship.

Blumsohn Should you include the placebo group in those studies?

Delmas Well, I think yes. I think the placebo group has to be included. You could perhaps criticize having as many patients on placebo as you have on active treatment, because that could be a bit unrealistic. But I think that, according to what you think is the compliance in real life, you may have 30% placebo-treated patients or 25% or 40% included in your analysis, because we know (and this argument has been made before) that these patients that we studied, because they were in clinical trials, were highly compliant, and we know that in real life they are not. So, that is why I think it is fair to include the placebo-treated patients, because compliance is much lower in the real world.

McClung Are there other comments on Dr Greenspan's question?

Seibel Aubrey (Blumsohn), did I understand you rightly when you said you cannot identify, on the basis of statistical reasons, by bone markers the change in BMD due to treatment?

Blumsohn Yes, not within the treatment group. Compliance is a complicated issue. Compliance is often not an all-or-nothing thing. Many patients who have been called non-compliant are in fact partially compliant patients and would not be the patients who would be in the placebo group. The second thing is that the vast majority of patients who are non-compliant tell you so. Another problem with detecting non-compliance is that you need to know exactly what you can do about non-compliance. If you confront a patient with a result and perhaps challenge them about possible non-compliance, does that improve compliance? Likewise, if you have a false apparent response to therapy, if you have a patient who knows that they are not taking their treatment and you confront them with the result and say that this is a very good response to treatment, what does that make them think about it?

Delmas Now we are not talking about science but clinical practice. I have been using the markers in my patients for the past 10 years, and while I could give you a lot of examples, I quoted this patient who was taking calcium with alendronate, and again I did not detect that before, it was actually a big glass of milk. The other thing is that when you monitor with BMD, and of course I do that like all of us, you have repeatedly the same question, which is: OK, it has increased by 5%, but it is still in the red, it is still a low value, I am still osteoporotic, and then if you want to look at the relationship between BMD and fracture injuries in treated patients—Good luck! I mean it is going to take you at least half an hour—at least it takes me half an hour—so it is impossible. But when you show the markers, it is much more gratifying, because you normalize the marker. So my patients are very happy when I show them the marker values. Maybe it is too simple, but I think it is much easier to handle a marker result than a BMD result under treatment, and I think that, although again, as I mentioned at the beginning of my talk, this has not been prospectively tested, when you have something to monitor osteoporosis treatment, you are more likely to monitor the compliance. The problem is that, whereas if you deal with hypertension, you can just take the blood pressure and are sure it is normalized, with osteoporosis, we really don't have much. So I think it is a valid approach and, regarding the non-compliance, I agree that there is partial compliance, and that is why I like to see studies where you have sub-optimal doses included. This is the case with the alendronate with 1 mg, and with our HRT when we had different doses between 25 and 75 mg, because many of the patients will take the drug 3 days per week, or who knows. So I think it reflects as much as possible the clinical situation.

Blumsohn OK, I'll stray outside the realms of science again. You are talking about giving nice numbers to patients that are going to make them compliant. We know that the marker responds in every patient, so why not just make up a number?

Delmas No, I don't think we should discuss that further.

Kleerekoper I strongly echo Pierre Delmas' concerns about waiting for the fracture data. I think that really has to be the outcome we need. I have a question for Dr Blumsohn to put into your model. As far as I understand it, when we look at markers of bone remodelling, we are looking at the effects of global skeletal remodelling. When we look at BMD in the lumbar spine, we are looking at a miniscule fraction of bone remodelling, even more minuscule when we look at the femoral neck or femoral trochanter or any other small segment that we measure by BMD. I am surprised that we can even begin to look for relationships between changes in global remodelling activity and changes in minuscule activity.

Blumsohn The aim of that model was to look at what sort of plausible results we might expect to get in studies looking at the relationship between changes in markers and changes in BMD, so I am not looking at the underlying biology of it. The model was asking what sort of plausible results we might expect from studies.

Kleerekoper Are there any data, and I think Pierre (Delmas) you published something on this, on total body bone mineral, for example, and markers?

Delmas We took the 24-h urinary excretion of bone resorption markers and, instead of correcting by creatinine, we corrected by total body bone mineral. The exercise was just to look at the fraction of the total skeleton which is resorbed per day.

Bauer This is a question for Dr Delmas. I agree with you that it seems that the Holy Grail here is the fracture data. But I would be interested in hearing your opinions a priori, that is before we know what the fracture data are, about how big an effect is going to be big enough to warrant routine use of markers to monitor therapy. Could you give me some idea here, and maybe some of the other clinicians on the panel could as well, about whether you are looking for an odds ratio of 0.5, or a 50% reduction, or some other meaningful reduction?

Delmas Of the fractures or the markers?

Bauer How strong a relationship between reduction in markers and fractures is going to be useful to clinicians to encourage them to use markers regularly?

Delmas Well, it depends how you express the data. What I would like to see is the relationship between the decrease in the marker and the probability of fracture. Now if you do that, of course, you don't integrate the severity of the disease, because you are just looking at the absolute probability, and we know that the effects of treatment in terms of probability of fracture are highly dependent on the prevalent fractures. Probably a 30% reduction, if you look at percentage reduction of fracture rate. But, ideally, the probability of fracture is a better endpoint. The problem is that it varies a lot in different trials, because of the severity of the disease.

Raisz One of the things that has not come up so far is the issue of non-absorption of bisphosphonates. Do we have any data on this? It always seems to me as one of the reasons I would do a marker, but I don't know how many people fail to absorb. I know that we have a large number of malabsorbing patients among our osteoporotics. Are there any data on malabsorption of bisphosphonates that we can use to guess whether this is an additional argument for markers?

Delmas I am not sure whether there are studies on that. What I have found, but again this is clinical experience, so probably this is irrelevant, is that when I don't see a decrease of the markers, especially if I repeat the marker and it is still there on a few occasions, I have ended up with patients who were talking calcium with the drug. So that is non-absorbers, but non-absorbers with bisphosphonates taken in the right way, I don't think we have the data.

Raisz We have the experience of patients who did not respond to oral bisphosphonates and who seemed to be tremendously careful in the way they did it, but responded to intravenous bisphosphonates. So I do think there is a subset, but I wish that I knew how big it was.

Delmas Did you look at the marker?

Raisz Yes, we showed that there were no marker responses to the oral drug on repeated occasions, but switched to intravenous and got a response.

Watts I've had the opposite experience published about a year ago, where we identified 25 patients who had failed to show an increase in bone density after oral etidronate, given cyclically. Not a single one of those 25 patients had an elevated resorption marker, yet most of them showed increases in bone density when changed to alendronate, so I think it may be a difference in bisphosphonates, rather than failure to absorb the drug.

Raisz I'm talking about failure to respond in terms of markers. Many people have low markers, and still respond to bisphosphonates.

Epstein I've got a question for Dr Blumsohn. I was fascinated by your model, because what we have heard is that you can have tertiles or quartiles of percentage changes in bone markers and judge a response on that. But what I think bothers us all is what the difference is between a 65% as opposed to a 75% as opposed to an 85% in BMD. It seems from your model, which you plotted, that you actually could look at changes from 70% to 80%, 60% to 70% etc., and when you plotted BMD on the vertical axis, it looked as though those data could be translated into increases in BMDs for the changes with those small ranges. Have you verified this model, looking at the alendronate data to see what the percentage change was in bone markers, and translated into increase in BMD?

Blumsohn I can't follow the question exactly.

Epstein In your last slide you showed a great relationship when you put your placebo and treated and HRT and you plotted percentage decrease in NTX and you showed increase in BMD.

Blumsohn That slide was just to make the point that if you do include your placebo group, then you would expect to get a correlation. Obviously, if you give a bisphosphonate with a much bigger response compared to the HRT, then you would expect a greater difference between your placebo group and your control, and therefore greater apparent prediction.

Epstein But it looks like plotted percentage changes from 80%, 60%, 40%, 20% and you had BMD and the slope was virtually a straight line. What I am asking is can you apply that in saying that if you have a 60% decrease of a bone marker that you could expect a certain amount of increase in BMD?

Blumsohn I think the critical thing is not the percentages and the marker. The critical thing is how much difference you get in the percentage decrease from one subject to another. That is, how much true difference you get after taking measurement noise into account. So I think that the actual response is irrelevant.

Epstein Has he tried his model then, and applied it to the existing data to see if that model holds true? That was really the question, to see if you could then use that. And the second question for the whole panel is: what is a fast loser? What is the definition of a fast loser?

Rosen I can just tell you, reviewing the literature, it's in the eyes of the beholder. It runs from 2% to 3.5% to . . .

Kanis The definition of a fast loser is Sol Epstein at the casino. The comment that I have relates to the fact that a lot of the issues seem to be black or white. Are you a responder or are you a non-responder? Do you have a least significant change or do you not? I just wonder, because this is such a grey area, where you have gradations, much the same as bone mineral, and you can't give a certificate to an individual that they will fracture on the basis of bone density—only say that the probability is higher. I wonder whether we look at the least significant change; for example, for the same reason people compute the 95% CI. Do we need to be 95% sure for all treatment decisions? For some decisions, maybe a 50% probability is adequate; for many clinical decisions, we take a 70% probability. And I just wonder whether for the least significant change for responders/non-responders we should be substituting the algorithm: what is the probability that somebody has had a change or somebody is a responder or non-responder to treatment? And I would be interested in your comments.

Blumsohn The least significant change could be calculated at any value you like, at a *P* value of 0.2 or 0.3 or 0.4. I think that, with any statistic, the trade-off is between sensitivity and specificity, and if you choose a *P* value of 0.2, then obviously you are going to detect almost everyone who is responding, but there are going to be a much larger number of people who are not responding who you are going to detect as responders. Where you draw that cut-off doesn't really depend on the statistics, but on the utility of false positives and false negatives.

Kanis I think that the point, that I am making is that the field seems to have chosen a 95% probability and I just can't see the logic of that.

Delmas I think it is a very good point, and that is why I said in principle I would like to see probabilities of fractures at the endpoint. The way to look at that in that type of analysis is really the positive predictive value, because that does not make a statement, it just tells us what is the probability. If we were to take 95% CI, we would never use DXA, for example.

Kanis Just to put that into application, if you have a given change in a marker, for example during treatment, what is the probability that that person is a responder or not a responder? That is something a clinician can relate to and take his own decisions on, depending on the clinical circumstances; and if this field is to survive for individual patients, then maybe this is one way to go.

Delmas Again, I think the only paper in which the data are expressed in that way is the paper of Patrick Garnero, with the alkaline phosphatase looking at the percentage, where you have probabilities and you can go from the 10% to 90% probability[†].

Rosen There are some data with probability of bone loss with baseline markers; both Ross and others expressed data as probabilities on the basis of baseline markers.

Chesnut III A quick update on calcitonin. In the 5-year trial at 1 year, the serum CTX demonstrated an approximate 23% reduction in the 200-unit dose. New data that we have from an ongoing trial, the so-called QUEST study, shows at 2 weeks an approximate 19% reduction in the serum NTX which is significantly different from baseline. This is at 2 weeks and is significantly different from the calcium control group. This raises the question, and I'd like to come back to a point from Susan Greenspan and Pierre Delmas as well, as to what does

[†] See Garnero P, Darte C, Delmas PD, A model to monitor the efficacy of alendronate treatment in women with osteoporosis using a biochemical marker of bone turnover. *Bone* 1999; 24: 603–9.

indeed reduce fracture risk. We have modest effects, I would say, on both markers and bone density (quantity) with calcitonin and raloxifen. In addition to these factors, we also need to consider an effect of therapies on bone quality to reduce fracture risk. Perhaps we can address this again in the subsequent session.

18

Assessment of fracture risk from bone mineral density and bone markers

Olof Johnell, Anders Odén, Alison Dawson and John A Kanis

Introduction • Bone mineral density • Biochemical markers • Application of risk factors • Long-term risk assessment • Conclusions • References

INTRODUCTION

The diagnosis of osteoporosis has been established by a WHO group as a bone mineral density (BMD) value that lies <−2.5 SD below the average value for young adult women.[1] The use of the T-score is most appropriate for dual energy X-ray absorptiometry (DXA) at the hip.[2] The use of the T-score alone for risk assessment is, however, problematic, since absolute risk for any T-score varies with age. Moreover, other risk factors also contribute to risk independently of age and BMD. Examples include the measurement of ultrasound and biochemical estimates of bone turnover. Other strong risk factors are age, body mass index, maternal family history of hip fractures and prior fragility fractures. A combination of these can be used to estimate long-term risks of hip fractures or other fractures to determine who should be treated. This chapter describes the principles of combining risk factors, with particular reference to the biochemical markers.

BONE MINERAL DENSITY

Several prospective studies have shown that BMD measurements predict fracture risk. The site with the highest predictive value is the hip, particularly for the prediction of hip fracture. A meta-analysis estimated that the risk of hip fracture had a 2.6-fold increase for each SD decrease in bone mass density at the femoral neck.[3] Bone mineral assessment at the hip to predict fracture risk is as good as blood pressure measurements to predict stroke and better than cholesterol to predict myocardial infarction in men.[1] Ultrasound has been shown to give information on fracture risk,[5] but the WHO criterion <−2.5 SD cannot be applied to ultrasound measurement.[2] Instead, other thresholds should be derived, preferably on the basis of fracture risk.

Despite these performance characteristics, assessment with ultrasound or with BMD has a high specificity but a relatively low sensitivity.[3,4] Thus, the majority of individuals who sustain a fracture during their lifetime will not be identified as being high-risk patients from a BMD measurement. This is one of the reasons why BMD assessment alone cannot be used for screening. In contrast, however, BMD assessment, particularly at the hip, is the cornerstone for a case-finding strategy.

BIOCHEMICAL MARKERS

A number of cross-sectional and prospective studies have shown that biochemical markers of bone formation and bone resorption have some utility in fracture prediction. For example, Ravn et al[6] found an association with high turnover and prevalence of spine fractures. Van Daele et al[7] found an association with pyridinoline and hip fracture risk. In a study by Garnero et al,[8] the odds ratios for hip fracture with a 1 SD increase were 1.4 (1.1–1.17) for free deoxypyridinoline (DPD) and 1.3 (1.0–1.6) for C-terminal cross-linking telopeptide of type I collagen (CTX). If instead the highest quartile was examined, the odds ratio for CTX was 2.1 (1.3–3.3), and that for free DPD was 1.5 (0.9–2.25).

The predictive value of markers appears to be independent of bone mass. Garnero et al[9] found that a combination of high CTX and low BMD had an odds ratio of 4.8 for hip fracture. Vergnaud et al[10] showed that undercarboxylated osteocalcin predicted hip fracture risk in the highest quartile with an odds ratio of 1.9 (1.2–3.0), a risk that persisted after adjustment for BMD (1.8 (1.1–3.0)). A combination of the lowest quartile of BMD and the highest quartile of undercarboxylated osteocalcin was associated with an odds ratio of 5.5 (2.7–11.2). In another study, Garnero et al[11] concluded that, if DXA is not available, the combination of history of fractures and urinary CTX might be used, since this performs as well as hip BMD in assessing hip fracture risk in elderly women. Ross et al[12] found an association between later spine and non-spine fracture and high serum bone alkaline phosphatase (ALP), with an odds ratio of 1.5–1.88 per SD change which persisted after adjustment for BMD (1.49–1.80). In the same study, urinary CTX predicted fracture with an odds ratio of 1.43–1.84 per SD change that also remained after adjustment for BMD (1.33–1.70/SD). These studies indicate that bone markers provide information on fracture risk independently of BMD and might add to the fracture risk assessment given by BMD alone. Other risk factors that might be used in an algorithm to identify high-risk patients are: low body weight, smoking, family history of fragility fracture and prior osteoporotic fractures. Several of these, e.g. prior fracture, the geometry of the hip and tests of balance and visual acuity, are also relatively independent of BMD[13] (Table 18.1).

It is important not only to focus on hip fractures, since they occur in the most elderly, but also to look at other osteoporotic fractures such as wrist fractures, spine fractures and shoulder fractures which are presented at an earlier age. Several of the risk factors are also associated with the other osteoporotic fractures.

APPLICATION OF RISK FACTORS

The assessment of a number of independent risk factors permits the identification of individuals at higher risk than is possible with the use of a single risk factor. In order to combine risks, the independent contributions of the factors need to be calculated. In the case of BMD at the hip, the risk of hip fracture increases 2.6-fold for each SD decrease in BMD. Thus an individual with a Z-score of -1 has a 2.6-fold higher risk than an individual of the same age and an average bone density. The risk has, however, to be adjusted to the population risk.[14] Hip fracture risk increases logarithmically with decreasing BMD, but BMD is normally distributed. Thus individuals with a BMD equal to the mean have a risk of hip fracture that is lower than the average risk for their age. It can be calculated[14] that the individual with a Z-score of -1 has a 1.65-fold increase in hip fracture risk compared to that of the general population.

This estimate can be used with other independent risks. As mentioned above, high urinary CTX is associated with a 1.9-fold increased

Table 18.1 Examples of relative risks of hip fracture in women with and without adjustment for bone mineral density (BMD).

	Relative risk	
	Crude	Adjusted
Hip BMD 1 SD below mean population value	2.6	
Non-carboxylated osteocalcin above normal	2.0	1.8
Biochemical index of bone resorption (above premenopausal range)	2.2	2.0
Prior fragility fracture after the age of 50 years	1.4	1.3
Body weight below 57.8 kg	1.8	1.4
Hip axis length 1 SD above population mean value	2.7	1.9
First-degree relative aged 50 years or more with a history of fragility fractures	1.7	1.5
Maternal family history of hip fracture	2.0	1.9
Current cigarette smoking	1.9	1.2
Poor visual acuity (<2/10)	2.0	2.0
Low gait speed (1 SD decrease)	1.4	1.3
Increase in body sway (1 SD)	1.9	1.7

risk of hip fracture after adjustment for BMD.[11] Among elderly women, 23% have CTX values higher than those of premenopausal women. The risk ratio of 1.9 compares those with high versus those with normal CTX. Thus the risk of hip fracture in the presence of high CTX compared to the risk of the general population is RR/CP $\times$ RR + $(1 - p)$, where p is the prevalence of the risk factor (23%) and RR is the unadjusted value (1.9). In this example, the population relative risk is 1.57. Thus the risk of hip fracture associated with a Z-score of -1 SD and a high CTX is $1.65 \times 1.57 = 2.59$, compared to the average risk for age.

An alternative approach is to determine the gradient of risk with combinations of risk factors; for example, if one risk factor alone gives a gradient of risk of, say, 1.5/SD of measurement, a second risk factor added to BMD gives a new distribution of risk where the gradient/SD exceeds 1.5.

The higher the gradient of risk, the higher the sensitivity of the combined test. The increase in sensitivity is, however, non-linear, and the increases in sensitivity with gradients of risk >3/SD are small. Thus, in practice, relatively few risk factors can be considered in risk algorithms.

LONG-TERM RISK ASSESSMENT

The assessment of absolute risk rather than relative risk is appropriate for the assessment of patients. In this context, long-term risks are useful, since they depend on age and life-expectancy as well as the current risk. Absolute risks of hip fracture have been well characterized in Sweden. The average lifetime risk of hip fracture is 22.7% at the age of 50 years and 18.9% at the age of 85 years.[15] Algorithms have been published[13] which determine the effect of

Table 18.2 Estimates of positive predictive value (PPV), sensitivity and specificity of measurements to predict hip fracture over 15 years or to death in women aged 50 years or 65 years, assuming that 15% of the population at highest risk would be identified.[13]

	Women aged 50 years			Women aged 65 years		
Gradient of risk (RR/SD)	PPV (%)	Sensitivity (%)	Specificity (%)	PPV (%)	Sensitivity (%)	Specificity (%)
1.5	2.2	26.5	85.1	12.8	25.7	85.9
2.0	3.1	36.4	85.3	17.0	34.1	86.5
2.5	3.8	44.5	85.4	20.1	40.4	87.0
3.0	4.3	51.1	85.5	22.3	44.8	87.4
4.0	5.1	60.6	85.6	24.7	49.7	87.8
5.0	5.6	66.1	85.7	25.6	51.4	87.9
6.0	5.8	69.0	85.7	25.6	51.4	87.9

increase in relative risk on lifetime risk (Table 18.2). They can be used to illustrate examples of combinations of tests from the EPIDOS study.[11] At the age of 81 years, the average lifetime risk of hip fractures is 21% in Swedish women. By selecting women within the lowest quintiles for BMD, the lifetime risk for all hip fractures at the age of 81 is 32%; for low ultrasound attenuation at the heel, the risk is 39%; and for CTX above the premenopausal range, the lifetime risk is 34%. There is a marked effect of combining the risk factors; for the combination of low hip BMD and high CTX, the lifetime risk is 45%. Similarly, the combination of CTX with a history of fracture has a lifetime risk of 52% at the age of 81 years. These data indicate that it is useful to add risk factors to get a higher lifetime risk of fracture and a high enough risk to indicate treatment.

It is important to recognize that long-term risks are slightly different in different countries, depending on absolute risk, mortality and mortality trends. It is also important to calculate the lifetime risk for other fractures alone and in combination; this has recently been done for Swedish women and men. The lifetime risk of a clinical spine fracture is about 18%, and the figure is similar for wrist fractures in women.

The use of lifetime risks may not be appropriate unless lifetime interventions are envisaged. A time frame of 10 years or so accommodates the duration of most treatments and the offset of effect that occurs once treatment is stopped.

The use of multiple risk factors can augment a case-finding strategy by identifying individuals at risk higher than that provided by a single risk indicator. An important implication of this finding is that intervention thresholds should be determined by the computation of absolute risk. Moreover, thresholds will need to be set according to total risk rather than any one threshold of measurement from a single test (e.g. BMD). For hip fracture, a 10% 10-year risk may be appropriate, but the challenge for the future will be to accommodate the many other types of fracture into these concepts.

CONCLUSIONS

The use of multiple risk factors substantially increases the sensitivity without a trade-off of specificity and thus gives better results than the use of BMD alone. This approach has been used in selecting individuals for intervention in cardiovascular disease. In Europe, a programme considering smoking, blood pressure, diabetes and serum cholesterol together could identify high-risk patients (>20% 5-year risk), whereas any one of these risk factors alone has a lower gradient of risk than these combined. It is also important to use the risk factors that are appropriate for the planned treatment. For example, falls are a strong risk factor for hip fracture, but patients identified by falling may not respond to pharmacological interventions.

It is likely that relatively few risk factors will be needed to optimize the risk assessment, and most of them have been identified today. If a 10% 10-year risk of hip fractures is considered to be unacceptable, then it is possible to calculate which combination of risk factors exceeds that threshold. This approach of assessing absolute fracture risk will be important to optimize the selection of individuals for treatment and may also, in future, be used in screening strategies.

REFERENCES

1. World Health Organization, *Assessment of fracture risk and its application to screening for postmenopausal osteoporosis.* WHO Technical Report Series, Geneva: WHO, 1994: 843.
2. Kanis JA, Glüer C-C, An update on the diagnosis and assessment of osteoporosis with densitometry. Committee of Scientific Advisors, International Osteoporosis Foundation. *Osteoporos Int* 2000; 11: 192–202.
3. Marshall D, Johnell O, Wdel H, Meta-analysis of how well measures of bone mineral density predict occurrence of osteoporotic fractures. *BMJ* 1996; 312: 1254–9.
4. Johnell O, Kanis JA, Odén A et al, Prediction of fracture from bone mineral density measurements overestimates risk. *J Bone Miner Res* 1998; 8(suppl 3): 29.
5. Hans D, Dargent P, Schott AM et al, Ultrasound heel measurements to predict hip fracture in elderly women: the EPIDOS prospective study, for the EPIDOS prospective study group. *Lancet* 1996; 348: 511–14.
6. Ravn P, Rix M, Andreassen H et al, High bone turnover is associated with low bone mass and spinal fracture in postmenopausal women. *Calcif Tissue Int* 1997; 60: 255–60.
7. van Daele PLA, Seibel MJ, Burger H et al, Case–control analysis of bone resorption markers, disability, and hip fracture risk: the Rotterdam study. *BMJ* 1996; 312: 482–3.
8. Garnero P, Hausherr E, Chapuy M-C et al, Markers of bone resorption predict hip fracture in elderly women: the EPIDOS prospective study. *J Bone Miner Res* 1996; 11: 1531–8.
9. Garnero P, Sornay E, Chapuy M-C, Delmas PD, Increased bone turnover in late postmenopausal women is a major determinant of osteoporosis. *J Bone Miner Res* 1996; 11: 337–49.
10. Vergnaud P, Garnero P, Meunier PJ et al, Undercarboxylated osteocalcin measured with a specific immunoassay predicts hip fracture in elderly women: the EPIDOS study. *J Clin Endocrinol Metab* 1997; 82: 719–24.
11. Garnero P, Dargent-Molina P, Hans D et al, Do markers of bone resorption add to bone mineral density and ultrasonographic heel measurement for the prediction of hip fracture in elderly women? The EPIDOS prospective study. *Osteoporos Int* 1998; 8: 563–9.
12. Ross PD, Kress BC, Parson RE et al, Serum bone alkaline phosphatase and calcaneus bone density predict fractures: a prospective study. *Osteoporos Int* 2000; 11: 7682.
13. Kanis JA, Johnell O, Odén A et al, Risk of hip fracture derived from relative risks: an analysis applied to the population of Sweden. *Osteoporos Int* 2000; 11: 120–7.
14. Kanis JA, Johnell O, Odén A et al, Combining bone mineral density measurement and other risk factors in osteoporosis. In: Aso T, Yanaihara T, Fujimoto S, eds. The Menopause at the Millennium. New York, London: Parthenon Publishing Group, 2000: 352–8.
15. Odén A, Dawson A, Dere W et al, Lifetime risk of hip fracture is underestimated. *Osteoporos Int* 1998; 8: 559–603.

19

Markers of bone turnover, endogenous hormones, rate of bone loss and fracture risk in the OFELY study

Patrick Garnero, Elisabeth Sornay-Rendu and Pierre D Delmas

Summary • Introduction • Materials and methods • Results • Discussion • References

SUMMARY

Increased bone turnover as assessed by biochemical markers has been suggested to be associated with increased bone loss and fracture risk in postmenopausal women. The mechanisms responsible for increased bone turnover in some women leading to increased fracture risk are however unclear. In a large prospective epidemiological study including 1039 women aged 30–89 years (OFELY study), we analysed the age-related changes in bone turnover and the relationships between baseline levels of a panel of biochemical markers, rate of postmenopausal bone loss and fracture risk. We also investigated whether levels of endogenous hormones mediated the association between increased bone turnover and fracture risk.

In a cohort of 305 untreated postmenopausal women, we found that increased bone formation and resorption markers were associated with faster forearm bone loss assessed prospectively by four repeated bone mineral density (BMD) measurements using dual X-ray absorptiometry over 4 years ($p < 0.0001$). Women with levels of bone markers at baseline 2 SD above the mean of premenopausal women, had a rate of bone loss that was 2–6-fold higher than women with a low turnover (p: 0.01–0.0001) according to the marker. During an average 5 years follow-up 58 incident fractures were recorded in 55 women. Markers of bone formation, except bone specific alkaline phosphatase, were not predictive of fracture risk. On the other hand, relative risks for fracture of women having their urinary and serum CTX in the highest quartile, were around 2 and remained significant after adjustment for BMD. Although low serum levels of estradiol and dehydroepiandrosterone sulfate were associated with increased fracture risk, adjustment of biochemical markers by hormone levels did not significantly alter the relationships between increased bone turnover and fracture risk.

We conclude that high levels of some biochemical markers of bone turnover are associated with faster bone loss and increased risk of osteoporotic fracture in postmenopausal women independently of BMD and of endogenous hormones.

INTRODUCTION

The mechanisms leading to increased bone loss and skeletal fragility in women with postmenopausal osteoporosis are still poorly understood. Increased bone resorption, low serum oestradiol and high serum sex-hormone-binding globulin (SHBG) have recently been reported as predictors of vertebral and hip fractures in elderly women.[1–5] However, no study has investigated in the same population the relationships between bone turnover, hormonal levels and fracture risk. In addition, most prospective studies looking at the determinants of fractures have been performed in elderly women (because of the high incidence of osteoporotic fractures), but the role of the residual secretion of steroids is more likely to play a role in younger postmenopausal women.

The OFELY study, which included a large population of healthy women from 30 to 89 years of age followed prospectively with annual measurements of bone mineral density (BMD) and registration of incident fracture, was used to test the relationships between baseline levels of bone turnover markers, rate of bone loss and fracture risk.

MATERIALS AND METHODS

Subjects

At baseline, the cohort of this study comprised 1039 women, 31–89 years of age, randomly selected from the regional section of a health insurance company (Mutuelle Générale de l'Education Nationale). According to the hypothesis tested, different subgroups of this population were analysed.

Age- and menopause-related changes in bone turnover

After exclusion of women with bone diseases or treatments that could interfere with bone metabolism, 766 healthy women could be analysed, including 243 premenopausal women characterized by normal serum follicle-stimulating hormone (FSH), 45 perimenopausal women characterized by increased FSH levels, and 434 postmenopausal women (absence of menses for at least 1 year). Details of the population are given elsewhere.[6]

Bone turnover and rate of bone loss

The rate of bone loss was assessed in 305 postmenopausal women (mean age 64 years; range 50–88 years) with no treatment or disease that could interfere with bone metabolism during the 4 years of follow-up. The rate of bone loss at the mid- and distal radius was assessed by four repeated BMD measurements using dual X-ray absorptiometry (DXA) performed at baseline, year 2, year 3 and year 4.[7]

Bone turnover markers, endogenous hormones and fracture risk

We analysed the relationships between baseline measurements of markers of bone turnover, endogenous hormones and the risk of fracture by comparing levels of biochemical markers in the 55 women who subsequently had a fracture (20 vertebral and 35 first peripheral fractures) to those in the 380 women who did not fracture during a 5-year follow-up.[7]

Measurement of bone turnover markers and endogenous hormones at baseline

For each woman, fasting blood samples were collected at baseline between 7:30 a.m. and 9:30 a.m. for measurements of hormones and serum bone turnover markers. First and second morning void and total 24-h urinary excretion were also collected at baseline without any preservative for measuring urinary resorption markers. Serum and urine samples were stored frozen at −80°C until assayed.

Bone formation markers

Serum osteocalcin was measured with a human immunoradiometric assay (IRMA) which uses two monoclonal antibodies recognizing, respectively, the 5–13 and 43–49 sequences of the molecule, and purified intact human bone osteocalcin as a standard (ELSA-OST-NAT, Cis Biointernational). Serum bone-specific alkaline phosphatase (bone ALP) was measured with a human-specific IRMA using two monoclonal antibodies directed against the human bone isoenzyme (Ostase, Hybritech). Serum C-terminal propeptide of type I collagen (serum PICP) was measured by a two-site enzyme-linked immunoassay (ELISA) (Prolagen-C, MetraBiosystems, Palo Alto, CA, USA). Serum intact N-terminal propeptide of type I collagen (PINP) was measured by a radioimmunoassay (Intact PINP, Farmos Diagnostica, Upsalla, Finland).

Bone resorption markers

Bone resorption markers included urinary type I collagen N-telopeptides (NTX, Osteomark, Ostex International, Inc., Seattle, WA, USA), isomerized type I collagen C-telopeptides (β CTX, CrossLaps ELISA, CIS Biointernational, Gif/Yvette, France) and urinary free deoxypyridinoline (urinary free D-Pyr, Pyrilinks-D, MetraBiosystem, USA). We also measured serum β isomerized type I collagen C-telopeptide (serum CTX) by two site immunoassays using monoclonal antibodies raised against an amino acid sequence specific for a part of the C-telopeptide of the α 1 chain of type I collagen (Glu-Lys-Ala-His-β Asp-Gly-Gly-Arg). Serum CTX levels were assessed by two different techniques: a manual ELISA (Serum CrossLaps one step, Osteometer Biotech, Ballerup, Denmark) and a recently developed automatic analyser (Elecsys, Roche Diagnostics).

Hormones

Endogenous hormones included serum total estrone, estradiol, testosterone and dehydroepiandrosterone sulfate (DHEA sulfate). These hormones were measured by radioimmunoassays after diethylether extraction or dilution (for DHEA sulfate). Serum intact parathyroid hormone (intact PTH) was measured with a two-site immunochemiluminometric assay (Magic Lite PTH, Ciba-Corning, USA) and 25 hydroxyvitamin D (25 (OH) D) level was assessed by a radiobinding assay kit (Bühlmann Laboratories AG, Switzerland).

Sex-hormone binding globulin was measured by IRMA (^{125}I SBP coatria, Bio-Mérieux, Marcy l'Etoile, France).

Bone mineral density

BMD of the spine (L1–L4), femoral neck, distal radius and whole body was determined by DXA on a QDR 2000 device (Hologic, USA) at baseline. The short-term in vivo precision of DXA was 0.9%, 1.0%, 0.6% and 1.0% respectively for lumbar spine, total hip, distal radius and whole body.

Statistical analysis

In calculating the rate of change in BMD, it was assumed that the expected change is linear over the 4-year follow-up period. BMD was regressed on time to yield a rate of change in BMD for each subject. Under this model, the annual percentage of change in BMD for each subject was then derived by dividing the regression slope by the intercept at time 0. We also calculated the absolute rate of BMD change by using raw values of the regression slope. Preliminary analyses showed that both expressions of the rate of change gave similar results, and data are thus presented using the annual percentage change. The mean +2 SD of premenopausal controls was used as a cut-off limit between low- and high-turnover groups. The premenopausal range was obtained from the values of 134 healthy premenopausal women, 35–55 years of age, drawn from the same Ofely cohort and described elsewhere.[6]

The relationships between baseline levels of bone turnover markers and hormones were analysed by logistic-regression analysis after adjustment for potential confounding variables such as age, body weight, prevalent fractures and physical activity. The upper limits of the premenopausal range for biochemical markers of bone turnover were determined from the mean and standard deviation (SD) values obtained from the baseline measured in the 134 healthy premenopausal women from the same cohort.[6]

All analyses were adjusted for age and performed using the Statistical Analysis Software (SAS, Cary, NC, USA).

RESULTS

Age- and menopause-related changes in serum CTX in healthy women

In the 254 premenopausal women with regular menses and normal FSH levels (mean age 39 years; range 30–58 years), no significant changes with age of bone resorption markers were observed (Fig. 19.1). Bone formation markers decreased slightly from 30 to 35 years and showed no change thereafter. In postmenopausal women, no significant changes with age or year since menopause were observed. When levels of serum CTX determined by Elecsys in premenopausal women, perimenopausal women ($N = 45$, characterized by increased FSH levels) and postmenopausal women (absence of menses for at least 1 year) were compared, we observed a 38% increase in levels in perimenopausal women ($P = 0.004$) and a 84% increase in levels in postmenopausal women ($P < 0.0001$) compared to premenopausal women. Such a pattern was also observed for U-NTX and U-CTX, and serum bone ALP.[6]

Prediction of rate of bone loss by bone markers

Baseline levels of all bone markers except serum bone ALP correlated significantly with the rate of BMD changes at both mid- and distal radius, high levels of markers being associated with faster bone loss (R values between -0.16 and -0.30; $P = 0.008$ to <0.0001). The regression and correlation coefficients were consistently higher in early postmenopausal women, in whom the rate of bone loss is greater (R: -0.3 to -0.53). For urinary resorption markers, the correlation was similar for all types of urine samples, i.e. first morning void, second morning void and 24-h samples. As shown in Table 19.1, when women were separated into high- and low-turnover groups with a cut-off between the two groups established as the mean + 2 SD of the premenopausal women of the same cohort, the 4-year rate of loss in women with high bone turnover at baseline was 2–6-fold faster than in women with levels within the premenopausal range.

Prediction of fracture risk

At baseline, women with fractures were slightly older (67 versus 64 years, $P = 0.004$), had a lower physical activity score (12.1 versus 13.8, $P = 0.004$), and had lower BMD at all skeletal sites. The proportion of women with prevalent fracture was also higher among those women who had an incident fracture (32% versus 14%, $P = 0.002$). Women with levels in the highest quartile of four bone resorption markers, including urinary free DPD (RR (95% CI): 1.8 (0.95–3.4)), (U-NTX) (1.7 (0.92–3.2)), urinary (2.3 (1.3–4.1)) and S-CTX by ELISA (2.1 (1.2–3.8)) and Elecsys (1.9 (1.02–3.2)) had about a 2-fold increased risk of fractures compared to women with levels within the premenopausal range. Among markers of bone formation, only serum

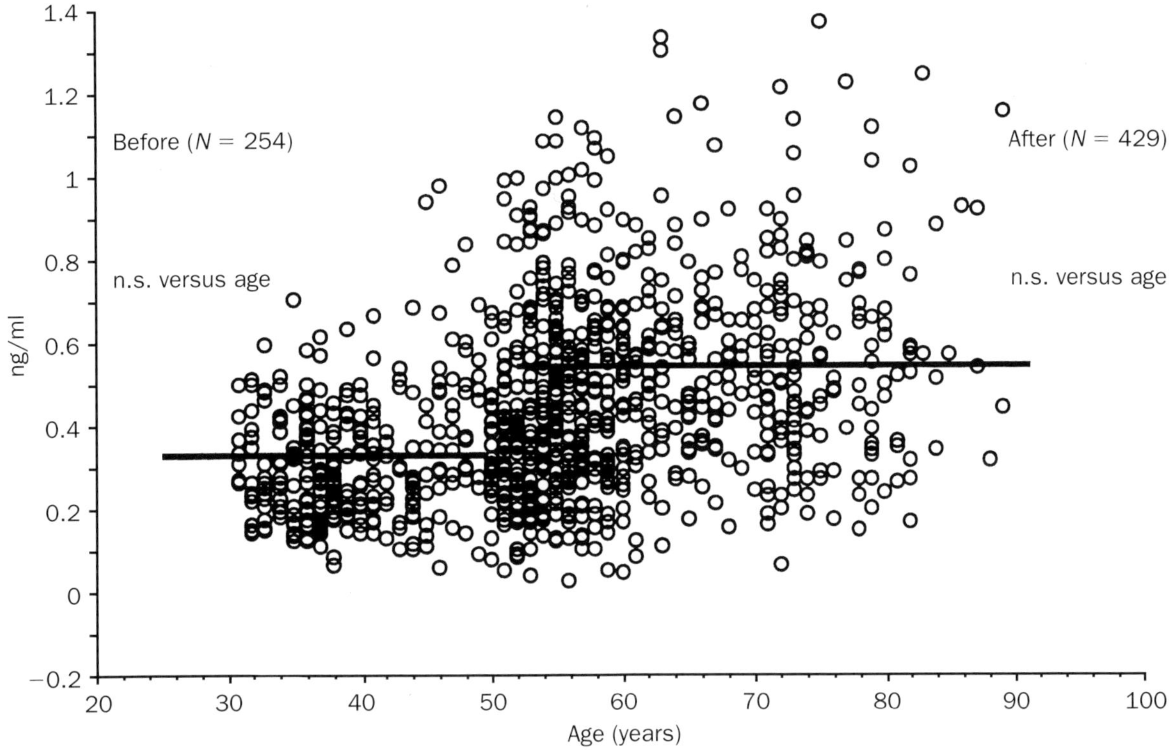

Fig. 19.1 Age-related changes in serum CTX levels measured by automated analyser in healthy women: the OFELY study.

bone ALP was predictive of fracture risk (RR (95% CI): 2.4 (1.3–4.2)).

Women with serum levels of oestradiol and (DHEA sulphate) in the lowest quartile had relative risks of fracture of 2.2 (1.2–4.0) and 2.1 (1.2–3.8) respectively. Increased levels of SHBG and intact PTH were moderately associated with an increased risk of fracture. Adjustment of biochemical markers by hormone levels did not significantly alter the results. Women with both high bone resorption markers and low oestradiol (or low DHEA sulphate) had a higher risk of fracture, with a relative risk of 3.0–3.3 ($P < 0.001$).

After adjustment for BMD of the hip, spine, radius or total body, bone markers and hormones were still predictive of fracture risk, with similar relative risks.

DISCUSSION

In this large, well-characterized population of healthy women, using a panel of sensitive and specific markers, we found the following:

- Bone turnover begins to increase before the menopause, during the perimenopausal period.
- Bone turnover is increased in post-menopausal women, and high levels are maintained in elderly women.
- Levels of bone resorption markers above the premenopausal range are associated with increased rate of bone loss and fracture risk independently of BMD.
- Low serum oestradiol, low DHEA sulphate, high SHBG and high PTH are associated with increased risk of osteoporotic fracture

Table 19.1 Four-year rate of bone loss in 305 healthy untreated postmenopausal women with high or low bone turnover at baseline.

Marker	% high	High turnover	Low turnover	*P*
Osteocalcin	24.7	−2.24 ± 4.1	−0.75 ± 3.2	0.0015
Bone ALP	27.5	−1.78 ± 4.14	−0.85 ± 3.23	0.06
PINP	14.3	−2.83 ± 4.54	−0.81 ± 3.2	0.0004
U-NTX	43.5	−2.06 ± 3.94	−0.38 ± 2.97	<0.0001
U-CTX	29	−2.11 ± 4.14	−0.71 ± 3.16	0.002
S-CTX (ELISA)	22	−2.78 ± 4.23	−0.63 ± 3.15	<0.0001
S-CTX (Elecsys)	39.7	−2.22 ± 4.00	−0.37 ± 2.96	<0.0001

High bone turnover was defined as levels 2 SD above the mean of premenopausal women from the same cohort. Results are the mean (±SD) of the 4-year rate of bone loss expressed in per cent.

in postmenopausal women. These hormones did not account significantly for the increase in bone turnover.

Both bone formation and bone resorption markers were associated with increased bone loss at the forearm with correlation coefficients which are in the same range as those previously reported in smaller studies investigating bone loss at the spine and radius.[8–10] The correlation coefficients were rather low, suggesting that biochemical markers would not be useful to accurately predict the magnitude of bone loss in a single individual. However, it should be pointed out that in these correlation analyses, changes in BMD by DXA is usually considered as the gold standard to estimate rate of bone loss, which may actually not be the case. Indeed, when the rate of bone loss is assessed over periods of 2–4 years, the amount of bone loss is in the same magnitude of the precision error of repeated BMD measurements in a single individual, precluding an accurate estimation of the rate of bone loss. This is actually illustrated by the fact that although osteoporosis is believed to be a generalized disease, rates of bone loss assessed by BMD changes at different skeletal sites usually do not exceed 0.3-0.4[11] i.e. in the same range as the correlation observed between bone markers and rate of bone loss. This technical limitation in estimating rate of bone loss from repeated DXA measurements, together with the variability of bone marker levels may account for the conflicting data which have been published. A more relevant approach would be, instead of considering bone markers as a surrogate for rate of bone loss, to use them as a prognostic factor for fracture risk per se. Indeed we found that women with levels of bone turnover markers 2 SDs above the premenopausal mean lost over the subsequent 4 years bone at the radius 2–6-fold more rapidly—a result in line with a retrospective 13 year study from Ross et al[12]—and had about a 2-fold higher risk of fracture than individuals with normal levels. Biochemical markers of bone resorption more consistently predicted fracture risk than bone formation ones. However increased levels of bone alkaline phosphatase was also associated with increased fracture risk in agreement with a recent prospective study[13] showing that this marker predicted the risk of fractures (33 vertebral and 25 non-spinal fractures) that occurred during 2.7 years follow-up. Of interest the predictive value of increased bone turnover markers was significant after adjustment for the level of BMD. Thus, an increased bone turnover, as detected by sensitive biochemical markers, is likely to be related to decreased bone strength as a result of two mechanisms. First, a sustained increase in bone turnover after the menopause will be associated with a greater bone loss as shown in that study and subsequently a lower level of bone mass, which is major determinant of reduced bone strength. Second, increased bone resorption above the upper limit of the normal premenopausal range threshold appears to be harmful for bone strength independently of bone mass, possibly by inducing perforation of trabeculae, a major component of bone strength.

In this study we reported for the first time that serum CTX, a marker of bone resorption, was predictive of fracture risk. Interestingly, provided that sampling is performed at standardized time in the morning in fasting state, the predictive value of serum CTX was similar to that urinary CTX. Measurements performed by a new analyser with improved precision and high output gave a similar predictive value as the manual ELISA, suggesting that this automated technique would be useful for the investigation of patients with osteoporosis in routine clinical laboratory.

The mechanisms explaining why some postmenopausal women have an increased bone turnover resulting in increased fracture risk remain unclear. In that study, we found that low serum estradiol is associated with increased fracture risk, findings which extend to younger postmenopausal women the results recently reported by Cummings et al[3] in elderly women. Amongst the other hormones investigated we also found that decreased levels of serum DHEA sulfate and to a lesser extent increased PTH and SHBG were also associated with increased fracture risk. However, after adjustments for these hormones, increased levels of biochemical markers of bone turnover still predict osteoporotic fractures with very similar relative risks. These findings suggest that the association between high bone turnover and increased risk of fracture may not be explained by the association between decreased hormone levels and increased fracture risk. Further studies would be needed to elucidate the determinants of increased bone turnover in a fraction of postmenopausal women.

In conclusion, data obtained in this large study using a panel a specific biochemical markers showed that an increased bone turnover in postmenopausal women is a risk factor for fracture which is independent of the level of bone mass. The role of bone markers in the assessment of fracture risk of individual postmenopausal women can be envisaged, probably for identifying patients at high risk together with other important risk factors such as low BMD, personal and maternal history of fracture and low body weight, although further studies are necessary to define the optimal strategy.

REFERENCES

1. Ettinger B, Pressman A, Sklarin P, Bauer D, Cauley JA, Cummings SR, Associations between low levels of serum estradiol, bone density, and fractures among elderly women: the study of osteoporotic fractures. *J Clin Endocrinol Metab* 1998; 83: 2239–43.
2. Stone K, Bauer DC, Black DM, Sklarin P, Ensrud KE, Cummings SR, Hormonal predictors of bone loss in elderly women: a prospective study. *J Bone Miner Res* 1998; 13: 1167–74.
3. Cummings SR, Browner WS, Bauer DB et al, Endogeneous hormones and the risk of hip and vertebral fractures among older women. *N Engl J Med* 1998; 339: 733–8.
4. Garnero P, Hausherr E, Chapuy M-C et al, Markers of bone resorption predict hip fracture risk in elderly women: the EPIDOS Prospective Study. *J Bone Miner Res* 1996; 11: 1531–8.
5. van Daele PLA, Seibel MJ, Burger H et al, Case–control analysis of bone-resorption markers, disability, and hip fracture: the Rotterdam study. *BMJ* 1996; 312: 482–3.
6. Garnero P, Sornay-Rendu E, Chapuy MC, Delmas PD, Increased bone turnover in late postmenopausal women is a major determinant of osteoporosis *J Bone Miner Res* 1996; 11: 827–34.
7. Garnero P, Sornay-Rendu E, Duboeuf F, Delmas PD, Markers of bone turnover predict postmenopausal forearm bone loss over 4 years: the Ofely Study. *J Bone Miner Res* 1999; 14: 1614–21.

8. Johansen JS, Riis BJ, Delmas PD et al, Plasma BGP: an indicator of spontaneous bone loss and effect of estrogen treatment in postmenopausal women. *Eur J Clin Invest* 1988; 18: 191–5.
9. Slemenda C, Longcope C, Peacock M, Hui S, Johnston CC, Sex steroids, bone mass, and bone loss: a prospective study of pre-, peri-, and postmenopausal women. *J Clin Invest* 1996; 97: 14–21.
10. Uebelhart D, Schlemmer A, Johansen J, Gineyts E, Christiansen C, Delmas PD, Effect of menopause and hormone replacement therapy on the urinary excretion of pyridinium crosslinks. *J Clin Endocrinol Metab* 1991; 72: 367–73.
11. Keen RW, Nguyen T, Sobnack et al, Can biochemical markers predict bone loss at the hip and spine? A 4-year prospective study of 141 early postmenopausal women. *Osteoporos Int* 1996; 6: 399–406.
12. Ross PD, Knowlton W, Rapid bone loss is associated with increased levels of biochemical markers. *J Bone Miner Res* 1998; 13: 297–302.
13. Ross PD, Kress BC, Parson RE, Wasnich RD, Armour KA, Mizrahi LA, Serum bone alkaline phosphatase and calcaneus bone density predict fractures: a prospective study. *Osteoporos Int* 2000; 11: 76–82.

20

Prediction of hip fracture with markers of bone turnover in the EPIDOS study

Roland D Chapurlat, Patrick Garnero, Pierre J Meunier, Gerard Bréart and Pierre D Delmas

Summary • Introduction • Methods • Results • Discussion • References

SUMMARY

Markers of bone turnover have been advocated as good predictors of the risk of osteoporotic fractures in elderly women. In particular, markers of bone turnover were studied in the EPIDOS cohort, which is a prospective study of risk factors for hip fracture in French healthy ambulatory elderly women. The predictive value of bone markers was studied within a nested case–control design, each fracture case being randomly assigned three controls matched for age and time of recruitment. Two analyses were conducted, the interim one with 109 fracture cases, and the second one with 212 fracture cases. Markers of bone formation, except undercarboxylated osteocalcin, were not predictors of hip fracture. On the other hand, relative risks of fracture for women with urinary C-terminal cross-linking telopeptide of type I collagen (CTX) and free deoxypyridinoline above the premenopausal range were around 2.0. Serum CTX also predicted the risk of hip fracture, if it was sampled in the afternoon. Bone resorption markers can be used to predict the risk of hip fracture in healthy elderly women, if sampling procedures are appropriate. In individuals, results should be compared to treatment thresholds, using likelihood ratios.

INTRODUCTION

Hip fracture is a major public health concern, as the lifetime risk of this fracture for a 50-year-old woman is about 18%,[1] and because its outcomes are poor. The total number of fractures and hence the cost to society will rise over the next 50 years as a result of an increase in the elderly population.[2] The validity of various risk factors for fracture, such as bone mineral density (BMD),[3] some clinical risk factors,[4,5] and markers of bone resorption,[6–8] has been established in the past two decades. Some of these risk factors are now used to screen for patients at risk of fractures.

Postmenopausal bone loss is associated with an increased bone turnover, reflected by an increase in biochemical markers of bone turnover.[9,10] Bone resorption markers have been shown to predict the risk of hip fracture in elderly women.[6–8] In contrast, markers of bone formation do not seem to predict fractures in elderly women, except in the particular case of undercarboxylated osteocalcin (ucOC).[6,11] Markers of bone turnover have also been advocated for monitoring of treatment, as they are markedly decreased after several months of antiresorptive therapy.[12–14]

The predictive value for fracture of markers has been studied in several large prospective cohorts (EPIDOS study, Rotterdam study and Study of Osteoporotic Fractures),[6–8,15,16] Some results are very consistent across these studies, and some others are not, but the explanation of these discrepancies is beyond the scope of this chapter. The purpose of this chapter is to review results of the EPIDOS study regarding markers, and draw a perspective for the potential use of markers as risk factors for hip fracture.

METHODS

Two analyses were conducted. The first one was done after 22 months of follow-up, with 109 hip fractures,[6,7,11] and the second one after 3.3 years, with 212 hip fractures.[8] These two analyses relied on the same design: a nested case–control study within the EPIDOS prospective study.

Elderly subjects

From January 1992 to January 1994, 7598 ambulatory healthy Caucasian female volunteers, aged 75 years and older (mean age: 81 years), were recruited into the EPIDOS study, which was a prospective study of the risk factors for hip fracture performed in five French cities. Participants were recruited from population-based registries such as voting or healthcare lists. The baseline investigation included a questionnaire, a clinical and functional examination, non-fasting blood and urine sampling, measurement of femoral neck BMD with a Lunar DPX-Plus, and ultrasound assessment of the calcaneus with the Lunar Achilles ultrasound system (Lunar Corporation, Madison, Wisconsin, USA). Blood was drawn between 7 and 11 a.m. or between 1 and 2 p.m. Morning samples were not controlled for the fasting/non-fasting state, while afternoon sampling was performed after lunch. First morning void urine samples were also collected. Serum and urine were stored at −80°C.

Each participant was contacted every 4 months by mail or telephone to get information about fractures. The diagnosis and classification of hip fracture were confirmed by reference to clinical and/or radiographic records in the medical files of the hospital where the patient was treated.

Women with diseases and treatments likely to affect bone metabolism were excluded from the analysis. We performed a nested case–control analysis by matching each fracture case with three controls randomly chosen from the same cohort according to age (±1 year) and the time of recruitment. None of the controls had incident osteoporotic fractures, and women undergoing treatment influencing bone metabolism or with a bone disease were excluded.

For the interim analysis,[6] after 22 months of follow-up 109 patients with hip fracture and 292 controls were analysed. In the final analysis,[8] during a mean 3.3-year follow-up period (maximum 4.9 years), 291 women had an osteoporotic hip fracture. Blood and urine samples were not available in 79 patients with subsequent hip fracture. Thus, in total, 848 women were studied, including 212 patients and 636 controls.

Premenopausal control subjects

A premenopausal reference group was used to define normal values of markers of bone resorption. These women belong to a prospective cohort, the OFELY study, which comprises 1039 healthy volunteers 31–89 years of age.[9]

These two studies (EPIDOS and OFELY) were approved by the local ethical committee, and written informed consent was obtained from all women. No compensation was provided to women in these studies.

Biochemistry

For the interim analysis,[6] serum total osteocalcin (total OC) was measured with a human-specific immunoradiometric assay (IRMA) (ELSA-OSTEO, Cis Biointernational, Bagnol/Ceze, France), which recognizes both the intact molecule and the N-terminal fragment.[17] Serum ucOC was measured with both a hydroxyapatite binding assay[11] and an ELISA (Biotechnology Research Laboratory, Takara Shuzo Co., Otsu, Shiga, Japan).[11] Serum bone-specific alkaline phosphatase (bone ALP) was measured with an IRMA (Ostase, Hybritech Inc., San Diego, CA, USA).

Urinary type I C-telopeptide breakdown products (urinary CTX) were measured in both studies by an ELISA (CrossLaps ELISA Osteometer, Biotech A/S, Herlev, Denmark) based on an immobilized synthetic peptide with an amino acid sequence specific for a part of the C-telopeptide of the α1 chain of type I collagen (Glu–Lys–Ala–His–Asp–Gly–Gly–Arg) (CrossLaps antigen).[18]

In the interim analysis, urinary free deoxypyridinoline (DPD) was measured by an ELISA that uses a monoclonal antibody (Pyrilinks-D, MetraBiosystem, Mountain View, CA, USA),[19] whereas, in the final analysis, it was measured with an automated device (Chiron Diagnostic), using the same type of monoclonal antibody.

Measurement of serum CTX was performed only in the final analysis. This assay is based upon a double-antibody sandwich technique, in which one biotinylated monoclonal antibody is specific for an epitope of an 8 amino acid sequence of the C-terminal portion of the telopeptide and the ruthenium-labelled partner monoclonal antibody for another epitope (serum CrossLaps Osteometer, Biotech A/S, Herlev, Denmark). Intra-assay CVs range from 1.2% to 4.1%, and inter-assay CVs are below 8%. The assays were performed with an automated device (Elecsys, Roche Diagnostics).

Statistical analysis

A conditional logistic model was used for the interim analysis, whereas a stratified Cox proportional-hazards model was fitted to analyse predictors of hip fracture, using the SAS 6.12 software (SAS Institute, Cary, NC, USA). This model gave the best goodness of fit in regard to this more prolonged follow-up. The proportionality assumption of this model was checked. Predictors were expressed as odds ratios in the first analysis, and as relative hazards (RHs) in the second analysis, both with 95% confidence intervals (CI). Relationships between serum CTX and the hip fracture risk were estimated on the whole group, and, to take into account the nycthemeral variations, on morning and on afternoon samples separately.

RESULTS

We briefly present the most significant results of these studies, as they have been published in detail elsewhere.[6–8,11] Characteristics of women in the final analysis are presented in Table 20.1. In the interim analysis, serum OC and bone ALP were not predictors of hip fracture, with ORs of 1.0 (0.6–1.6) and 1.1 (0.7–1.7), respectively, for women having these markers above the premenopausal range, compared with all the others. In contrast, ucOC predicted the risk of hip fracture with an OR of 2.0 (1.3–3.3), when higher than the premenopausal range. On the other hand, two markers of bone resorption, urinary CTX and urinary free DPD, were independent predictors of the risk of hip fracture. The OR for urinary DPD was 2.2 (1.3–3.6), and it was 1.9 (1.1–3.2) for urinary CTX, when women with markers above the premenopausal range were compared with all the others.

In the final analysis, only three markers of

Table 20.1 Baseline characteristics of hip fracture patients and age-matched controls.

	Controls	Hip fractures	*P*-value
Age (years)	82.5 ± 4.5	82.4 ± 4.4	0.65
Weight (kg)	60.1 ± 10.5	57.7 ± 9.9	0.001
Height (cm)	152.8 ± 62.8	152.9 ± 62.7	0.8
Urinary CTX	307.8 ± 199.2	330.46 ± 221.9	0.15
Serum CTX (entire group)	0.342 ± 0.197	0.347 ± 0.194	0.83
Morning samples	0.379 ± 0.214	0.381 ± 0.211	0.9
Afternoon samples	0.304 ± 0.170	0.321 ± 0.177	0.02
Urinary free DPD	8.9 ± 3.9	9.9 ± 3.6	0.001
Femoral neck BMD (g/cm^2)	0.719 ± 0.116	0.647 ± 0.101	<0.001
Gait speed (m/s)	0.81 ± 0.22	0.73 ± 0.2	<0.001

CTX, type I collagen breakdown product; BMD, bone mineral density; DPD, deoxypyridinoline.

bone resorption were measured: urinary CTX and free DPD, as well as serum CTX. Women with baseline urinary excretion of free DPD and CTX above the upper limit of the premenopausal range had a relative hazard (RH) of 2.07 (1.49–2.9) for DPD, and of 1.67 (1.19–2.32) for urinary CTX.[8] For women with serum CTX above the upper limit of the premenopausal range, the RH was 1.21 (0.8–1.83).[8] Because of the diurnal variation of bone resorption, women with hip fractures and their respective controls were separated into two categories—morning or afternoon—according to the hour of serum sampling. One hundred and fifteen hip fracture cases and 293 controls had blood samples taken during the early afternoon. For women whose samples were drawn in the morning, the RH of women with serum CTX above the premenopausal range to sustain a hip fracture was of 0.9 (0.44–1.87). For women whose samples were drawn in the afternoon, the RH was 1.86 (1.01–3.76).[8] Results of the final analysis are summarized in Fig. 20.1.

DISCUSSION

In French healthy elderly women over the age of 75, three bone resorption markers predicted the risk of hip fracture: urinary CTX and free DPD, as well as serum CTX if sampled after lunch. In contrast, serum OC and bone ALP were not associated with an increased risk of hip fracture in these women, whereas ucOC predicted the risk of hip fracture in these elderly women.

Our results show that elderly women with markers of resorption (urinary and serum CTX and urinary free DPD) above the upper limit of the premenopausal range have an increased risk of hip fracture, with a relative risk around 1.7–2.0.[6,8] Only women with their urinary markers above the premenopausal range or in the highest quartile had an increased risk for hip fracture, and that susceptibility to hip fracture was not proportionally associated with the increase in absolute levels of bone resorption markers. These results suggest that bone resorption becomes harmful for bone strength only when it exceeds a normal physiological threshold, i.e. the mean + 2 SD of the premenopausal population. Increased bone resorption persists

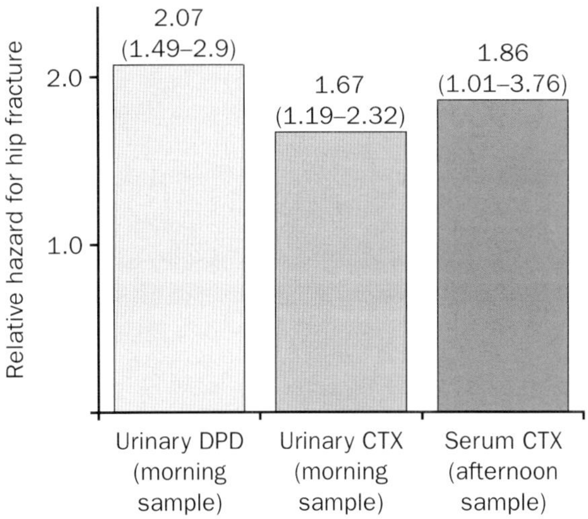

Fig. 20.1 Prediction of hip fracture by markers of bone resorption increased above the premenopausal range (*T* score > 2). The prediction of hip fracture by urinary markers (morning urine) is of the same magnitude as that of serum CTX (afternoon sample). CTX, collagen C-telopeptide breakdown product; DPD, deoxypyridinoline.

in elderly women and explains in part the rise in fracture incidence in elderly women, as the association between bone resorption and hip fracture risk is independent of BMD. In combination with BMD and/or ultrasound measurement, these two markers can be used to detect elderly women who are at highest risk of hip fracture.[7]

Data obtained with serum CTX deserve further comment. The inability of morning serum CTX to predict hip fracture is likely to be due to the lack of control of the fasting/non-fasting state and to the sampling time spanning from 7 to 11 a.m. During this period, bone resorption decreases steeply. In another cohort (the OFELY study), it has been found that, in postmenopausal women, fasting morning serum CTX collected under controlled conditions is able to predict the risk of osteoporotic fractures as well as urinary CTX.[20] In that study, blood was drawn after an overnight fast, between 8 and 9.30 a.m. A circadian rhythm of serum CTX has been recently reported in a Danish randomized crossover study of the diurnal variation of serum CTX in postmenopausal women.[21] In non-fasting women, the highest level of serum CTX was observed during the night, followed by a steep decrease in the morning, followed by a nadir at the onset of the afternoon. In addition, this circadian rhythm was attenuated when women were fasting, and the day-to-day variation of serum CTX sampled in the morning was markedly higher when women were non-fasting than when they were fasting (S. Christgau et al, personal communication). Thus, the variability of the measurement is increased by circadian variation of serum CTX and non-fasting state and precludes its ability to predict the risk of fracture. However, in the final analysis of the EPIDOS study, all women with afternoon samples were not fasting, but they were at the nadir of the circadian rhythm, and thus had a reduced variability as compared with those sampled in the morning.

Measurement of urinary free DPD was changed between the interim and final analyses. In the interim analysis, the assay was an ELISA, whereas an automated device was used for final analysis, using the same type of monoclonal antibody. This modification resulted in an improvement in the analytical performance of the assay. As prediction of hip fracture remains of the same magnitude, this automated technique could now be used on a larger scale, such as in routine laboratories.

Undercarboxylated osteocalcin was able to predict the risk of hip fracture in the EPIDOS cohort,[11] confirming earlier data obtained in a cohort of elderly institutionalized women.[22] The relative risk is around 2.0 with the traditional assay using hydroxyapatite or with the ELISA. The mechanism by which ucOC is linked to bone fragility is not yet understood. However, increased levels of serum ucOC probably reflect the low level of both vitamin D and vitamin K,[22] so that it could be considered as an integrated index of deficiency in these vitamins. It should be stressed that measurement of ucOC needs to be performed under carefully controlled conditions to provide meaningful results.[23] In this context, the use of ucOC is not yet possible in clinical practice. However, data regarding the relationship between fractures and the level of ucOC give an interesting insight into the pathophysiology of bone fragility of elderly women.

The main issue that needs additional study is the practical use of these markers of bone resorption in screening women at risk of osteoporotic fractures. Sensitivity and specificity are of the same order of magnitude as that of BMD.[7] They can be increased by combining of two or three tests. However, as we are dealing with continuous risk factors, there is a large overlap between the distributions of these risk factors (markers and BMD) for the populations of women who will fracture and for the population of women who will not. Therefore, the increase in sensitivity must be achieved at the expense of specificity, and vice versa. Moreover, it seems impossible, given this large overlap and the low incidence of fracture, to obtain a very good trade-off between sensitivity and specificity, so that it might be preferable to calculate the posterior probability of fracture, using the likelihood ratio for a positive test. The combination of several tests (e.g. a marker and BMD) is likely to increase likelihood ratios, and so posterior probability of fracture. This posterior probability should be compared to the treatment threshold defined for specific age groups, specific treatments and specific populations. These treatment thresholds should be obtained by appropriate decision analysis and cost-effectiveness analysis, but these types of result have not been available for treatments of osteoporosis so far.

In conclusion, markers of bone resorption predict hip fracture in healthy elderly women. These tests could be used to generate posterior probabilities necessary to improve therapeutic decisions, given predefined thresholds.

REFERENCES

1. Delmas PD, Biochemical markers of bone turnover. *J Bone Miner Res* 1993; 2(suppl): 549–55.
2. Garnero P, Delmas PD, Clinical usefulness of markers of bone remodelling in osteoporosis. In: Meunier PJ, ed. *Osteoporosis: diagnosis and management*. London: Martin Dunitz, 1998: 79–101.
3. Cummings SR, Black DM, Nevitt MC et al, Bone densitometry at various sites for prediction of hip fractures. *Lancet* 1993; 341: 72–5.
4. Dargent-Molina P, Favier F, Grandjean H et al, Fall-related factors and risk for hip fracture: the EPIDOS prospective study. *Lancet* 1996; 348: 145–9.
5. Cummings SR, Nevitt MC, Browner WS et al, Risk factors for hip fracture in white women. *N Engl J Med* 1995; 332: 767–73.
6. Garnero P, Hausher E, Chapuy MC et al, Markers of bone resorption predict fractures in elderly women: the EPIDOS prospective study. *J Bone Miner Res* 1996; 11: 1531–8.
7. Garnero P, Dargent-Molina P, Hans D et al, Do markers of bone resorption add to bone mineral density and ultrasonographic heel measurement for the prediction of hip fracture in elderly women? The EPIDOS prospective study. *Osteoporos Int* 1998; 8: 563–9.
8. Chapurlat RD, Garnero P, Breart G, Meunier PJ, Delmas PD, Serum type I collagen breakdown products (serum CTX) predicts hip fracture risk in elderly women: the EPIDOS study. *Bone* 2000; 27: 283–6.
9. Garnero P, Sornay-Rendu E, Chapuy MC, Delmas PD, Increased bone turnover in late postmenopausal women is a major determinant of osteoporosis. *J Bone Miner Res* 1996; 11: 337–49.
10. Garnero P, Gineyts E, Riou JP, Delmas PD, Assessment of bone resorption with a new marker of collagen degradation in patients with metabolic bone disease. *J Clin Endocrinol Metab* 1994; 79: 780–5.
11. Vergnaud P, Garnero P, Meunier PJ, Breart G, Kamihagi K, Delmas PD, Undercarboxylated osteocalcin measured with a specific immunoassay predicts hip fracture in elderly women: the EPIDOS Study. *J Clin Endocrinol Metab* 1997; 82: 719–24.
12. Garnero P, Shih WJ, Gineyts E et al, Comparison of new biochemical markers of bone turnover in late postmenopausal osteoporotic women in response to alendronate treatment. *J Clin Endocrinol Metab* 1994; 79: 1693–700.
13. Rosen HN, Moses AC, Garber J et al, Serum CTX: a new marker of bone resorption that shows treatment effect more often than other markers

because of low coefficient of variability and large changes with bisphosphonates. *Calcif Tissue Int* 2000; 66: 100–3.
14. Rosen HN, Moses AC, Garber J, Ross DS, Lee SL, Greenspan SL, Utility of biochemical markers of bone turnover in the follow-up of patients treated with bisphosphonates. *Calcif Tissue Int* 1998; 63: 363–8.
15. van Daele PLA, Seibel MJ, Burger H et al, Case–control analysis of bone resorption markers, disability and hip fracture risk: the Rotterdam study. *BMJ* 1996; 312: 482–3.
16. Bauer DC, Black DM, Ensrud K, Qvist P, Williams EN, Serum markers of bone turnover and fractures of the hip and spine: a prospective study. *J Bone Miner Res* 1999; 14(suppl 1): S147.
17. Garnero P, Grimaux M, Demiaux B, Preaudat C, Seguin P, Delmas PD, Measurement of serum osteocalcin with a human-specific two-site radioimmunoassay. *J Bone Miner Res* 1992; 7: 1389–97.
18. Bonde M, Qvist P, Fledelius C, Riis BJ, Christiansen C, Immunoassay for quantifying type I collagen degradation products in urine evaluated. *Clin Chem* 1994; 40: 2022–5.
19. Robins SP, Woitge H, Hesley R, Ju J, Seyedin S, Seibel MJ, A direct enzyme-linked immunoassay for urinary deoxypyridinoline as a specific marker for measuring bone resorption. *J Bone Miner Res* 1994; 9: 1643–9.
20. Garnero P, Sornay-Rendu E, Claustrat B, Delmas PD, Markers of bone turnover, endogenous hormones and the risk of fractures in postmenopausal women. *J Bone Miner Res* 1999; 14(suppl 1): S171.
21. Christgau S, Bitsch-Jensen O, Hanover-Bjarnason N et al, Serum CrossLaps for monitoring the response in individuals undergoing antiresorptive therapy. *Bone* 2000; 26: 505–11.
22. Szulc P, Chapuy MC, Meunier PJ, Delmas PD, Serum undercarboxylated osteocalcin is a marker of the risk of hip fracture in elderly women. *J Clin Invest* 1993; 91: 1769–74.
23. Gundberg CM, Nieman SD, Abrams S, Rosen H, Vitamin K status and bone health: an analysis of methods for determination of undercarboxylated osteocalcin. *J Clin Endocrinol Metab* 1998; 83: 3258–66.

21

Biochemical markers of bone turnover: the Study of Osteoporotic Fracture

Douglas C Bauer

The Study of Osteoporotic Fracture • Markers and bone loss • Markers and fracture • Discussion • References

THE STUDY OF OSTEOPOROTIC FRACTURE

The Study of Osteoporotic Fracture (SOF) is a multicentre study of risk factors for fracture in 9704 non-black women, 65 years of age or older, recruited from population-based listings at four clinical centres: The Kaiser-Pemanente Center for Health Research, Portland, Oregon; the University of Minnesota, Minneapolis, Minnesota; the University of Maryland, Baltimore, Maryland; and the Monangehela Valley, Pennsylvania.

MARKERS AND BONE LOSS

Fasting serum and 2-h morning urine were collected and stored at −190°C in a consecutive sample of 501 women (approximately 125 from each clinical centre) between April and July 1989. Bone loss was measured by hip dual X-ray absorptiometry (DXA) (Hologic QDR 1000) at baseline and after a mean follow-up of 3.8 years (range 3.3–5.1 years). For these analyses, we excluded 89 women who reported oral oestrogen use at baseline. Of the 412 non-users who had an initial bone mineral density (BMD) measurement, 295 (77% of survivors) had a follow-up DXA measurement, 31 died prior to the second measurement, 4 did not return, and 82 attended both SOF visits but did not undergo a second DXA.

Biochemical markers of bone formation

Serum total osteocalcin (OC) (ELSA-OSTEO, Cis Biointernational, Baglos/Ceze, France) and serum bone-specific alkaline phosphatase (bone ALP) (Ostase, Hybritech, Inc., San Diego, CA, USA) were measured with standard assays.

Biochemical markers of bone resorption

Urinary type I collagen cross-linked N-telopeptides (NTX) (Osteomark, Ostex International, Inc., Seattle, WA, USA), urinary type I

C-telopeptide breakdown products (CTX) (CrossLaps ELSA, Osteometer Biotech A/S, Herlev, Denmark), urinary free deoxypyridinoline (DPD) (Pyrilinks-D, Metra Biosystems, Mountain View, CA, USA) and urinary free pyridinolines (PYD) (Pyrilinks, Metra Biosystems, Mountain View, CA, USA) were measured by an ELISA and were corrected by the urinary creatinine concentration.

Bone mineral density

BMD of the total hip and three subregions was measured by DXA at two visits, a mean of 3.8 years apart, using Hologic QDR 1000 bone densitometers.

Statistical analysis

We compared the mean rate of change in BMD among women above and below the median for each marker, and compared those in the highest quartile of marker level to those below the highest quartile. Bone loss models were examined with and without baseline BMD as a co-variate. All regression models were adjusted for age.

To determine the sensitivity, specificity, positive predictive value and negative predictive value of normal and elevated marker levels for rapid bone loss, we divided the cohort into tertiles of bone loss and calculated the ability of elevated baseline markers (above the median or the highest quartile) to predict the highest tertile of bone loss. The highest tertile of bone loss was chosen, as there is no generally accepted clinical cut-point that defines excessive bone loss in older untreated women.

Bone loss results

The mean age of our participants was 73 years (range 67–89). Mean bone loss at the total hip was 0.6% per year. Correlations between baseline marker levels and rate of change in BMD ranged between −0.01 and 0.16. By comparison, correlations between age and rates of change in BMD ranged from 0.16 for the femoral neck to 0.30 for the total hip (Table 21.1).

After adjusting for age, the mean rate of bone loss from the total hip significantly increased across quartiles of each urine marker (Table 21.2). Trends were similar and remained significant after further adjustment for baseline BMD (data not shown). Conversely, although there was a trend for increasing mean rate of bone loss with increasing concentration of serum markers, these relationships did not reach statistical significance (Table 21.2). Women who had baseline levels of NTX, CTX, PYD, DPD and OC that were above the median had significantly higher levels of bone loss than did women whose marker levels were below the median, even after adjusting for age. The age-

Table 21.1 Correlation coefficients of age and biochemical markers of bone turnover with annual percentage change in BMD.

	Age	CTX	DPD	NTX	PYD	Bone ALP	OC
Total hip	0.30	0.08	0.13	0.15	0.09	0.08	0.12
Femoral neck	0.16	−0.01	0.06	0.04	0.04	0.00	0.06
Trochanter	0.22	0.07	0.11	0.09	0.11	0.05	0.10
Intertrochanter	0.28	0.11	0.16	0.19	0.10	0.12	0.15

Table 21.2 Comparison of mean age-adjusted total hip bone loss in women with marker levels in and below the fourth quartile and above and below the median. Reproduced with permission from *J Bone Miner Res* 1999; 14: 1404–10.

	Mean %/year loss of hip BMD[a]						
Marker	**Q1**	**Q2**	**Q3**	**Q4**	**Test for trend**	**Q3–4 versus Q1–Q2**	**Q4 versus Q1–3**
CTX	0.39	0.49	0.90	0.75	$P = 0.025$	$P = 0.007$	$P = 0.43$
DPD	0.47	0.54	0.83	0.82	$P = 0.026$	$P = 0.017$	$P = 0.19$
NTX	0.39	0.49	0.77	0.89	$P = 0.004$	$P = 0.005$	$P = 0.04$
PYD	0.33	0.58	0.94	0.77	$P = 0.007$	$P = 0.004$	$P = 0.40$
Bone ALP	0.57	0.63	0.64	0.81	$P = 0.259$	$P = 0.423$	$P = 0.22$
OC	0.59	0.47	0.71	0.86	$P = 0.056$	$P = 0.044$	$P = 0.67$

[a]Least square mean, adjusted for age.

adjusted difference in mean bone loss between women above and below the median level was not statistically significant for bone ALP (Table 21.2). The age-adjusted difference in bone loss rates between women with marker levels in the highest quartile and those in the lowest three was statistically significant for NTX only (Table 21.2).

The sensitivity, specificity, positive predictive value and negative predictive value of marker levels above and below the median and fourth quartile levels for identifying women in the highest tertile of total hip bone loss (loss > 1.1%/year) are listed in Table 21.3. Using the median marker level as a cut-point, the sensitivity for identifying women who subsequently lost more than 1.1% of total hip BMD per year varied from 56% for DPD to 72% for NTX. Thus, among those women with more than 1.1% bone loss per year, 56% had DPD levels above the median (6.8 nmol/mmol Creatinine (Cr), and 72% had NTX levels above the median (41.8 nmol BCE/mmol Cr). The corresponding sensitivities using the upper quartile of marker as a cut-point ranged from 28% for DPD to 38% for both NTX and OC. Using the median marker level as a cut-point, the specificity for identifying women who did not lose more than 1.1% per year ranged from 46% for bone ALP to 56% for PYD. The corresponding specificities using the upper quartile of marker levels as a cut-point ranged from 74% for CTX and bone ALP to 79% for PYD.

From a clinical standpoint, the sensitivity and specificity of a test are less important than the positive predictive value, which describes the probability of excessive bone loss among women with elevated marker, and negative predictive value, which describes the probability of not having excessive bone loss among women with normal markers. With a pretest probability of 33% (excessive bone loss was defined as those in the upper tertile, >1.1% per year), the positive predictive values of elevated marker levels for excessive bone loss were between 35% and 42% when the median was used as a cut-point, and between 36% and 46% when the highest quartile was used as a cut-point. Thus, among those women with elevated marker levels using either the median or highest quartile cut-point, less than half lost more than 1.1% per year of bone mass.

MARKERS AND FRACTURE

During the baseline examination in 1986–88, participants underwent a detailed interview and examination. Lateral X-rays of thoracic and lumbar spine were obtained using a standard protocol, and, after a modified fast, serum was collected and stored at −190°C. Serum was collected between 8 a.m. and 2 p.m., and the time of collection was recorded at three of the four SOF clinics.

Fracture outcomes

After the baseline examination, participants were queried every 3 months for the occurrence of hip fractures. All self-reported fractures were confirmed centrally. To detect incident vertebral fractures, spine X-rays were repeated after a mean follow-up of 3.7 years. Incident fractures were defined morphometrically using a standard definition: at least a 20% and 4-mm reduction in the anterior, mid- or posterior height. Approximately 8 years after the initial baseline examinations, we selected baseline serum from 700 participants for biochemical analysis. From the entire study of 9704 women, we identified all women with documented new hip and spine fractures and randomly selected 150 women with each fracture type. We also identified a randomly selected control pool of 400. Biochemical studies were performed on baseline sera from all 700 women. A small number of women in the control pool had experienced a hip or spine fracture during follow-up: these individuals were analysed as cases for that fracture type. Women with other non-hip and non-spine fractures during follow-up were not excluded from the control pool.

Table 21.3 Sensitivity, specificity, positive predictive value (PPV) and negative predictive value (NPV) of marker levels above and below the median and fourth quartile levels for identification of women who lost more than 1.1% BMD per year from the total hip (highest tertile). Reproduced with permission from *J Bone Miner Res* 1999; 14: 1404–10.

	Above median				Fourth quartile			
Marker	Sensitivity (%)[a]	Specificity (%)[b]	PPV[c]	NPV[d]	Sensitivity (%)[a]	Specificity (%)[b]	PPV[c]	NPV[d]
CTX	68	49	39	76	33	74	38	69
DPD	56	55	38	72	28	78	38	68
NTX	72	50	42	79	38	78	46	71
PYD	63	56	42	71	30	79	41	69
Bone ALP	59	46	35	69	29	74	36	68
OC	63	47	37	72	38	76	44	71

[a]Proportion of women with excessive bone loss (>1.1% per year at the total hip), with marker levels above the cut-point.
[b]Proportion of women with normal bone loss (<1.1% per year), with marker levels below the cut-point.
[c]Probability of excessive bone loss of women among those with marker levels above the cut-point.
[d]Probability of normal bone loss among those with marker levels below the cut-point.

Biochemical markers

Thawed serum from the fracture cases and control pool were assayed for the following biochemical markers. Total OC (ELSA-OSTEO, Cis Biointernational, Baglos/Ceze, France) and serum bone ALP (Ostase, Hybritech, Inc., San Diego, CA, USA) were measured at an independent research laboratory. Serum CTX was measured using a one-step ELISA assay (CrossLaps ELISA, Osteometer Biotech A/S, Herlev, Denmark). All assays were performed without knowledge of fracture outcomes. All analyses were adjusted for age, weight and oestrogen use. Serum CTX levels obtained between 10 and 12 a.m. and between 12 and 2 p.m. were approximately 25% lower than those drawn between 8 and 10 a.m. Therefore, all subsequent serum CTX analyses are also adjusted for the time of collection.

Fracture results

The mean age of our participants was 72, and approximately 10% were current users of oestrogen. The means of SDs of each marker among the hip fracture cases, vertebral fracture cases and their respective controls did not differ. Women in the lowest quintile of each marker tended to have lower fracture rates, but these relationships were not significant ($P > 0.05$). In analyses adjusted for age and oestrogen use, we found no evidence that marker levels, analysed per SD, lowest quintile versus other quintiles, or highest quintile versus other quintiles, were associated with the risk of hip or vertebral fractures.

Despite the lack of significant associations with fracture, we did find that markers were strongly affected by oestrogen use in this cohort. The mean levels of OC and bone ALP were 40% lower among oestrogen users than in non-users, and serum CTX levels were 60% lower among the oestrogen users. These differences were highly significant.

DISCUSSION

Prospective analyses of SOF have demonstrated that higher levels of urine NTX, CTX, PYD and DPD, and serum OC, are associated with somewhat faster total hip bone loss in elderly women. However, the predictive value of these markers for bone loss in an individual woman is quite limited, and therefore their clinical utility remains uncertain. Our data do not support the widespread use of currently available markers to identify older women at risk of rapid bone loss. It is likely that future assays for markers of bone resorption and formation will have improved biological and analytical precision and perhaps will better reflect bone turnover and bone loss in older women.

In the prospective analyses of markers and fracture, we observed a suggestive trend that women with lower levels of OC, bone ALP and serum CTX had fewer fractures, but there were no strong or independent relationships with fracture risk. Our inferences about the utility of serum CTX must be tempered by the unexpectedly low levels in our cohort, and additional data are needed on population means and the effects of fasting and time of collection. Nonetheless, our data do not support the routine use of serum markers of bone resorption and formation to predict fracture risk among older women.

REFERENCES

1. Bauer DC, Sklarin PM, Stone KL et al, Biochemical markers of bone turnover and prediction of hip bone loss in older women: The Study of Osteoporotic Fractures. *J Bone Miner Res* 1999; 14: 1404–10.
2. Bauer DC, Black DM, Ensrud K et al, Serum markers of bone turnover and fractures of the hip and spine: a prospective study. Oral presentation at the national meeting of the American Society for Bone.

22

The association between bone turnover and fracture risk (Sheffield Osteoporosis Study)

Diana M Greenfield, Rosemary A Hannon and Richard Eastell

Summary • Introduction • Materials and methods • Results • Discussion • Acknowledgements • References

SUMMARY

The aims of this study were to estimate the relative risks of bone turnover markers (BTMs) for vertebral deformities and non-vertebral fractures and to compare the predictive value of BTM with that of bone mineral density (BMD). Three hundred and seventy-five women aged 50–85 years (mean 64 years) were selected from three general practices and observed over 5 years. Nine BTMs were measured. The urine markers were: total pyridinoline (PYD), free and total deoxypyridinoline (ifDPD, DPD) and N-terminal telopeptide (NTX). The serum markers included: tartrate-resistant acid phosphatase (TRAP), bone alkaline phosphatase activity (bone ALP), immunoreactive bone alkaline phosphatase (bone iALP), procollagen type I C-terminal propeptide (PICP) and osteocalcin (OC). BMD of the lumbar spine (LS), proximal femur (FN) and total body (TB) was measured at baseline by dual X-ray absorptiometry (DXA). Radiographs of the thoracic and lumbar spine were taken at 0, 2 and 5 years and were examined for vertebral deformities by two observers using a consensus method after triage with semi-quantitative and quantitative–morphometric techniques. Fifteen per cent of subjects were identified as having prevalent vertebral deformities. Of the BTMs, all resorption markers and one formation marker (bone ALP) were raised in subjects with prevalent vertebral deformities compared to those with no deformities. At 2 and 5 years, the rates of incident vertebral deformity were 3.6% and 5% respectively. In the multivariate model, prevalent vertebral deformities, PYD/creatinine and TB BMD were found to be significant risk factors for incident vertebral deformities. Thirty-nine women incurred incident non-vertebral fractures over 5 years. BMD (FN or TB) is a strong risk factor for incident non-vertebral fractures. No BTMs remained as risk factors for non-vertebral fracture. We conclude that: (1) high BTM levels may be associated with increased risk of vertebral fracture; and (2) BMD is a stronger risk factor than BTMs.

INTRODUCTION

In cross-sectional studies, vertebral fractures have been found to be associated with increased bone turnover, as measured by biochemical markers.[1,2] This association could reflect an increase in bone turnover markers (BTMs) as a result of the fracture, rather than high BTM levels being indicators of underlying mechanisms responsible for the fracture. The association between high bone turnover and fracture has also been observed in older cohorts of subjects, particularly in those with hip fractures. The aims of this study were to see if the same association could be observed in a younger cohort of women with other fractures, including vertebral deformities, and to compare the predictive abilities of the different BTMs and BMD and to estimate their relative risks.

Table 22.1 Recruitment figures for population-based study.

	Number
Subjects randomly selected from GP lists	816
Agreed to participate	375
Refused	310
Withdrawn by GP	116
Did not respond	15

MATERIALS AND METHODS

Subjects

From July 1990 to October 1991, we recruited 375 women aged 50–85 years (mean 64.5 years, 9.1 SD) by age-stratified randomization from three general practice populations in Sheffield, UK. Strata were half decades of age. Women were excluded if they were too ill to take part (e.g. because of terminal illness) or if they were unable or unwilling to give informed consent ($N = 307$). Among those invited to participate, the uptake rate was 55%. Subjects returned at 2 ($N = 310$) and 5 ($N = 242$) years, the recall rates being 83% and 65% respectively. The recruitment figures are given in Table 22.1. By 5 years, 27 women had died and 28 were lost to follow-up.

Subjects closely represented the Sheffield social class population distribution according to the 1991 population census. The distribution of social class published by HMSO (Her Majesty's Stationery Office) is shown in Table 22.2.

Table 22.2 Social class distribution of the population-based cohort compared with the entire Sheffield population.

Social class	Study population (%) (*N*)	City average (%)
I	2.4 (9)	4.6
II	21.8 (82)	20.4
IIIM	38.7 (146)	37.4
IIIN	19.1 (72)	11.2
IV	13.3 (50)	19
V	4.8 (18)	5.7

Fracture history

Radiographs of the thoracic and lumbar spine, including both anteroposterior (AP) and lateral views, were taken at the diagnostic imaging department at 0, 2 and 5 years. The film-to-focus distance was 100 cm. The radiographs of the AP thoracic, AP lumbar, lateral thoracic and lateral lumbar spine were centred over the usual landmark, namely 1 inch below the xyphisternum, T7, lower costal margin and L3, respectively. Baseline radiographs were examined for prevalent vertebral deformities by a single radiologist (Dr Guirong Jiang) using both a semi-quantitative and a quantitative morphometric technique.[3,4] These methods were further modified by Dr Jiang to reduce the number of false positives and false negatives. Any disagreements between techniques were resolved by consensus with a second observer (Professor Richard Eastell). Any subject noted to have vertebral deformity due to osteoporosis was followed up at the metabolic bone clinic. Incident vertebral deformities were assessed from radiographs taken at 2 and 5 years, using the same techniques, and, in addition, height ratios were compared between visits.

Incident non-vertebral fractures were reported at years 2 and 5 for returners and by reviewing medical notes for non-returners. In the case of the deceased, medical notes were reviewed at the local health authority with the general practitioner's permission. Reported incident non-vertebral fractures were verified by radiology report. Fracture information was not available in subjects who were lost to follow-up ($N = 28$).

Bone density measurements

Bone mineral density (BMD) of the lumbar spine (LS), proximal femur (FN, WT, TR) and total body (TB) was measured at baseline by dual energy X-ray absorptiometry (DXA) using a Lunar DPX densitometer (Lunar Corp., Madison, WI, USA). The in vivo coefficients of variation (CV) for LS, FN and TB BMD measurements were 1%, 3% and 1% respectively. All scans were analysed using software version 3.6y.

Biochemical markers of bone turnover

Blood samples were collected at baseline between 9.00 a.m. and 9.45 a.m. after an overnight fast, centrifuged at 2000 *g* for 30 min, and frozen within 60 min. Subjects also provided a 24-h urine sample. Serum and urine were stored at −70°C and −20°C respectively, and all assays were performed in duplicate. The biochemical markers of bone formation measured in serum were as follows. Procollagen type I C-terminal propeptide (PICP) was measured by radioimmunoassay (Orion Diagnostica, Turku, Finland); the intra-assay CV was 4% and the inter-assay CV was 7%. Osteocalcin (OC) was measured by radioimmunoassay (Nichols Institute Diagnostics, San Juan Capistrano, CA, USA); the intra-assay CV was 4% and the inter-assay CV was 7%. The bone isoform of alkaline phosphatase (bone ALP and immunoreactive bone ALP (bone iALP)) were measured by both a wheat germ lectin precipitation method based on that of Rosalki and Foo[5] and by ELISA (Alkphase-B) (Metra Biosystems, Mountain View, CA, USA); the intra-assay CVs were 5% and the interassay CVs were 7%. The biochemical markers of bone resorption measured in urine were as follows. Total urinary pyridinoline (PYD) and deoxypyridinoline (DPD) were measured by HPLC following acid hydrolysis according to the method of Colwell et al;[6] the intra-assay CVs for PYD and DPD were <10% and 13% respectively. Free urinary deoxypyridinoline (ifDPD) was measured by Pyrilinks-D ELISA (Metra Biosystems, Mountain View, CA, USA); the intra-assay CV was 5% and the inter-assay CV was 8%. Urinary cross-linked N-telopeptides of type I collagen (NTX) were measured by the Osteomark ELISA (Ostex, Seattle, WA, USA); the intra-assay CV was 5% and the inter-assay CV was 7%. Results of urinary markers were expressed as a ratio to urinary creatinine. Urinary creatinine was measured by a kinetic Jaffe method. The biochemical marker of bone resorption measured in serum was tartrate-resistant acid phosphatase (TRAP) by a standard kinetic method (bioMérieux, Marcy-Étoile, France); the intra-assay CV was 3% and the inter-assay CV was 5%.

Anthropometric measurements

Height and body weight were measured at baseline with subjects wearing light clothing. Height was measured to the nearest millimetre at baseline using a wall-mounted Harpenden stadiometer. Weight was measured to the nearest 0.1 kg. Body mass index (BMI) was calculated as body weight (kg)/height $(m)^2$.

Medical and lifestyle history

A standardized medical and lifestyle questionnaire was administered by interview at baseline.[7] Two further non-standardized in-house questionnaires were administered by interviews at 2 and 5 years. A history of previous fractures was obtained at baseline.

Data analysis

Comparisons of characteristics between those with and without prevalent vertebral deformities were made at baseline by Student's *t*-tests or by Mann-Whitney tests. Further comparisons were made at 5 years between those with and without incident vertebral deformities. Data were analysed by Cox's proportional-hazards model using forward stepwise Wald analysis (SPSS Inc., Chicago, IL, USA). Relative risks (RR) and 95% confidence intervals were estimated for those variables that remained in the model.

Ethical approval

The study was approved by the North Sheffield Local Research Ethics Committee, and all subjects gave written informed consent.

RESULTS

Non-vertebral fractures

Characteristics of the 375 women are shown in Table 22.3. During the five years of follow-up, 48 women reported 49 non-vertebral fractures. Nine fractures were excluded either because radiographs were not taken to confirm the presence of fracture ($N = 7$) or because they were pathological ($N = 2$). In the subject with more than one fracture, time to event of the first fracture was noted, so details of the second fracture were excluded from the analysis. Sites and frequencies of incident fractures are shown in Table 22.4. Information regarding trauma level was not available in the subjects where fracture details were obtained from general practitioner case notes or radiograph reports, so analysis included all non-vertebral fractures regardless of the degree of trauma.

Relative risks for incident non-vertebral fractures are shown in Table 22.5. These were calculated for each variable individually, excluding all other variables. Other factors that provided a significant fit ($P \leq 0.05$) were: increased age, decreased height and late menarche. When all variables were entered into Cox's proportional-hazards model, only the BMD was significant (at any measured site). This is attributed to high correlations between factors.

Prevalent vertebral deformities

At baseline, 56 (14.9%) subjects were found to have 85 prevalent deformities. The characteristics of subjects with vertebral deformities are compared with those with no deformities in Table 22.6.

Incident vertebral deformities

Of the 310 women who returned at 2 years, 11 subjects were found to have 17 incident vertebral deformities (3.6%); 6 of these subjects already had vertebral deformities at baseline. At

Table 22.3 Characteristics of the 375 women studied.

Characteristic	Value[a]
Age (years)	64.5 (9.1)
Height (m)	1.58 (0.06)
Weight (kg)	66.1 (12.2)
BMI (kg/m^2)	26.3 (4.6)
Age at last menstrual period (years)	48.2 (5.1)
Still menstruating (%)	4
Biochemical markers of bone resorption	
PYD/Cr (nmol/mmol)	46.9 (16.3)
DPD/Cr (nmol/mmol)	14.6 (5.6)
ifDPD/Cr (nmol/mmol)	5.1 (1.6)
TRAP (IU/l)	3.70 (0.58)
NTX/Cr (nmol BCE/mmol)	49.3 (19.2)
Biochemical markers of bone formation	
PICP (μg/l)	104 (35)
Bone ALP (IU/l)	38.7 (16.3)
Bone iALP (IU/l)	14.4 (6.0)
OC (μg/l)	6.60 (3.08)
Bone mineral density (g/cm^2)	
Lumbar spine	1.069 (0.189)
Femoral neck	0.845 (0.134)
Trochanteric region	0.742 (0.134)

[a]Mean (SD). Cr: creatinine; PYD: pyridinoline; DPD: total deoxypyridinoline; ifDPD: free deoxypyridinoline; TRAP: tartrate-resistant acid phosphatase; NTX: N-terminal telopeptide; PICP: procollagen type I C-terminal propeptide; ALP: alkaline phosphatase; iALP: immunoreactive alkaline phosphatase.

5 years, 242 women returned and 12 subjects were found to have 16 incident vertebral deformities (5%); 7 of these subjects already had vertebral deformities at baseline, and 3 had deformities between baseline and 2 years.

Prediction of incident vertebral deformities

Relative risks for incident vertebral deformities are shown in Table 22.7. These were calculated for each variable individually, excluding all other variables. Other factors that provided a significant fit ($P \leq 0.05$) were: decreased height, TR BMD and WT BMD. All variables were entered into Cox's proportional-hazards model, using forward stepwise variable selection with Wald analysis. Prevalent vertebral deformities, PYD and TB BMD remained in the model, giving two acceptable models. There were no statistical differences between the models by chi-square score. The RR and 95% CI are given in Table 22.8. Therefore, our results indicate that prevalent vertebral deformity with decreased TB BMD and prevalent vertebral deformity with increased PYD/Cr are equally competent risk factors for incident vertebral deformities in older women. This is a reasonable result, since

Table 22.4 Description of incident fractures in 39 women over 5 years.

Site of fracture	Frequency
Wrist	13
Humerus	6
Foot	4
Hip	3
Ribs	3
Big toe	2
Scaphoid	2
Ankle	2
Hand	1
Clavicle	1
Elbow	1
Patella	1

PYD/Cr may be independent of BMD, with an increased risk indicating an increase in turnover per se, e.g. increase in trabecular perforations. Non-significant variables included weight or BMI, bone ALP or bone iALP, OC, DPD/Cr or ifDPD/Cr, PICP, TRAP and NTX.

DISCUSSION

PYD was the only BTM that was significantly associated with incident non-vertebral fractures. However, PYD did not remain in the multivariate model that included BMD. This may be because there is an inverse relationship between BMD and BTMs.[8]

In subjects with prevalent vertebral deformity, all markers of bone resorption were raised compared with those without vertebral deformities. In a previous study by Eastell, those with vertebral osteoporosis had raised bone resorption markers as measured by pyridinium cross-links.[2] Moreover, a comparison of bone formation markers identified our subjects with vertebral deformity as having raised levels of bone ALP (44.9 versus 37.7). None of the other markers of bone formation was significantly different; OC was raised in those with deformity, but did not reach statistical significance. Similarly, Eastell et al found raised levels of OC in those with vertebral osteoporosis.[2] PICP and bone iALP were not significantly different between the two groups. Our findings contrast with those of Åkesson et al, who found that in women aged 40–80 years a decrease of 1 SD of PICP and ICTP gave odds ratios for any future fracture of 1.8 and 1.9 respectively, independent of BMC and age.[9] However, our findings are in keeping with other researchers, who found that

Table 22.5 Risk factors significantly associated with incident non-vertebral fractures.

Risk factor	Relative risk	95% CI
Previous fracture	2.2	1.2, 4.0
PYD/Cr (in upper quartile)	1.2	1.2, 1.3
Bone mineral density (per SD decrease)		
Lumbar spine	1.7	1.2, 2.4
Femoral neck	2.8	1.9, 4.1
Total body	2.5	1.8, 3.6

PYD: pyridinoline; Cr: creatinine.

Table 22.6 Comparisons of characteristics between those with and without vertebral deformities

Characteristic	Mean (SD)	
	No deformities	Deformities
–	319	56
Age (years)	63.8 (8.8)	68.4 (9.4)[a]
Weight (kg)	66.7 (12.6)	63.1 (9.5)[b]
BMI (kg/m^2)	26.5 (4.7)	25.4 (3.8)
Biochemical markers of bone resorption		
PYD/Cr (nmol/mmol)	46.3 (16.6)	50.7 (13.2)[b]
DPD/Cr (nmol/mmol)	14.5 (5.7)	15.6 (4.3)[b]
ifDPD/Cr (nmol/mmol)	5.0 (1.6)	5.4 (1.5)[b]
TRAP (IU/l)	3.7 (0.6)	3.9 (0.6)[b]
NTX/Cr (nmol BCE/mmol)	52.9 (32.1)	70.0 (53.0)[b]
Biochemical markers of bone formation		
PICP (μg/l)	103 (34)	108 (41)
Bone ALP (IU/l)	37.7 (14.6)	44.9 (22.8)[b]
Bone iALP (IU/l)	14.1 (5.6)	16.1 (7.6)
OC (μg/l)	6.5 (3.1)	7.1 (3.1)
Bone mineral density (g/cm^2)		
Lumbar spine	1.09 (0.19)	0.98 (0.19)[a]
Femoral neck	0.90 (0.14)	0.79 (0.12)[a]
Total body	1.08 (0.10)	1.01 (0.11)[a]

[a]$P < 0.001$, [b]$P < 0.05$, by Student's *t*-tests or Mann-Whitney tests. Cr: creatinine; PYD: pyridinoline; DPD: total deoxypyridinoline; ifDPD: free deoxypyridinoline; TRAP: tartrate-resistant acid phosphatase; NTX: N-terminal telopeptide; PICP: procollagen type I C-terminal propeptide; ALP: alkaline phosphatase; iALP: immunoreactive alkaline phosphatase; OC: osteocalcin.

elevated levels of resorption markers are predictive of fracture risk.[1,10] Garnero et al have shown that levels of free DPD and CTX above the premenopausal range predict hip fracture in women over the age of 75 years with odds ratios of 1.7 and 2.0, respectively. These ratios were independent of FN BMD.[10]

Levels of NTX and markers of bone formation did not predict hip fracture. Similarly, in a nested case–control study, van Daele et al found that both PYD and DPD were associated with an increased risk of hip fracture.[11] Furthermore, in a recent retrospective study of osteoporotic fracture, PYD was the only marker from a comprehensive panel of markers to predict a history of fracture independently in a multivariate model.[1] Although DPD is more specific for bone than PYD, it was PYD that was found to be significantly associated with fracture in the longitudinal study. It is possible that PYD may be a stronger indicator of fracture than DPD because its peak on the chromatograph is larger and so the measurement is more precise.

It is clear from our cross-sectional analysis of BTMs in subjects with and without vertebral deformities that a differentiation of BTMs between those with and without osteoporosis may be an unachievable ideal. High BTM levels

Table 22.7 Risk factors significantly associated with incident vertebral deformities ($N = 20$).

Risk factor	Relative risk	95% CI
Prevalent vertebral fracture	7.7	3.1, 19.0
PYD/Cr (in upper quartile)	1.3	1.1, 1.5
NTX/Cr (in upper quartile)	1.2	1.0, 1.5
Bone mineral density (per SD decrease)		
Lumbar spine	1.8	1.2, 3.5
Femoral neck	2.3	1.4, 4.0
Total body	2.7	1.6, 4.6

Cr: creatinine; PYD: pyridinoline; NTX: N-terminal telopeptide.

Table 22.8 Relative risks of factors remaining in the multivariate models for incident vertebral deformities.

Model	Relative risk	95% CI
Prevalent vertebral deformity with 1 SD decrease in TB BMD	9.4	2.1, 43.4
Prevalent vertebral deformity with PYD/Cr in upper quartile	7.8	2.6, 23.2

TB BMD: total body bone mineral density; PYD: pyridinoline; Cr: creatinine.

are associated with other conditions, such as Paget's disease and osteomalacia, as well as risk of fracture. The drawback to those studies (e.g. the EPIDOS study) that have incorporated the use of BTMs to distinguish between subjects with high or low fracture risks is that they are self-limiting, in that subject exclusion includes those taking medications or with specific medical conditions that affect bone metabolism.[10] The advantage of our study sample was that it included a normal rather than a healthy population. This allows us to consider a representative clinical scenario rather than a pre-selected subsample. Clearly, in this situation, a high turnover state, as shown by the BTMs, should be useful in a differential diagnosis of osteoporosis, Paget's disease, osteomalacia and so on.

Our study had several limitations. The standardized questionnaire used at baseline was one designed for the European Vertebral Osteoporosis Study (EVOS).[7] This questionnaire was modified slightly after the first 100 subjects, so that some of the data were non-comparable, particularly with regard to previous fracture information. The number of incident non-vertebral fractures was small (39), resulting in wide confidence intervals of our estimates of relative risk.

The number of people with incident vertebral

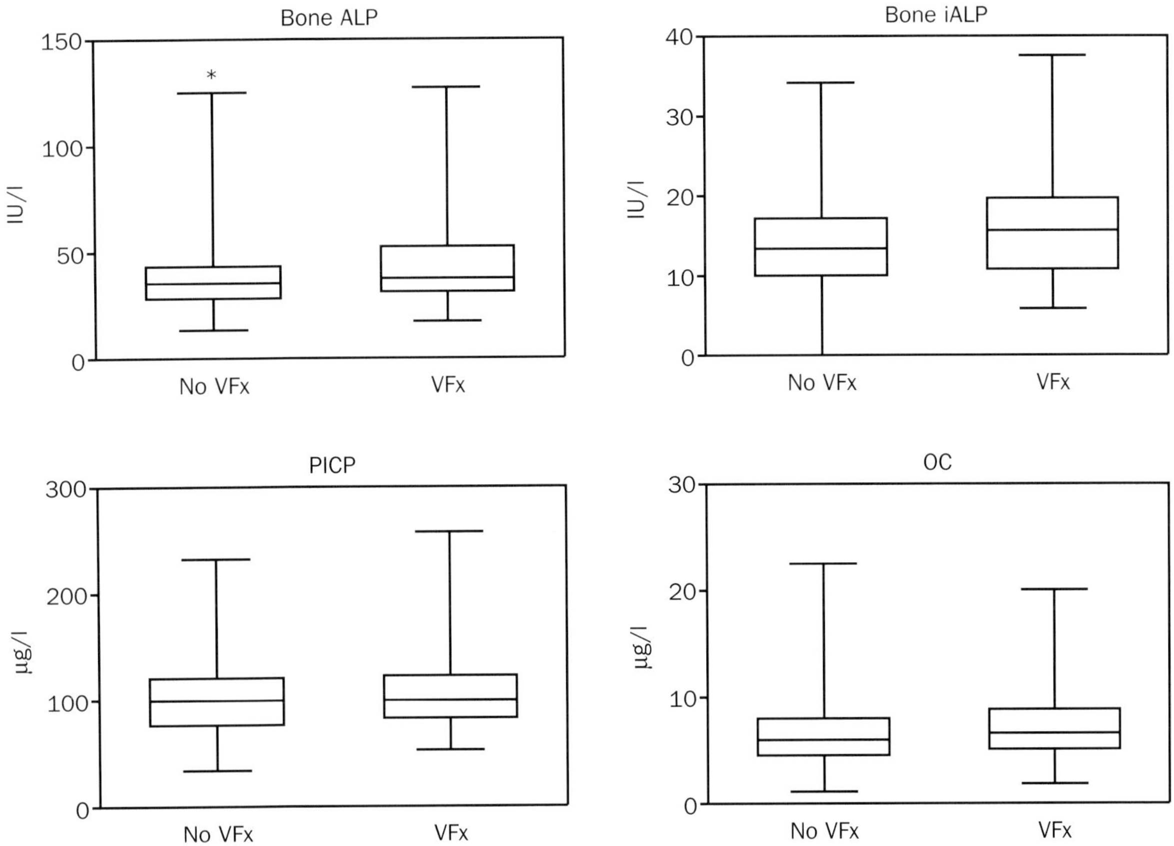

Fig. 22.1 Bone formation markers: cross-sectional results comparing subjects with and without prevalent vertebral deformities at baseline. The box gives the inter-quartile range, the line through the middle is the median, and the whiskers give the range. BTMs compared by Mann-Whitney tests, $*P < 0.05$.
ALP: alkaline phosphatase; iALP: immunoreactive alkaline phosphatase; PICP: procollagen type I C-terminal propeptide; OC: osteocalcin; VFx: vertebral deformity.

deformities was also small ($N = 20$). Thus, results may not represent a larger cohort. Additionally, over the 5 years of follow-up, 133 subjects did not return, either because of death, ill-health, being lost to follow-up, or dropout. Since we were unable to radiograph the spines of non-returners, it was not possible to establish the incidence of vertebral deformities in the whole cohort. It could be speculated that the rate is higher in the non-returners. This is highly probable, since the mean age at baseline was higher in non-returners (67.8 versus 62.7, $P < 0.001$), and the prevalence of vertebral deformities rises with increasing age. Furthermore, subjects identified as having vertebral deformities, both prevalent and incident, were investigated at a metabolic bone clinic and, where appropriate, were offered treatment to prevent further bone loss. Treatments included hormone replacement therapy, bisphosphonates and calcium supplements. Prevention of bone loss in these individuals may have weakened the link between prevalent and incident vertebral fractures.

In summary, our study has found that FN and TB BMD are both competent risk factors of non-vertebral fracture and are rather better than LS BMD. Of all the biochemical markers

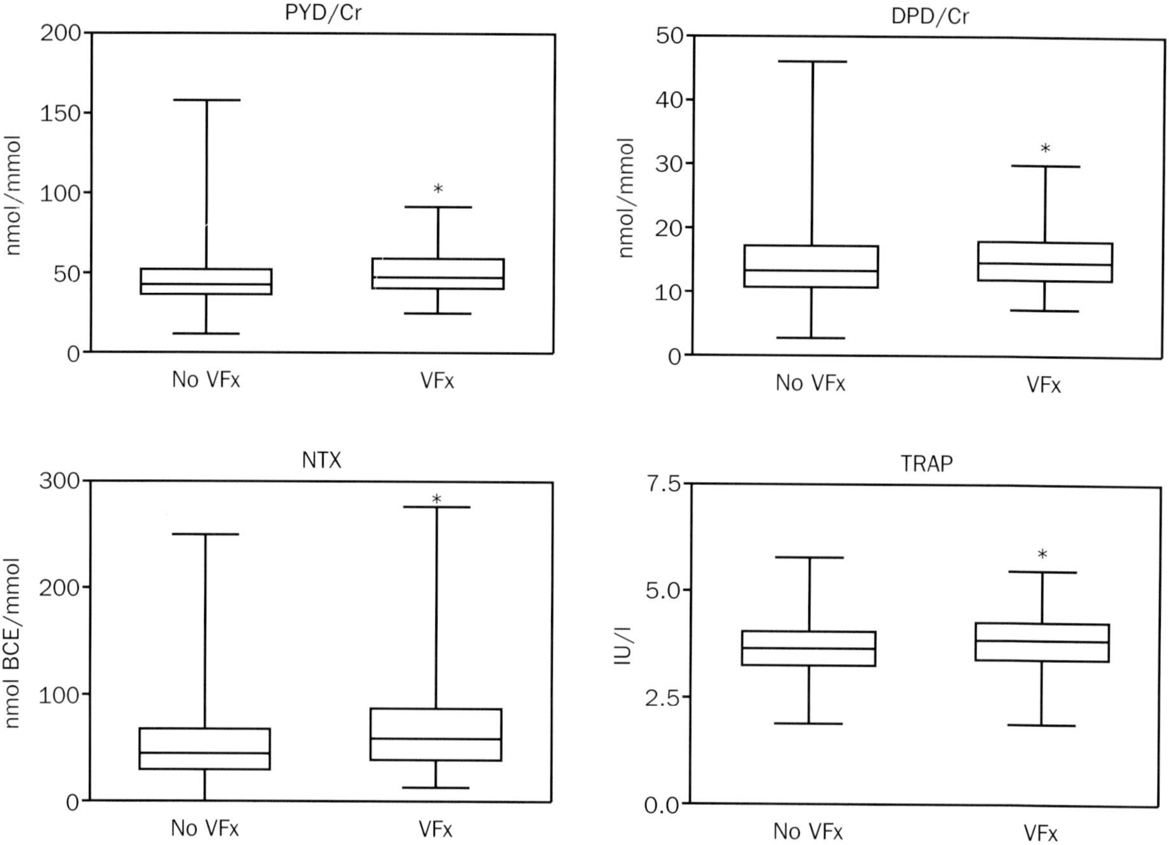

Fig. 22.2 Bone resorption markers: cross-sectional results comparing subjects with and without prevalent vertebral deformities at baseline. The box gives the inter-quartile range, the line through the middle is the median, and the whiskers give the range. BTMs compared by Mann-Whitney tests, $*P < 0.05$.
PYD: pyridinoline; DPD: total deoxypyridinoline; Cr: creatinine; NTX: N-terminal telopeptide; TRAP: tartrate-resistant acid phosphatase; VFx: vertebral deformity.

measured, only PYD was weakly predictive of incident non-vertebral fractures, and this was not independent of BMD. In terms of risk factors for incident vertebral deformity, prevalent vertebral deformity, TB BMD and PYD/Cr were competent risk factors. However, the fact that some markers appear to be non-significant statistically may not mean that they are non-significant clinically. It may be because the sample size and the number of events are insufficient. With a larger sample or a sample accruing a larger number of events (e.g. studied for a further 5 or 10 years), some of the markers may have reached statistical significance.

Finally, in terms of risk factors for vertebral deformities, our prevalence results give similar results to the incident study. High BTM levels are associated with both vertebral deformities and non-vertebral fractures. However, high BTM levels may indicate other conditions, so it would be misleading to expect them to be used in isolation for predicting fractures. Hence, high BTM levels may be an indication for further investigation. In the context of this study, therefore, we conclude that BTMs do not predict fracture. They do provide relative risks, but these must be taken in combination with other factors such as age, BMI, BMD, medical history

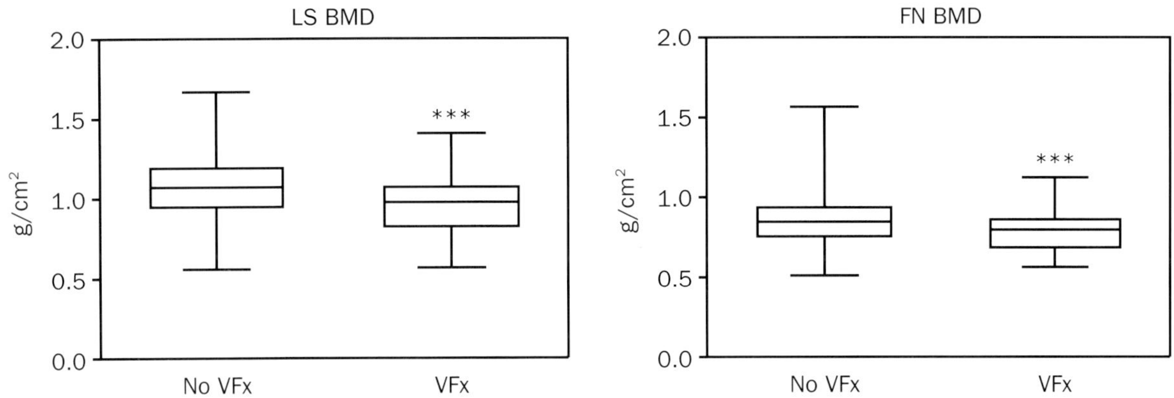

Fig. 22.3 Bone mineral density: cross-sectional results comparing subjects with and without prevalent vertebral deformities at baseline. The box gives the inter-quartile range, the line through the middle is the median, and the whiskers give the range. BMDs compared by Student's *t*-tests, $***P < 0.001$.
LS BMD: Lumbar spine bone mineral density; FN: proximal femur; VFx: vertebral deformity.

and so on. A larger and longer data set will provide the opportunity for more refined data analysis and the development of predictive models.

ACKNOWLEDGEMENTS

We thank all the staff and students of the Bone Metabolism Group, both in the Clinical Sciences Centre and the Osteoporosis Centre, for their assistance in data collection, project management and measurement of biochemical markers. This work was partly funded by a programme grant awarded by the Arthritis Research Council.

REFERENCES

1. Melton LJ3, Khosla S, Atkinson EJ, O'Fallon WM, Riggs BL, Relationship of bone turnover to bone density and fractures. *J Bone Miner Res* 1997; 12: 1083–91.
2. Eastell R, Robins SP, Colwell A, Assiri AMA, Riggs BL, Russell RGG, Evaluation of bone turnover in type I osteoporosis using biochemical markers specific for both bone formation and bone resorption. *Osteoporosis Int* 1993; 3: 255–60.
3. Genant HK, Wu CY, Vankuijk C, Nevitt MC, Vertebral fracture assessment using a semiquantitative technique. *J Bone Miner Res* 1993; 8: 1137–48.
4. Eastell R, Cedel SL, Wahner HW, Riggs BL, Melton LJI, Classification of vertebral fractures. *J Bone Miner Res* 1991; 6: 207–15.
5. Rosalki SB, Foo AY, Two new methods for separating and quantifying bone and liver alkaline phosphatase isoenzymes in plasma. *Clin Chem* 1984; 30: 1182–6.
6. Colwell A, Russell RGG, Eastell R, Factors affecting the assay of urinary 3-hydroxypyridinium crosslinks of collagen as markers of bone resorption. *Eur J Clin Invest* 1993; 23: 341–9.
7. O'Neill TW, Cooper C, Cannata JB et al, Reproducibility of a questionnaire on risk factors for osteoporosis in a multicentre prevalence survey: the European Vertebral Osteoporosis Study. *Int J Epidemiol* 1994; 23: 559–65.
8. Mole PA, Walkinshaw MH, Robins SP, Paterson CR, Can urinary pyridinium crosslinks and urinary oestrogens predict bone mass and rate of bone loss after the menopause? *Eur J Clin Invest* 1992; 22: 767–71.
9. Akesson K, Ljunghall S, Jonsson B et al, Assessment of biochemical markers of bone metabolism in relation to the occurrence of fracture: a retrospective and prospective population-

based study of women. *J Bone Miner Res* 1995; 10: 1823–9.

10. Garnero P, Hausherr E, Chapuy MC et al, Markers of bone resorption predict hip fracture in elderly women: the EPIDOS prospective study. *J Bone Miner Res* 1996; 11: 1531–8.
11. van Daele PL, Seibel MJ, Burger H et al, Case–control analysis of bone resorption markers, disability, and hip fracture risk: the Rotterdam study. *BMJ* 1996; 312: 482–3.

Discussion paper D

Panel: *Douglas C Bauer, Roland D Chapurlat, Pierre D Delmas, Peter R Ebeling, Patrick Garnero, Diana Greenfield, Olof Johnell, John A Kanis and Michael R McClung*

Delmas We now have some questions that Richard (Eastell) wants to be addressed by the panel. The first one is: is high bone turnover a risk factor for fracture? Who wants to address that? Nobody! I think actually that the data are pretty consistent for that, but probably the most important question is if and how it can be used clinically. The criticism that has been made of the markers is that they are significant for the group value but that may not be true at individual level. I guess the question is, if you consider increased turnover as a risk factor in the same way as other risk factors, then is this an issue? Who wants to address that?

Chapurlat I think that the way to address the issue is again to calculate posterior probabilities—we need to define treatment thresholds. Let's say that a treatment threshold is an incidence of fracture of 10%. If the prior probability is 5% and the marker gives a likelihood ratio of 2 or 2.5, then these prior probabilities double and the posterior probability is above the treatment threshold. So markers could be used as other risk factors.

Johnell It depends if the result would be universal from the EPIDOS centre and from OFELY. But the question is, why are all studies not the same? I think that with the magnitude of the risk time from EPIDOS and OFELY, you have sufficient for risk factors to be used clinically, and in a combination of risk factors. I think we can all agree on that, but why are all studies not the same?

Bauer I would ask the question why we see such heterogeneity between the studies before you can answer the question of who you apply the diagnostic test to. I think actually that I am more confused than I was even before I heard the panel presentations, because I am not sure that I see total agreement among the groups. I would be interested in what other people think. In SOF the serum CTX results appear to be an outlier, perhaps related to the lower mean levels compared to other population-based studies, as well as the difficulty with fasting. So now we actually see some other discrepancy and a fairly consistent lack of effect with the formation markers, and yet at least two of the four studies show a strong effect with resorption markers. I don't have a good explanation for this—does anyone else?

Delmas Let's make it more lively by having questions from the floor. Larry?

Raisz I want to talk about a problem that has been bothering me for a long time. Old people are frail and have less muscle, and their creatinine excretions are lower; all of these data for the urine are corrected for creatinine. Are we measuring the frailty of our population in the urine samples?

Delmas Yes, that is a very good point, Larry, as usual.

Raisz So what are the data on creatinine levels in the blood or excretion in the urine and prediction of fracture?

Delmas That's a very good point which we have actually considered. We had the same reasoning that they have a low muscle mass so they have a low urinary excretion of creatinine, and maybe what we measure is only the predictive value of urinary creatinine. Patrick (Garnero), maybe you can comment on that.

Garnero In the EPIDOS study we corrected the odds ratio for serum creatinine and also for disability status. We found that the odds ratio were very similar, and therefore believe that this odds ratio reflects not only diminution of excretion of creatinine but really an increase in bone resorption.

Delmas I seem to remember that there was also a correction for GFR evaluated from serum and urinary creatinine which did not change the results either.

Garnero Yes, we did correct for this.

Johnell We looked at serum creatinine and we couldn't find it as a predictor in the 50-year-old men and women for hip fractures. Did you look at serum creatinine by itself as a risk factor?

Garnero Yes; there was no association.

Fraser I wonder if some of the differences that we've seen in the SOF study, against other studies in serum CTX, relate to the length of time for which the samples were stored before analysis? One of the things that we have noticed in a small sub-study is that EDTA seems to protect CTX, whereas with serum storage, even at −50°C and at higher temperatures, the CTX degrades. Does that explain why you've got different mean values in your samples compared to the European studies?

Delmas Especially because they were much lower.

Fraser Especially since they are almost three times lower. In fact, they are a lot lower than the subjects that we've studied—significantly lower. For all of the studies that I presented this morning, we measure those samples within months, whereas you have stored the SOF samples for a significant period of time.

Delmas That's a good point.

Bauer I agree with you, but they were stored at −90°C to −100°C. The other thing is that we saw the expected relationships with oestrogens. We saw a very dramatic difference between oestrogen users and non-users; you could argue that perhaps there could be some selective drop-out, but again it would have been more convincing to me if we had seen a lack of other expected associations with weight or with oestrogen.

Fraser But oestrogen is an extremely stable steroid hormone, as opposed to CTX, which is isomerizing.

Bauer No, I am referring to self-reported oestrogen use—which was associated with a 60% reduction in serum CTX levels in the cohort.

Delmas The point is well made.

Eastell I wanted to make a comment and ask a question. The comment was that in the Sheffield study, the bone resorption markers were all 24-h excretion, there was no creatinine correction, and these did show prediction of fractures. The question I was going to ask was from the French study, particularly in the OFELY study. One of the problems we encountered, Patrick (Garnero), was do you or don't you exclude people who you may consider might have disease or be taking drugs that you know affect bone turnover? Obviously, you might exclude HRT and bisphosphonates, but we have also found other things that affect bone turnover, like thiazide diuretics and cod-liver oil—there are lots of reasons for exclusion. In fact, when Diana went through all these cases and thought about the exclusion, she ended up excluding about 85% of the people. I know that in your EPIDOS case–control you excluded several people, but in your OFELY study, had you excluded people and how did that affect your analysis?

Garnero We excluded, obviously, those women taking oestrogens, bisphosphonate or corticosteroids at baseline, but we kept patients on thiazide treatment. We also made the analysis by adjusting for thiazide use, and there was no effect. Therefore we only excluded HRT, bisphosphonates and some metabolic bone diseases.

Delmas Which were obvious like Paget's or myeloma or primary hyperparathyroidism.

Garnero But the bottom line is that, of the patients who were excluded from the analysis, most were HRT users.

Chesnut III One quick comment on the stability of serum CTX in our PROOF calcitonin study, some of the samples for serum CTX had indeed been stored at −70°C for up to 7 years, and Dave Baylink did confirm stability of the serum CTX over that long period of time at that particular temperature. There were aliquots removed during that time and checked, and there were essentially the same levels throughout. I can show you the data.

Ebeling I would like to ask whether, as men and women get older, do you think that the predictive value of these markers might be improved if you corrected the bone marker you are looking at for total bone mineral content? Have you tried to do that in your analyses?

Raisz I may have misunderstood something. Patrick (Garnero), you had 12.5% fractures in your control group population, is that right? You have a cut-off of 10% for treatment. Well then, shouldn't we be treating everybody, or do I misunderstand these numbers?

Garnero The mean age was 64.

Raisz Sixty-four is young. Your number for 64 was way below 10%.

Johnell But we showed you hip fractures; for all fractures it would probably be much higher. It would be important to create some sort of hip fracture equivalents to calculate the other fractures back to hip fractures and add them on, because hip fractures are the most important to be able to incorporate all osteoporotic fractures.

Raisz Your hip fracture number would be much lower.

Kanis A couple of questions for Patrick (Garnero). One of the things that we didn't hear about, but that we have heard from EPIDOS is about undercarboxylated osteocalcin—so that's one question. The second question is that you seem to have shown independent contributions of various hormones, for example IGF-1, and presumably you have already calculated what the relative risks would be with a combination of hormones, markers and BMD?

Garnero Regarding the first point, we did not measure undercarboxylated osteocalcin in the OFELY study. With regard to the second point, we did actually combine, for example, not IGF-1 but oestradiol and markers of resorption like serum CTX, and we see that relative risk increases from about 2 to about 4.

Ebeling My comment doesn't really relate to prediction of fracture, but to a study that we published about 4 years ago. At that time we were looking at the perimenopause and the changes in bone markers that occurred during the perimenopause. You may recall that we found that the markers went up mainly in the postmenopausal phase, but in perimenopausal women the NTX marker was increased. So we've now looked at these women for a total of 6 years, and then in a prospective fashion looked at the changes in bone density to see where bone density really changes in relation to the stage of the menopause that they were in over that time. The Melbourne Mid-life Women's Health Study comprises about 350 women who have been followed up for 6 years with about three bone density scans, and they're in about four different groups. The first group remained pre- or early perimenopausal; the second group became late perimenopausal, which was 3–11 months of amenorrhoea; the third group became truly postmenopausal—that's 12 months or more of amenorrhoea; and the fourth group were those women who remained postmenopausal. None of these women were on hormone replacement therapy. So what we found when we looked at the analyses of the different bone resorption markers was that the total pyridinoline and deoxypyridinoline were increased in groups 3 and 4 at baseline—this was at entry to the study, when they were all aged between 45 and 55 years. Interestingly, the free deoxypyridinoline was increased in the group that went from premenopause to late perimenopause, which was the earliest of the menopause transition stages. The rates of spinal bone loss in each of those groups were: no change in group 1, 1% per year change in group 2, 3% change in group 3, and 1% change per year in group 4. Thus the largest changes in bone density occurred in the transition from pre- or peri- to postmenopausal. With the telopeptide cross-links, surprisingly we found that the NTX values weren't increased in group 3, whereas the CTX values were increased in both groups 3 and 4. By contrast, in the cross-sectional study, it looked like the NTX was a more sensitive marker in this sort of situation. So really that's all I wanted to show you today. We've got to go on and correlate the markers with FSH, in particular, and oestradiol levels to see what the important factors were, but these changes in bone resorption occurred before the progression of women through the menopause, indicating that bone resorption increases very early.

Delmas Well, thank you very much. Although that is a bit far from the fracture, it is definitely a clinically important question, because we see these perimenopausal women that do not tolerate HRT and those that start with a low BMD. If they have an increased rate of bone loss, that could clearly put them in the osteoporosis level very quickly, and today we have no regimen that is fully approved. So I think it is an area that deserves more studies. Questions from the floor or panelists?

Kleerekoper Can I follow up on Larry's question and perhaps ask from each of the studies, do you feel the markers you are measuring are better than clinical markers of frailty in the elderly in predicting fractures? Are they as good as, better than, or additive to?

Delmas Well, that has been looked at, I know, at least in the EPIDOS study, so Patrick (Garnero) do you want to comment?

Garnero In the EPIDOS, bone resorption markers also predict the risk of fracture independently of frailty assessed by the gait speed, and we don't combine the two factors, but maybe they could be combined. It is additive; the high resorption, at least in the EPIDOS study, is not related to frailty, not just assessed by the speed gait, but also by the body weight. I know there was an issue in your first report—it was associated, and when you had more cases it became independent.

Ebeling I would like to ask the panel something pertaining to osteocalcin. I have several patients who are on glucocorticoids, and if they come to me and I measure their osteocalcin and it's low, does that mean that I can reassure them that they are not going to have a fracture, because they have low bone turnover? In the setting of glucocorticoid use, does having a low osteocalcin level increase or decrease their risk of having a subsequent fracture?

Bauer In SOF they were not on corticosteroids, but osteocalcin didn't affect the risk at all—either high or low.

Garnero Simply, there is no association, as you saw, between osteocalcin and fracture rate. But obviously, if you treat with corticosteroids, this should be taken into account.

Ebeling So we don't know, is that the answer?

Chapurlat In corticosteroid-treated patients, bone turnover is really different compared to postmenopausal women, and there is a non-coupling, so things are really different. I think we cannot speculate from results of these prospective studies of healthy people.

Delmas I don't think there are any long-term data actually looking at the value of the markers in steroid-treated patients to predict fractures.

Johnell In 1 or 2 days after you started your glucocorticoid treatment, you got a very low osteocalcin value, but we don't know what is happening after 5 or 6 months—how much it returns. Do you know that Pierre (Delmas)?

Delmas In the only really longitudinal study in which the first measurement was done prior to the steroid treatment, they don't last for very long—the longest one I think was 12 months. What happens is that very often the dose of steroids is tapered, so the increase could reflect the fact that the dose is decreased. Now our Chairman wants us to address the question about how to use that—these markers—if we want to predict the risk of fracture. Would there be a consensus that perhaps the most conservative use would be to see increased resorption as a risk factor that you will add onto the information provided by BMD measurement? So, I have one Yes and Doug probably No.

Bauer Could I just make one clarification point? Just because two predictors are independent in a multivariate model, it doesn't necessarily mean they give more information on the overall likelihood of a fracture. So what I am wondering, and Patrick (Garnero) I think you have done this before, when you plot ROC curves, is the area under the curve different when you add markers to bone mineral density to just BMD alone?

Delmas You mean, if our relative risk goes from 2 with BMD alone to 4, this is not meaningful.

Bauer No, I think the best example is, for instance, quantitative ultrasound and BMD, where in populations they are both independently associated with hip fractures, but ultrasound doesn't appear to add much additional information to BMD alone.

Garnero When we combined CTX with ultrasound, the area under the curve increased a little bit. When we analysed the data for a given sensitivity, we see that actually, by combining CTX and BMD, we could significantly increase the specificity. So maybe it is adequate not to look at the area under the curve, but to look at a specific cut-off. Because for bone resorption markers we have seen that the increase in the risk is only for a value above a certain level, we have to look to analyse the data for this cut-off and then not use area under the curve.

Kanis Yes, if you have independent risk factors, you can combine them in a variety of ways. But if you combine it with BMD, you can convert the new combination into a normal distribution. If you then look at the combined risk factor, compared to either risk factor alone, then the specificity in application to the population as a whole remains the same, but, as Olof (Johnell) showed in a microsecond on one of his slides, the sensitivity improves. Thus you can increase sensitivity in our assessment without trading off specificity—that is a very important principle in terms of adding risk factors together. The other situation where the combination is important is along the lines that Olof was trying to develop. If you think that you want to take intervention decisions above a given risk, and Olof was suggesting a 10-year probability of 10% for hip fracture, then this is a mechanism by which you can attain that, not in the whole population, but in a segment of the population, whereas a single test alone would not be able to achieve that. And it is probably true in the field of bone disease at the moment that no single test will allow us to treat health—economically a segment of the population using that test alone. On the other hand, with a combination of factors, if we can improve the sensitivity of the test, even marginally—even by 50–60%—then we stand a chance of being able to identify people above a threshold of risk which is justifiable from a healthcare purchaser point of view or from a clinical perspective. So I think it's a very important issue and I just want to know (the question that Pierre was asking) what is the consensus view on whether the combination of markers and bone mineral density improves the predictive value?

Delmas So those who think yes, raise their hands—so let's say who disagrees based on the data presented? Just for fun—who disagrees that you cannot improve your prediction by combining bone markers with a BMD? Who is sceptical about the use in that way that was so nicely described, as usual, by John?

Raisz The problem is that we are dealing with multiple studies in multiple populations, and until we have seen a wider distribution of population, none of us can disagree because we have some evidence but we can't agree because we don't have enough data. I know it seems like a lot of data, but to me the number of fractures that we've talked about has not been astounding. I'd like to see data on 1000 fractures.

Kanis Larry, there have been no studies of the relationship of BMD and fracture risk in Madagascar—do you think that we should reserve judgement on the relationship?

Raisz The number of fractures is not that astounding in my view; maybe other people feel that there are plenty of fractures here.

Delmas Two hundred hip fractures.

Raisz Two hundred hip fractures is approximately half the number that happen in Hartford, Connecticut.

Delmas Let me challenge that; do you use drugs in your clinical practice that have shown efficacy in hip fracture on a larger sample?

Raisz On the number of fractures, no—I agree, and we have concern about that as well. If you want votes of confidence, they are going to be partial when the data are somewhat limited in my view, that's all.

Delmas Larry, I think what was nice was that in one hour we have seen data from five different cohorts, and if you exclude the SOF study, which was the only negative one, all the other cohorts have shown exactly the same trend—at least for resorption markers and in a population that varies according to age. The OFELY study starts at 50 years of age; the EPIDOS starts at 75; the other cohorts are in between. So for me, what is reassuring is to see the same trend in cohorts that are quite different in terms of age and other characteristics.

Raisz I agree, it's in that direction, I just think we need more data.

McClung One comment. Regardless of the consensus of whether we can add or combine risk factors, let me point out that there are at least two separate uses of clinical risk factors. One is in predicting fracture risk, which is what the consensus is about, but deciding whether therapeutic intervention is indicated and can be assessed by adding or combining risk factors is a different question. Not all risk factors bear the same information or weight in terms of predicting response to therapy. You are aware that we have just completed a large study in which patients were identified at risk for hip fracture on the basis of frailty, in which antiresorptive agents were not effective despite the fact that those patients were clearly at high fracture risk based on clinical frailty-type risk factors. The only patients in whom we have demonstrated a reduction in fracture risk with our osteoporosis drugs are patients with previous fractures or patients with very low bone density. It would be interesting in clinical trials to look at the subset of patients with low or high markers in any bone density group and to see if that gives us more information about the therapeutic efficacy, so we can decide whether we can use that piece of information to decide whether to treat or not to treat. But mostly, I want to emphasize that deciding to use risk factors for treatment is a different issue for predicting fracture risk.

Delmas Absolutely, and I think it is a very good point; again, what the study you are referring to really demonstrated is that selecting patients for treatment based on clinical risk factors that are independent on osteoporosis does not make any sense. All the data we have seen today suggest that this would not be the case for markers because, clearly, increased resorption is associated with bone metabolism. Any comment?

Johnell I actually had that on a slide; most of the risk factors suggested are tested in some of the trials.

Kanis There is quite lot of information that has been published, and I would have thought that there are enough studies published now for someone who is good at this to be able to undertake a meta-analysis and come to some conclusion as to what the overall effect and what the magnitude of the effect might be. Can I just finally add that I am gratified to see that this is the first time that lack of evidence has stimulated Larry Raisz not to have an opinion.

Delmas Thank you, I think it is time to close this round-table. Richard (Eastell), do you want to say a few words?

Eastell I think today we have heard about the different candidates for bone markers and how we really have a nice selection of these now. Where the future lies is perhaps having markers of osteocyte action, or of the different envelopes of bone. I think there is still a lot of work to do on finding new markers, but we have plenty to keep us going, proven in clinical practice to be useful. We had a discussion about variability and emphasized the importance of two particular factors, namely the circadian rhythm and day-to-day variability, and addressed this issue by making some provisional recommendations about having preferably more than one marker, those of us who can afford it, and having the sample at a particular time of day—probably in the morning, fasting. We then went on to talk about the issue of monitoring and had a really lively debate on that topic, and I think basically people in general felt that this was a good way of identifying non-compliers and non-responders, and thresholds are now being developed and quite a few of the markers have proved useful in the setting of monitoring. There are some questions that still remain, like what is the relationship between changes in bone turnover and fracture risk. Finally, in this section here on prediction, I think people are coming towards the idea of using bone markers in fracture prediction models, although there are still some concerns about inconsistency across studies, and I think it would be good to see more data—the meta-analysis that was proposed is a good idea, and the idea that was proposed of looking at the clinical trials to see whether bone turnover predicted efficacy is also really important for future research. So I think today we have heard some really positive things about bone markers, that really there are some situations where they are likely to prove useful to us and that we currently have markers that will be useful in this way, but there are still things that we would like to learn from more extensive application of these to our clinical practice. So I would like to end up by thanking today all of the Chairpeople, the panelists, the speakers and also the audience for all the comments and interesting debate we've had.

Index

Page numbers in *italic* indicate references to figures; number in **bold** indicate references to tables.

achondroplasia 12
acid phosphatase *see* tartrate-resistant acid phosphase (TRACP)
adolescence *see* childhood
adrenal steroids *see* steroid hormones
adulthood, bone turnover 97–8, 99, 100
age effects
  bone mineral density (BMD) **220**
  bone turnover 203
  bone turnover markers (BTM) 95–9
  C-telopeptide cross-links (CTX) 206
  reference ranges 137
alendronate
  in antiresorptive therapy 20, 22, 150, 151, 152
  bone mineral density (BMD) 124, 171
  C-telopeptide cross-links (CTX) 168, *169*
  deoxypyridinoline (DPD) 91
  diurnal variation/variability 120
  N-telopeptide cross-links (NTX) 168, *169*
  response 168, *169*
  see also antiresorptive therapy
algorithm
  fracture prediction 198, 199–200
  monitoring 160
alkaline phosphatase (bone)
  age-related changes 95, 97
  assay 30–1, 58–60, 61–2, 205, 213, 227
  and bone mineral density (BMD) 153
  as bone turnover marker (BTM) *6*, 57–62, 93
  cross-reactivity 59, 60, 62
  diurnal variation/variability 119–20
  ethnic differences 100, *101*
  genetics 57
  and immobilization 127
  isoenzymes 28–9, 57, 58
  localization 28
  and metabolic bone disease 129
  metabolic function 57
  monitoring 2, 61
  oestrogens 151
  osteoblasts 58
  osteoclasts 29
  physiological peaks 58
  post-translational modification 57
  predictive value 154, 203, 208–9, 230
  radioimmunoassay 59
Alkphase B 59, 60
among-day variation/variability 133
anabolic agents 22–3, 183–4
analytical variation/variability 44, 131, 132, 135
androgens 17
ankylosing spondylitis 85
anthropometric measurements 228
antiresorptive therapy
  bone mineral density (BMD) changes 152–3, 180
  bone turnover markers (BTM) 33, 151–2, 159–65, 167–77, 212
  C-telopeptide cross-links (CTX) 39, 45
  clinical trials 154–5, 182–3
  compliance 45, 61, 124, 150, 192, 193
  cut-offs 150, 191–2
  monitoring 34–5, 45, 61, 149–55, 162–3, 166–7, 190, 212
  patient selection 34, 45
  preventive 34
  response prediction 154–5, 175, 183, 191
apoptosis 8, 22
Arg-Gly-Asp (RGD) sequence 10, 83, 84, 86
aspartyl isomers 51

assays
  accuracy 144, 147
  alkaline phosphatase (bone) 30–1, 58–60, 61–2, 205, 213, 227
  automated 31, 39, 42, 45, 67, 205, 213
  C-telopeptide cross-links (CTX) 32, 39, 40–6, 109, 205, 213, 223
  deoxypyridinoline (DPD) 31, 49–50, 109, 213, 227
  N-telopeptide cross-links (NTX) 32, 227
  osteocalcin 67–9, 70, 205, 213, 227
  point-of-care 27, *31*, 32, 39, 40, 132, 179
  procollagen C-terminal propeptide (PICP) 73–80
  procollagen N-terminal propeptide (PINP) 73–80, 205
  pyridinoline cross-links (PD) 49–50, 227
  standardization 53, 61–2, 138, 174
  techniques 31–2, 40–2

bed rest *see* immobilization
benign vertebral osteoporosis (OPO) 86
biochemical monitoring of bone metabolism 2, 3
biological variation/variability *see* variation/variability
biomechanical forces *see* mechanical forces
bisphosphonates
  antiresorptive therapy 2, 20–2
  bone sialoprotein (BSP) 85, 86
  bone turnover markers (BTM) 33, 119, 121, 151–2
  chemical structures *21*
  malabsorption 150, 171, 194
  osteoclast regulation 8
  procollagen C-terminal propeptide (PICP) 151
body weight 198
bone
  cells 3–10, 14–15
  densitometry *see* dual-energy X-ray absorptiometry (DXA)
  differentiation 3
  disease
    bone turnover markers (BTM) 123–9
    diminished resorption 17
    increased resorption 17–18
    non-osteoporotic 127–9
    pathogenesis 1
    therapies 18–23
  formation
    markers 28–9
      see also alkaline phosphatase (bone), procollagen C-terminal propeptide (PICP); osteocalcin; procollagen N-terminal propeptide (PINP)
  growth and development 3
  histomorphometry 28, 32
  loss
    and bone turnover 204
    and bone turnover markers (BTM) 220–1
    inflammatory 18
    menopause 98–9
    prediction 34, 45
    preventive treatment 233
  markers *see* bone turnover markers (BTM)
  matrix proteins **4**, 10, 28
  metabolism 1–23, 28
  metastases
    bisphosphonates 20
    bone sialoprotein (BSP) expression 84
    bone turnover markers (BTM) 33, 127, 129
    breast cancer 2, 18, 86, 129
  mineral density (BMD)
    age-related changes **220**
    alendronate 124, 171
    antiresorptive therapy 33, 152–3, 171, 180
    assessment, hip 197–198
    bone turnover markers (BTM) 98–9, 152–5, 215, 225
    C-telopeptide cross-links (CTX) 39, 153, *172*
    ethnic differences 100
    least significant change 150
    monitoring 34–5
    negative correlation with ucOC 66
    predictive value 34, 153–5, 191, 192, 193, 197–201, 204, 221
    risk factor 233
    stepwise stratification 138
    time scale 171
  mineralization 84
  morphogenetic proteins (BMPs) 3, 13
  oestrogen anabolic effect 16
  remodelling 1, 9–10, 27–8, 125–6
  resorption 17–18, 214–15
    markers 29–31
      see also C-telopeptide cross-links (CTX), deoxypyridinoline (DPD); N-telopeptide cross-links (NTX)
  sialoprotein (BSP) 83–7
  status, clinical assessment 17
  turnover
    age-related changes 97–8, 203, 204, 207
    and apoptosis 8
    baseline 152
    and bone loss 204
    and fracture risk 225–35, 237
    markers (BTM)
      age-related changes 240
      anabolic therapy 183–4
      antiresorptive therapy 33, 151–2, 159–65
      assays *see* individual markers
      baseline levels 172, 179, 180–1, 184, 204–5
      biological variation/variability 95–102, 119–21, 131–8, 180–1
      bisphosphonates 119, 121, 151–2
      and bone disease 123–4
      and bone loss 220–1
      and bone mineral density (BMD) 98–9, 152–5, 215, 225

and calcitonin 152
clinical utility 33–4, 45–6, 117
comparisons 51
cut-offs 149, 168
and densitometry 132
deterioration during storage and collection 134
diurnal variation/variability 92–3, 107–18, 119–20, 132–3, 187–8
dose-related changes 168
elevation 33, 144
gender effects 99
hyperparathyroidism 123
and immobilization 119, 126–7
least significant change 154, 172
in monitoring 61, 149–55
new 243
non-osteoporotic bone disease 127–9
oestrogen 119, 121, 151, 174–5
osteomalacia 123
osteoporosis 132, 149–55
Paget's disease 123
predictive value *see* individual markers
procollagen N-terminal propeptide (PINP) 74
and raloxifene 152
reference ranges 100–2
reproducibility 161
thyrotoxicosis 114
tissue specificity 50–1
urine 50, 108
validation 32–3
rate 1, 3, 28
and vertebral fracture 226
breast cancer, bone metastases 2, 18, 86, 129

C-telopeptide cross-links (CTX)
age-related changes 206
alendronate 168, *169*
antiresorptive therapy 39, 45, 167
assays 32, 39, 40–6, 109, 205, 213, 223
baseline 174
bone mineral density (BMD) 39, 153, *172*
calcitonin 196
diurnal variation/variability 44, 107, 109, *110*, *111*, *112*, 120, 215
gender effects 31
ibandronate *170*
least significant change 176
low-end detection limit 43
menopause-related changes 206
and oestrogens 151
old age 199
osteoporosis 114, *115*, 117
point-of-care 40
possible improvements **46**
predictive value 154, 175, 176, 198–9, 203, 209, 213–14, 231
sensitivity and specificity 167, 172
serum 39, 42, 45
structure 28
thyrotoxicosis 117
tissue specificity 42
and ultrasound 241
urine 39, 45, 108
C-telopeptides 39–46
cachectin 10
calcification 84
calcitonin 1, 2, 15, 22, 150, 152, 182, 196
calcitriol 1, 2, 10, 15
calcium
intake 119, 121, 137, 141, 180
metabolism 1–3
PROOF study 239
regulating hormones 1, 2
supplementation 120
in urine 30
cancer
secondary bone changes 2
see also bone metastases
cartilage morphogenetic factors (CMPs) 3
cartilage-derived morphogenetic proteins (CDMPs) 13
case studies 124–5
cathepsin K 17, 27, 30, 46
cbfa1 gene 1, 3
change
expected 162–3
not important 163
relative and absolute 189–90
unexpected 162
childhood
bone turnover markers (BTM) 96–7
ethnic differences 100
chondrodysplasias 3
circadian rhythm *see* diurnal variation/variability
cleidocranial dysplasia **4**
clinical trials
antiresorptive therapy 154–5, 182–3
bone turnover markers (BTM) 181–2
heterogeneity 237
population-based statistics 165, 168
time scale 163, 180
treatment response monitoring 168–71
clinical utility
bone turnover markers (BTM) 117
and least significant change (LSC) 43–4
clodronate 20, 22
collagen
in bone and connective tissue 73, 78–9
C-terminal propeptide *see* procollagen C-terminal propeptide (PICP)
degradation products 2, 17, 28, 39–40, 108
dietary intake 27
metabolism in bone 28

collagen *cont.*
N-telopeptide cross-links *see* N-telopeptide cross-links (NTX)
N-terminal propeptide *see* procollagen N-terminal propeptide (PINP)
propeptides 2, 10
telopeptide assay epitopes *40*
collagenase 10
colony-stimulating factors (CSFs) 6, 7, *8*, 11
compliance 45, 124, 150, 171–2, 192, 193, 243
computer simulations 163–5
corticosteroids 8, 10, 13, 241
coupling index 125, *125*
creatinine 108, 134, 227, 238
cross-link assays *see* pyridinium cross-link assays
CrossLaps 32, 39, 40, 187–8, 190, 205, 213
cut-offs
antiresorptive therapy 150, 191–2
bone turnover markers (BTM) 149, 168
deoxypyridinoline (DPD) 144
determining 153, 154, 195
N-telopeptide cross-links (NTX) 182
stepped 137–8
cyclooxygenase-2 10, 13–14
cytokines 1, 3, **4**, 6, 7, 8, *9*, 14, 15–17

deoxypyridinoline (DPD)
accurate parameter 52
age-related changes 95, 96, 97
and alendronate 91
among-day variation/variability 13
assays 31, 49–50, 109, 213, 227
and bone mineral density (BMD) 153
bone specific 28
bone turnover markers (BTM) 198
collagen degradation product 108
cut-offs 144
diurnal rhythm 107, 109, *110*, *111*, *112*
excretion *51*
gold standard 172
and immobilization 126–7
in liver and kidney 27
and oestrogens 151
predictive value 198, 213–14, 231
proportion in collagen 91
reproducibility 161
structure 28, *50*
urine 30
derived index *see* least significant change
dietary factors 119, 134, 137
diurnal variation/variability
and alendronate 120
alkaline phosphatase 119–20
bone resorption 214
bone turnover markers (BTM) 92–3, 107–18, 119–20, 132–3, 187–8
C-telopeptide cross-links (CTX) 44, 107, 109, *110*, *111*, *112*, 120, 215
deoxypyridinoline (DPD) 107, 109, *110*, *111*, *112*
and fasting 44, 187, 215
gender differences 120
N-telopeptide cross-links (NTX) 109, *110*, 120
osteocalcin 66, 119–20
urine 116, 133, 136–7
drug design and development 23, *24*
dual-energy X-ray absorptiometry (DXA) 17, 109, 149, 150, 203, 204, 205, 208, 219, 220, 227
gold standard 153, 180, 208
T-score 197
dwarfism 12, 13

Early Postmenopausal Intervention Cohort Study 189
eicosanoids 14
elderly *see* old age
electrophoresis 58–9
ELISA
C-telopeptide cross-links (CTX) 39, 42, 205
osteocalcin 67
procollagen C-terminal propeptide (PICP) 78, 205
procollagen N-terminal propeptide (PINP) 74
pyridinoline cross-links (PD) and deoxypyridinoline (DPD) 50
enzyme activity measurements 59
EPIC trial 182, 183
epidermal growth factors (EGFs) 12
EPIDOS study 35, 45, 66, 188, 200, 211–16, 239
estr- *see* oestr-
ethnic factors
bone mineral density (BMD) 100
bone turnover markers (BTM) 99–100
osteocalcin 146
etidronate 20, 22, 152
European Vertebral Osteoporosis Study (EVOS) 232
exercise 119, 120, 134, 137, 141

familial expansile osteolysis (FEO) 18
familial factors 198
fasting 32, 44, 147, 187, 209, 212, 215, 243
fibroblast growth factors (FGFs) 3, 12
fluoride 22, 191
follow-up 175
fracture
and bone mineral density (BMD) 191, 192, 193, 204
and bone remodelling 125–6
and bone turnover markers (BTM) 98–9, 123–9, 190
history 227
normal healing 124
and oestrogen 223
prediction 34, 66
alkaline phosphatase (bone) 208–9
bone turnover markers (BTM) 155, 198, 208, 243–4
C-telopeptide cross-links (CTX) 198, 209, 213–14

deoxypyridinoline (DPD) 198, 213–14
ucOC 213
predictive factors 204
reduction 149
risk 180
absolute and relative 198, 199–200
assessment 197–201
and bone mineral density (BMD) 34
and bone turnover 225–35, 237
and bone turnover markers (BTM) 197–201, 203, 222–3
C-telopeptide cross-links (CTX) 198–9
factors 198–9, 211, 216, 228
and oestrogens 207–8, 209
prediction 206–7
scoring 144
ultrasound 197
Fracture Intervention Trial (FIT) 180, 183
frailty 238, 240, 243

galactosyl-hydroxylysine *6*, 96
gender effects
bone turnover markers (BTM) 99
diurnal variation/variability 108, 120
in infancy 99
N-telopeptide cross-links (NTX) levels *98*
reference ranges 136, 137
gene polymorphisms
alkaline phosphatase (bone) 57
bone sialoprotein (BSP) 84, 85
cytokines 14
genetic defects **4**, *9*, 12
genetic factors 1, 3, *19*
geographical variation/variability 200
gestational age 95
Gla protein 10
glucocorticoids *see* corticosteroid therapy
gonadal steroids *see* steroid hormones
growth factors 1, 3, 14, 16, 23
growth plates, closure 97, 143
growth velocity 96, 97

Hawaii Osteoporosis Study 181
heat-inactivation methods 58
height, as risk factor 228, 229
heterogeneity *see* variation/variability
high-performance liquid chromatography (HPLC) 31, 49, 227
hip fracture
lifetime risk 200, 211
prediction 194, 211–16, 231
see also fracture
hormone replacement therapy (HRT) 119, 121, 151, 171
see also oestrogen
hormones
calcium-regulating 1, 2
endogenous 205

humoral hypercalcaemia of malignancy 129
hydroxyproline *6*, 30, 31
hydroxypyridinium cross-links, assay 31
1,25-hydroxyvitamin D *see* calcitriol
hyperparathyroidism
alkaline phosphatase assay 59
bone sialoprotein (BSP) levels 85
bone turnover markers (BTM) 33, 123, 127
and fracture risk 180
monitoring 61
procollagen N-terminal propeptide (PINP) 74
hypogonadism 17, 126
hypophosphatasia **4**

ibandronate 151, *170*
IBSP gene 84, 85
ICTP 32, 39, 46, 96, 153, 161
immobilization 119, 120, 126–7, 134
immune system 15
immunoassays 31, 32, 39, 40, 59, 61, 85
immunoelectrochemical assay 74, 109
immunoradiometric assays 205, 213
index of individuality 136
infancy 58
inflammatory bone loss 18
insulin-like growth factors (IGFs) 3, 12
integrin receptors 10, 84, 86
interferon-gamma 7, *8*, 11–12
interleukins 6, 7, *8*, 10, 11, 15
intra-individual variation/variability 132–3, 136–7
ipriflavone 22

kidney
calcium uptake 2
failure 30, 46, 134
osteocalcin uptake 66–7
knockout animals *see* transgenic animals

laboratory measurements, advantages 132
latency-associated protein (LAP) 13
least significant change (LSC)
bone mineral density (BMD) 150
bone turnover markers (BTM) 154, 172
C-telopeptide cross-links (CTX) 176
calculating 195
and clinical usefulness 43–4
definition 131–2, 135–6
derived index 161
leptin 92
leukotrienes 14
lifestyle factors 119, 120, 121, 134, 137, 228
liver disease 58, 61
liver enzyme, alkaline phosphatase cross-reactivity 59, 60, 62
lymphotoxin 10
lysyl pyridinoline *see* deoxypyridinoline (DPD)

Marfan syndrome **4**
matrix proteins *see* bone, matrix proteins
mechanical loading 1, 3, 7, 27–8
Melbourne Mid-life Women's Health Study 240
menopause
  bone loss 98–9
  bone turnover 97, 204, 207, 240
  C-telopeptide cross-links (CTX) 206
menstrual cycle 66, 133
meta-analysis 243
metalloproteinases 10, 11, 12–13
metaphyseal dysplasia, Jansen type 13
mevalonate pathway *21*, 22
MHC antigens 11–12
minimum significant change (MSC) *see* least significant change (LSC)
molecular defects **4**
monitoring
  aims 160
  types 162–3
monoclonal gammopathy of undetermined significance (MGUS) 86
MORE study 155
myeloma 2, 18, 86, 129, 146

N-telopeptide cross-links (NTX)
  age-related changes 95, 96, 97
  alendronate 168, *169*
  in antiresorptive therapy monitoring 167
  assays 32, 227
  baseline 174
  and bone mineral density (BMD) 153
  calcitonin 196
  cut-off 182
  diurnal variation/variability 107, 109, 116–17, 120
  gender factors 31, *98*
  and immobilization 126–7
  inter-subject variation/variability 120
  in liver and kidney 30
  and oestrogens 151
  predictive value 154, 176
  sensitivity and specificity 167, 172, 182
  serum 27
  structure 28
  urine 27, 32, 33, 108
nitric oxide 1, 7, 14
nitric oxide synthase (NOS) 7, 14
non-compliance *see* compliance
non-fasting *see* fasting
non-steroidal antiinflammatory drugs (NSAIDs) 14
norethisterone acetate (NETA) 151
NTX *see* N-telopeptide cross-links (NTX)

obesity, and osteoporosis 91–2
oestrogen
  alkaline phosphatase (bone) 151
  anabolic effect 16
  bone turnover markers (BTM) 33, 174–5
  C-telopeptide cross-links (CTX) 151
  chemical structures *21*
  and cytokines 15–17
  deficiency 180
  deoxypyridinoline (DPD) 151
  and fracture risk 207–8, 209, 223
  and growth factors 16
  and interleukin-6 11
  N-telopeptide cross-links (NTX) 151
  osteocalcin 151
  osteoclast regulation 8
  procollagen N-terminal propeptide (PINP) 151
  receptors 15, 19–20
  therapy 2, 18–20
OFELY study 35, 188, 203–9, 212–13, 215, 239
old age
  bone resorption 214–15
  bone turnover 97–8, 207
  C-telopeptide cross-links (CTX) 199
  EPIDOS study 212
  gender differences 99
Ostase assays 59–60
osteitis fibrosa, bone turnover markers (BTM) 127
osteoblasts
  alkaline phosphatase (bone) 58
  in bone formation and remodelling 1, 3–6
  cellular lineages *5*
  differentiation 1, 3
osteocalcin
  aberrant levels 66, 182
  age-related changes 96, 97
  assays 67–9, 70, 205, 213, 227
  as bone formation marker 2, *6*, 28–9, 65
  and bone mineral density (BMD) 153
  diurnal variation/variability 66, 119–20
  ethnic differences 100, *101*, 146
  fragments 67, 69–70
  function 65
  gender effects 31
  and immobilization 127
  kidney uptake 66
  menstrual cycle 66
  metabolic function 10
  oestrogens 151
  Paget's disease 33
  predictive value 154
  in rheumatoid arthritis 45
  serum levels 65–6
  structure 65
  and thyrotoxicosis 124
  undercarboxylated (ucOC) 66, 212, 213, 215–16
osteoclasts
  activating factors (OAFs) 10, 18
  alkaline phosphatase (bone) 29
  in bone remodelling 1
  in bone resorption 6–7

differentiation 7, 8
gene defects *9*
regulation *9*
osteocytes, and mechanical loading 7
osteogenesis imperfecta **4**
osteomalacia 85, 123, 232
Osteomark 32
osteonectin 28
osteopenia 15
osteopetrosis **4**, *9*, 17
osteopontin 10, 28, 84
osteoporosis
alkaline phosphatase (bone) 59
biological variation/variability 32
bone sialoprotein (BSP) levels 85
bone turnover markers (BTM) 132
C-telopeptide cross-links (CTX) 114, *115*, 117
clinical trials 167–77
clinical value of bone turnover markers (BTM) 33–5
diagnosis 33–4
diurnal rhythm blunted 116
ethnic effects 99–100
genetic factors *19*
importance 18
mechanism 204
in men 17, 18, 184
monitoring 61, 149–55
multifactorial 15–17
and obesity 91–2
pathogenesis *19*
physiological basis 1
prevalance 1
procollagen N-terminal propeptide (PINP) 74
risk assessment 174
treatment 1–2
WHO criteria 197
osteoporotic fracture *see* fracture
osteoprotogerin *8*

Paget's disease
acquired osteoclast abnormality 9
alkaline phosphatase (bone) 58, 59, 67
aspartyl isomers 51
bisphosphonate therapy 20
bone resorption increased 2, 17–18
bone sialoprotein (BSP) 85
bone turnover markers (BTM) 33, 123, 127, 129, 232
monitoring 61
procollagen N-terminal propeptide (PINP) 74, 94
viral origin 17–18
pamidronate 20, 152
parathyroid hormone (PTH)
assay 205
and bone remodelling 1, 2, 6, 10, 23, 117, 183–4
circadian rhythm 116
predicting response 183–4
parathyroid hormone-related peptides (PTHrP) 3, 13, 18
peptide hormones, anabolic effects 23
periodontal disease 7
physical activity *see* exercise
phyto-oestrogens 22
pituitary hormones 1, 3
plasminogen activator 10, 12
plasminogen activator inhibitor 12
point-of-care assays 27, *31*, 32, 39, 40, 132, 179
population-based methods 181
posterior probabilities 216, 237
postmenopausal women
cytokine production 15
elevation of bone turnover markers (BTM) 33
see also old age
prediction, in individual patients 153–5, 171–6
pregnancy 85
prior fractures, as risk factor 198, 225, 228, 229
procollagen
C-terminal propeptide (PICP)
age-related effects 96, 98
assays 73–80
and bisphosphonates 151
bone turnover marker (BTM) 79, 96
ELISA 205
gender effects 31
molecular heterogeneity 73–80
and procollagen N-terminal propeptide (PINP) 78, 80
serum concentration 29, 78
structure 79
N-terminal propeptide (PINP)
age-related effects 98
antigen profile 75
assays 73–80, 205
blood levels 77
as bone turnover marker (BTM) 29, 74, 79
molecular heterogeneity 73–80
no diurnal variation/variability 93
oestrogens 151
Paget's disease 74, 94
and procollagen C-terminal propeptide (PICP) 78, 80
size chromatography 74, 75
stability 94
structure 74
peptides *6*, 28
PROOF study 239
prostacyclin 14
prostaglandins 13–14, 17
puberty 58, *96*, 99
see also childhood
pycnodysostosis **4**, 17, 30
pyridinoline cross-links (PD)
age-related changes 95, 96, 97
among-day variation/variability 133
assays 49–50, 227
and bone mineral density (BMD) 153

pyridinoline cross-links (PD) *cont.*
  and bone sialoprotein (BSP) 85
  diurnal variation/variability 107
  predictive value 184, 225, 230, 231, 234
  reproducibility 161
  structure 28, *50*
  tissue specificity 50–1
  urine 30, 108
Pyrilinks-D 31, 213

quality control 146
questionnaire 228, 232

race *see* ethnic effects
radioimmunoassays 46, 59, 67, 78, 109, 205
raloxifene 19, *21*, 150, 152, 155, 190, 191, 196
RANK/RANK-ligand system 1, 3, 6, 7, *8*, *9*, 17, 18
reactive oxygen species (ROS) 18
reference methods 138
reference ranges 100–2, 136, 137, 141–3
regulation
  bone metabolism 10–14
  calcium metabolism 1, 2–3
retinoids 10
RGD *see* Arg-Asp-Gly
rheumatoid arthritis 7, 10, 45–6, 85
risendronate 20, 22, 152
risk
  absolute and relative 199–200
  factors combined 200–1, 216, 234–5, 237, 242, 243
  see also fracture, risk
rocket electrophoresis 74, 75

sample collection, timing and handling 32, 39, 44, 45, 46, 121, 132, 133–4, 136, 144–5, 209, 212, 227, 238, 243
screening tests 132
seasonal factors 141, 142
selective oestrogen-receptor modulators (SERMs) 2, 19–20, 155
septic shock 10
serum, sample handling 44
sex-hormone binding globulin (SHBG) 204, 205
Sheffield Osteoporosis Study 225–35
sialoproteins *see* bone sialoprotein (BSP)
signal-to-noise ratio 135, 136, 145, 146, 149, 161, 190
size chromatography 74, 75
skeletal, skeleton *see* bone
smoking 119, 120, 121, 198
standardization of assays 53, 61–2, 138, 174
statistical analysis 109, 154, 159–65, 205–6, 213, 220, 228, 229
steroid hormones 1, 3
stress 134
stromelysin 10, 12
strontium 23
Study of Osteoporotic Fracture (SOF) 181, 188, 219–22, 237, 238

T and Z scores 143–4, 163, 197
tamoxifen *21*
Tandem-MP Ostase 59
Tandem-R Ostase 59, 60
tartrate-resistant acid phosphatase (TRACP) 6, 29, 96, 100, 228
technetium-99m-labelling studies 152
tests, sensitivity and specificity 221
thiazide 239
thyroid hormones 1, 3, 33, 114, 117, 124
tissue inhibitors of MMPs (TIMPs) 12–13
transforming growth factors (TGFs) 3, *8*, 12–13
transgenic animals 15, 17, 65, 85
tumour necrosis factors (TNFs) *8*, 10–11

ucOC *see* osteocalcin, undercarboxylated (ucOC)
ultrasound 197, 215, 241
urine
  bone turnover markers (BTM) 50
  C-telopeptide cross-links (CTX) 39
  collection procedures and sample handling 32, 39, 44, 134, 136, 243
  confounding factors 136–7
  creatinine 30, 108, 134
  N-telopeptide cross-links (NTX) 32, 33
  pyridinium cross-links 31
  variation/variability 32, 116, 133, 136–7, 161

variation/variability
  among-day 133
  analytical 44, 131, 132, 135
  between-person 160
  biological 44
    bone turnover markers (BTM) 95–102, 119–21, 131–8
    and response to therapy 161
    within-person 160
  components 160
  geographical 200
  intra-individual 132–3, 136–7
  measurement 160
  methodological 160
  pre-analytical 131, 132, 160
  and specimen collection 133–4
  total 135, 147
  urine 32, 161
  see also diurnal variation/variability
vertebral fractures
  and antiresorptive therapy 155
  and bone turnover 226
  case study 125
  incident 222, 228–9, 232–3
  in men 184
  prediction 229–30
vitamin D deficiency 74, 126, 129, 141, 180, 215
vitamin K 65, 66, 215